Q427333

WITHDRAWN

METALLOGRAPHIC SPECIMEN PREPARATION
Optical and Electron Microscopy

METALLOGRAPHIC SPECIMEN PREPARATION
Optical and Electron Microscopy

Edited by
James L. McCall
Battelle Columbus Laboratories
Columbus, Ohio

and
William M. Mueller
Director of Education
American Society for Metals
Metals Park, Ohio

PLENUM PRESS · NEW YORK AND LONDON

Library of Congress Cataloging in Publication Data

Main entry under title:

Metallographic specimen preparation.

Proceedings of a conference sponsored by the International Metallographic Society and the American Society for Metals and held in Beverly Hills, Calif., Sept. 23-24, 1973.
Includes bibliographical references.
1. Metallographic specimens — Congresses. 2. Electron metallography — Congresses. I. McCall, James L., ed. II. Mueller, William M., ed. III. International Metallographic Society. IV. American Society for Metals.
TN690.7.M47   669'.95'028   74-8391
ISBN 0-306-30791-X

Proceedings of the 1973 Metallographic Symposium held in Beverly Hills, California, September 23-24, 1973

© 1974 Plenum Press, New York
A Division of Plenum Publishing Corporation
227 West 17th Street, New York, N.Y. 10011

United Kingdom edition published by Plenum Press, London
A Division of Plenum Publishing Company, Ltd.
4a Lower John Street, London W1R 3PD, England

All rights reserved

No part of this book may be reproduced, stored in a retrieval system, or transmitted in any form or by any means, electronic, mechanical, photocopying, microfilming, recording, or otherwise, without written permission from the Publisher

Printed in the United States of America

PREFACE

Metallography is much more than taking striking pictures at high magnifications or polishing and etching specimens in such a way that no scratches can be seen. Basically, metallography is the physical metallurgist's most useful and most used tool for studying metals. Although it is perhaps his oldest tool, it certainly is not likely to become obsolete. Rather, the continued demands that have been placed upon materials have required more detailed characterizations of their microstructures and this, in turn, has required the metallographer to develop new techniques to make these characterizations.

Not too many years ago, the metallographer had only optical microscopes with which to examine his specimens. Now he has electron microscopes, scanning electron microscopes, and a whole host of instruments which were unknown to him only a relatively few years ago. This has forced him to learn not only how to use these new instruments and how to interpret the information that they provide but it also has made him develop new techniques for preparing the samples for examination.

Metallography frequently has been described as a combination of both art and science, with the science contributing to the interpretation and the art primarily entering into the specimen preparation function. Hardly any metallographer, regardless of how long he has been plying his trade, fails to become thrilled when he examines a specimen he has prepared and sees the microstructure revealed. This thrill seems to exist regardless of whether the metallographer prepared the specimen using only a polishing wheel, a cloth, and an abrasive or if he placed the specimen in some type of "black box", turned the crank, and out came his prepared specimen. Essentially both of these two extremes are in use today.

In view of the many developments that have been made in methods for preparing metallographic samples, and because of the unique demands of specimen preparation required by some of the new tools that are now available to metallographers, the time seemed appropriate to hold a symposium where metallographers could discuss and exchange the techniques which they have found to be most useful to them in their work. The International Metallographic Society and the American Society for Metals generously agreed to co-sponsor this symposium, which was held in Beverly Hills, California, on September 23 and 24, 1973. In planning this 2-day symposium, the topics selected were based upon an expected audience of metallographers and physical metallurgists involved to one extent or another in preparing or examining prepared metallographic specimens. It was expected that the exchange of preparation methods would benefit each attendee in his work, whether it be a rather repetitious quality control function or an ever-changing research operation.

The content of the papers presented was found to be of such widespread value to the participants and attendees that it was decided to publish them in this way. Hopefully, this will serve as a useful reference book for metallographers everywhere.

A symposium of this type would not be possible without the assistance of many people. Mr. Art Calabra of IMS and Mr. Allan Ray Putnam, ASM, provided the full support of their respective societies. John Dresty and Russel Staub assisted tremendously by serving as session chairmen. James H. Richardson, perhaps more than anyone else, made the symposium a reality by handling all the arrangements in Beverly Hills and to him a very special thanks is given. Lastly, we thank Connie McCall, who came out of "retirement" to type the entire manuscript in a rather uniform style from the many forms that were given to her. We hope a useful, easily readable reference book has resulted.

James L. McCall
Battelle-Columbus Laboratories

William M. Mueller
American Society for Metals

CONTENTS

A Review of Specimen Mounting Methods
    for Metallography   1
  Delmar V. Miley and Arthur E. Calabra

Abrasive Cutting in Metallography   41
  James A. Nelson and Robert M. Westrich

The Use of Wire Saws for Metallographic
    Sectioning   55
  H. B. McLaughlin

Electric Discharge Machining   69
  Joseph Barrett

Diamond Compounds for Polishing Metallographic
    Specimens.   77
  R. D. Buchheit and J. L. McCall

Linde Alumina Abrasives for Metallographic
    Polishing   95
  F. R. Charvat, P. C. Warren, and
  E. D. Albrecht

The Evaluation of Final Polishing Abrasives 109
  William C. Coons

Mounting, Lapping, and Polishing Large Sizes
    and Quantities of Metallographic Specimens 121
  N. J. Gendron

Techniques of Specimen Preparation for Metallography
    Using a New and Unique Automated Polishing
    Process 129
  K. H. Roth and D. M. R. Taplin

Mechanical Preparation of $Gd_3Ga_5O_{12}$ and Various
 Electronic Materials by Vibratory
 Polishing Techniques 143
 D. Medellin, M. F. Ehman, and G. W. Johnson

Magnetic Etching with Ferrofluid 155
 R. J. Gray

Decorative Etching 179
 R. J. Gray, R. S. Crouse, and B. C. Leslie

Practical Applications of Cathodic Vacuum Etching 207
 F. M. Cain, Jr.

Metallographic Methods Applied to Ultrathinning
 Lunar Rocks, Meteorites, Fossils, and
 Other Brittle Materials for Optical
 Microscopy 233
 R. H. Beauchamp and J. F. Williford

The Metallographic Sample Preparation of
 Fiber-Reinforced Composites 251
 T. J. Bertone

Plutonium Metallography 275
 E. M. Cramer

On the Use of Electron Beams to Characterize
 the Microstructure of Radioactive
 Materials 283
 H. S. Rosenbaum

Specimen Preparation Techniques for Scanning
 Electron Microscopy 297
 N. M. Hodgkin

Electropolishing of Thin Metal Foils 307
 Charles Hays

Preparation of Samples for Electron Microscopy
 by Ion Micromilling 333
 Ram Kossowsky and G. P. Sabol

Author Index 349

Subject Index 355

A REVIEW OF SPECIMEN MOUNTING METHODS FOR METALLOGRAPHY

Delmar V. Miley and Arthur E. Calabra

Dow Chemical U.S.A.
Rocky Flats Division
Golden, Colorado

INTRODUCTION

Metallography is one of the oldest and also one of the most modern branches of metallurgy. It has been variously described as an art and as a science, and undoubtedly it has been considered by some to be a "black art". Its recognition as a distinct area of study goes back at least as far as 1916 when the ASTM Committee on Metallography was first established (1).

No matter how sophisticated our new techniques become though, there will always be a need to perform optical metallography, either solely or in conjunction with other techniques. And of initial interest to the optical metallographer will be selection of an appropriate method of mounting the sample so it may be properly prepared. It is the purpose of this paper to briefly review the development of mounting methods, and to discuss the more common methods now being used.

HISTORY OF MOUNTING METHODS

Before beginning our discussion of the more modern mounting techniques, it might be interesting to take a brief look back and see how the art of specimen mounting has progressed.

Prepared Under Contract AT (29-1)-1106 for the U.S. Atomic Energy Commission

As might be expected, the very earliest "mounts" were simply hand-held specimens, normally for examination of fracture surfaces. The use of the microscope to study fracture surfaces was reported before 1800 (2). In the years immediately following, there were some studies of ground-and-etched metals, and quite a bit of interest in the microscopic examination of meteorites (2).

Henry Clifton Sorby, considered by most to be the father of modern metallography, was a wealthy Englishman who conducted studies in many areas of the natural sciences (3). One of his areas of interest was petrography, and he is credited with preparing the first petrographic specimens in 1849 (4). He used Canada balsam to mount 1-in. square specimens on glass, and it is interesting to note that this is still one of the principal methods of petrographic preparation almost 125 years later.

By the early 1860's, Sorby's interests had expanded to include the examination of metallic structures. It seems only natural that he would adapt his earlier petrographic methods for use with metals. He cut his samples with a saw, then ground or filed them to a convenient thickness. They were then affixed to a piece of glass with Canada balsam, and the exposed surface was prepared by polishing and etching. A thin cover glass was then applied over the prepared surface, again using Canada balsam (3). Sorby's historic first publication of a metal microstructure appeared in 1864 in the Proceedings of the Sheffield Literary and Philosophical Society.

At about the same time in Germany, Adolph Martens was also developing similar techniques. He was apparently unaware of Sorby's 1864 publication, and his first papers appeared in 1878 (2). According to Smith (3), Martens was not strong on metallurgical theory, but he was a devoted technical metallographer. He also mounted his specimens on glass for preparation.

Similar work was also going on in France during the late 1800's, where Osmond used Canada balsam to mount thin metal slices to glass for preparation (3).

The turn of the century saw the beginning of the plastic age with the production of the first synthetic plastic, phenolic (5). The phenolic resins exhibited many desirable properties, and entered the commercial market in 1909 (6). The widely-used phenolic, bakelite, is named for its developer, Dr. Leo H. Baekeland.

It was apparently some time, however, before bakelite became significant as a metallographic mounting medium. Some of the more common materials used during the early 1900's included paraffin wax, sealing wax, sulfur, various plasters and cements, and low-melting point alloys (7-11). (See Appendix I for compositions of various low-melting point alloys.) A paper by Ragatz in 1929 (9) gave a comprehensive review of some of the mounting materials in use at that time including Lipowitz alloy, iron cement, litharge and glycerin cement, plastic magnesia cement, and sulfur. He also referred to a 1927 paper on the use of glass for mounting of fine tungsten filaments. (A tungsten-in-glass technique was reported more recently in 1959 (12).)

All of these early mounting materials were generally undesirable for one reason or another. They resulted in difficulties such as rapid clogging of grinding papers, undesirable etching effects, corrosion of the sample, inadequate hardness, poor grinding and polishing characteristics, water absorption, and poor adherence. In spite of these problems some of the early materials, e.g. low-melting alloys, are still used in specific applications (13,14).

By the early 1930's, the use of bakelite as a mounting material was becoming more common, and one paper described it as "the most satisfactory mounting material to be had" (15). And by standards of that time, it probably was, since it avoided most of the difficulties mentioned above.

Bakelite gained wider acceptance and usage during the 1930's, and pressure-molding techniques were developed (8,15). Also during this decade, more synthetic plastics became commercially available. These included the vinyls, polystyrene, polyethylene, and the acrylates (8,16). Some of these plastics were evaluated for use as mounting materials by Wyman (16), and although he mentions the use of casting resins, the emphasis at that time remained on pressure-molding techniques.

Even though the epoxies were developed in the mid-1940's (5), their use in metallography did not occur for some years. During the '40's, the mainstay materials included bakelite, lucite, and tenite. However, Kehl and Church (17) recognized the value of using a castable resin rather than pressure molding for some types of samples, and they devised a rather involved technique for casting of methyl methacrylate (lucite). The development of many castable resin systems and epoxies followed during the 1950's and 1960's, bringing us up to the commonly-used products of today.

Most of the important aspects of high-quality metallographic preparation were recognized by the metallographers of the early 1900's. One of the primary subjects of concern then, as now, was that of edge retention and preservation of surface features. References to the use of electroplating for edge retention go back as far as 1921 (see ref. 9), and this has remained one of the most acceptable techniques. Additions of various filler materials to the mounting medium in order to control mount hardness and polishing characteristics has also been widely used. These additions include such diverse materials as wood and mineral flours (16), iron grit (19), ground glass (20), and pelletized alumina (21, 22).

And so we see that metallographic mounting methods have been extremely varied over the years, making use of some materials and techniques which today seem rather primitive. Nevertheless, the basic principles and objectives of metallographic mounting were established using these early techniques, and with all of our modern technology, we still don't have one universal mounting method for all applications - we still resort to a variety of techniques. And that is what we shall attempt to cover in the rest of this presentation.

MODERN MOUNTING TECHNIQUES

Requirements

As an introduction to our discussion of the commonly-used mounting techniques, it might be worthwhile to briefly consider some of the requirements that should be kept in mind. The desirable characteristics of mounting materials have been discussed by several authors (6,7,11,16,17,22,23), and it is from these sources that the following remarks are taken.

The primary reason for mounting samples is simply to facilitate ease in handling. Sharp corners are eliminated, thereby increasing the safety to the metallographer and avoiding damage to the papers and cloths used for preparation. Also it is obviously very convenient to have all samples of a uniform size and shape, for automatic preparation equipment, viewing systems, and storage facilities.

Standard mounts usually measure 1-, 1 1/4- or 1 1/2-inch in diameter. Smaller samples will tend to rock during grinding and polishing, and larger samples reduce the abrasion rates. It is often necessary to use a much larger mount, however. The mount thickness should be about one half the mount diameter. Thin mounts are harder to handle, and very thick mounts tend to rock during preparation.

The first concern in selecting a mounting medium has to be protection and preservation of the sample. Both physical damage to fragile or delicate specimens and structural changes due to excessive heat or pressure must be considered. Also, it may be necessary to retain components in a specific relation to one another; surface scales, coatings, etc. must be preserved.

The mount should have sufficient hardness, and it should, ideally, have grinding and polishing characteristics similar to the specimen. (Hardness is not always an indication of abrasion characteristics.) It should resist physical distortion due to the heat generated during grinding, polishing, etching, and washing, and it should not clog grinding papers rapidly. The mount should also be chemically inert to the variety of lubricants, solvents, and etchants which are used.

The mount material should be capable of penetrating small pores, crevices, convolutions, and other surface irregularities. It should exhibit good adhesion to the sample surface, and not form a shrinkage gap between the sample and the cured mount. It should retain the specimen edge without rounding.

Depending on the type of the sample, the mount material may need to be either electrically conductive or insulating. The mount material must not cause unwanted etching effects from chemical or electrochemical reactions.

The mounting medium should be simple and fast to use and convenient to store. It should not be prone to formation of defects in the cured mount, such as cracks or voids. Often it is advantageous for the mount to be transparent. The mount material should present no health hazards, and it should be readily available at a reasonable cost.

This is a pretty formidable list of requirements, and no one mounting technique will fulfill them all. However, by judicious selection of materials and methods, it should be possible to obtain a mount which does achieve the most critical requirements for a given situation.

Mechanical Mounts

Numerous references to the mounting of samples by mechanical devices can be found in the literature (e.g., 9,11,18,23-25). Normally these techniques will be employed for specimens of regular shape and sturdy character. Although steps can be taken to achieve a degree of edge retention, this is not normally an advantage to mechanical methods. These techniques can be quite rapid, and if for some reason plastic mounts cannot be used, mechanical means may be the only alternative. In cases where subsequent examinations require an unmounted sample, removal from a mechanical clamp is obviously much easier than removal from a plastic mount.

Normally the specimen, or a stack of specimens, is clamped between suitable end plates (Fig. 1a & b) or is held within a cylinder

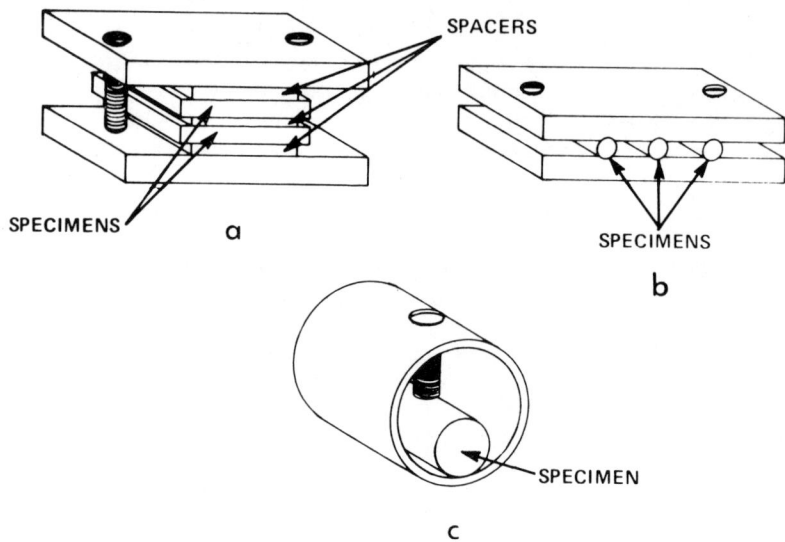

Figure 1. Examples of mechanical mounts.

or other convenient shape by a set-screw (Fig. 1c). Ideally, the end plates or surrounding member should have abrasion and polishing characteristics similar to the specimen material. Also, the mechanical mount should either be inert to the solvents and etchants used or have the same reactivity as the specimen.

The clamping pressure used to hold the sample within the mechanical device can be very critical. If insufficient pressure is used, there can be gaps remaining which may trap grit particles or cause staining due to seepage of lubricants and etchants. If the clamping pressure is excessive, physical damage to the specimen may result.

Various materials may be used as spacers if a stack of samples is being prepared. For maximum edge retention, a spacer with the same or slightly lower abrasion and polishing rates than the specimen would probably be selected. However, care must be exercised in selecting the spacer material so that undesirable galvanic reactions are avoided. If edge staining is a problem, the use of softer spacers, such as a soft metal or plastic sheet, would help. Coating of the individual samples with a thick layer of lacquer or phenolic or epoxy resin can also be effective. Immersion in molten paraffin is another method of filling pores and reducing the absorption of reagents. Reagent absorption and seepage can also be reduced by using alternate layers of dense blotting paper, but the paper may pick up abrasive particles, thus causing more problems than it cures.

Pressure Mounts

The various plastics used for metallographic mounting may be classed in several different ways, according to the technique used and the properties of the material. It is often convenient to divide these plastics into one group which requires the application of heat and pressure and another group which is castable at room temperature, and this is the basis for classification that we shall use here. Thermosetting and thermoplastic materials are in the class which requires the application of heat and pressure for curing. Let us start this discussion with some remarks which are applicable to pressure molding of both the thermosetting and the thermoplastic resins (7,11,16,18,23,24,26,27).

It was mentioned earlier that pressure-molding techniques using various plastic resins were developed during the 1930's. For about three decades, these methods were the most widely used, since they overcame many of the difficulties inherent in the earlier techniques. Even with the advent of the castable resins with their advantages, the pressure-molding methods have remained very popular. Good quality mounts can be prepared rapidly and simply, and they are characterized by relatively high hardness and good chemical resistance to a wide variety of reagents and solvents. These materials are fairly versatile and are available in transparent form as well as in a wide range of colors. They are all sold as uncured powder, and bakelite can also be purchased as a preformed pellet. Filler materials may be added to improve abrasion resistance or to achieve electrical conductivity. Several pressure mounts are illustrated in Figure 2.

Mounting of the thermoplastic and thermosetting resins is accomplished in a heated press, which may range from a simple, home-made design to an elaborate automatic unit. Three examples of commercially-available units are shown in Figure 3. Temperatures up to about 300°F and pressures up to 4200 psi are required. The press should have a high capacity heater surrounding the specimen mold so that curing times are held to a minimum. A means of rapidly cooling the mold is also required for the thermoplastic materials. The press should also have an automatic temperature controller, a mold ejection device, and a timer. Standard mold diameters are 1", 1-1/4", and 1-1/2".

Proper mounting technique starts with a sample which is clean and dry, and at least 1/2 in. smaller than the mold diameter. Unless they are necessary for examination of the sample, all sharp corners and edges should be eliminated. In order to avoid early partial curing of the resin, the die set should be reasonably cool when the resin is added. Alternatively, pressure must be applied without delay. Sufficient plastic must be used to cover the specimen to prevent damage to the top die and to maintain sufficient

Figure 2. Several types of pressure mounts.

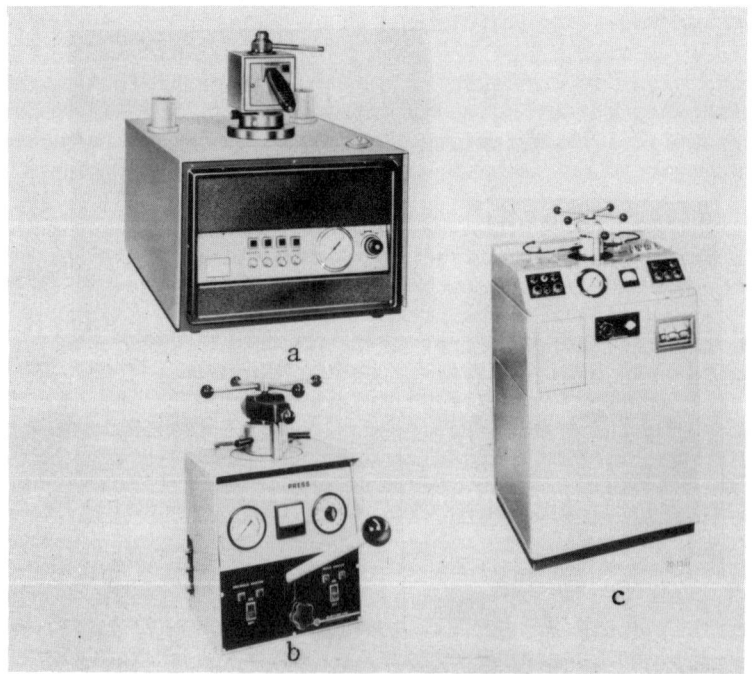

Figure 3. Commercial mounting presses:
    a. Pneumatic press (courtesy Leco Corp.)
    b. Hydraulic press (courtesy Buehler Ltd.)
    c. Mechanically-driven automatic press (courtesy Buehler Ltd.)

molding pressure. Pressure control is not overly critical, as long as the required minimum for the resin is exceeded. Excessive pressure may damage some samples. Temperature control is more critical; the required minimum must be reached without exceeding the maximum. If temperature is too high, the thermosetting resins may char and the thermoplastic resins may become so fluid that they are squeezed out through the mold parts. Good surface finish and close tolerances on the mold parts are important.

With these general comments in mind, let's look now at some of the differences between the thermosetting and thermoplastic resins.

Thermosetting Resins

The two most widely used thermosetting resins are bakelite and diallyl pthalate. Melamine has also been used, but is rather brittle when used alone. These resins actually undergo a chemical change at the molding temperature and pressure, and after sufficient time at the appropriate conditions they are fully cured. They may be ejected from the mold while hot. Several types of molding defects are common with these resins, and they are illustrated in Figure 4.

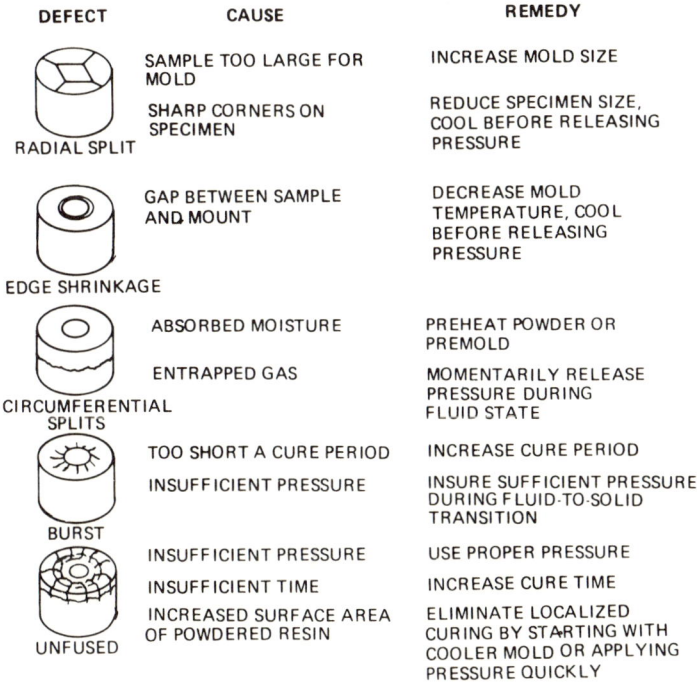

Figure 4. Mold defects in mounts made with thermosetting resins (Refs. 7,23,26).

Table I summarizes typical properties of bakelite and diallyl phthalate (11,18,23,24,26). The data from Table I show that both of these resins have similar curing cycles, so one mounting press can be used for either type. The heat distortion temperature listed in the table probably has more meaning in relation to the curing cycle than to subsequent distortion after curing. The thermosetting resins are normally considered to undergo no actual softening at high temperatures, but rather to char if the temperature is too high.

The coefficient of thermal expansion of these resins is relatively high compared to most metals, which are normally about $1-3 \times 10^{-5}$ in/in/°C. According to Samuels (23) this large difference is desirable in most cases, since the greater shrinkage of the plastic during cooling tends to hold the sample more tightly. This is an advantage because with these resins there is little or no actual adhesion to the sample surface. However, for certain sample shapes, this large difference in shrinkage can be a big disadvantage. For example, a tube section may be gripped tightly on the outside diameter, but on the inside, the shrinkage tends to pull the mount material away from the sample. This is demonstrated in Figure 5. The differential in thermal expansion can be altered somewhat by adding wood or mineral fillers to the powdered resin. These resins are normally purchased with the desired filler already added.

The abrasion and polishing rates of the thermosetting resins can also be altered by changing the type of filler material. For example, glass fibers are often added to diallyl phthalate and this lowers the abrasion rate and improves edge retention.

The thermosetting resins are decomposed by strong oxidizing acids and strong alkalies. They are resistant to strong reducing acids, organic acids, weak alkalies, mild etchants, alcohols, and organic solvents (16,23).

The major advantages of the thermosetting resins are that they are inexpensive, quickly and simply prepared, reasonably hard, and chemically resistant to most metallographic reagents. Their disadvantages include possible damage or alteration to fragile or heat-sensitive samples, gaps between sample and mount due to lack of adhesion and differential thermal contraction, and usually less than optimum edge retention properties.

Thermoplastic Resins

Several types of molding resins are classed as thermoplastics. These include polystyrene, polyvinyl chloride (PVC), polyvinyl formal (formvar), and methyl methacrylate (lucite, transoptic). Of these, methyl methacrylate is probably most widely used. The thermo-

TABLE I. TYPICAL PROPERTIES OF THERMOSETTING MOLDING RESINS (1)

Resin	Molding Conditions			Heat Distortion Temp.-°C (2)	Coeff. of Thermal Expansion in/in/°C	Abrasion Rate μm/min (3)	Polishing Rate μm/min (4)	Transparency	Chemical Resistance
	Temp. °C	Press. psi	Time min						
Bakelite (wood filled)	135-170	2500-4200	5-12	140	$3.0-4.5 \times 10^{-5}$	100	2.9	Opaque	Attacked by strong acids & alkalis
Diallyl phthalate (asbestos-filled)	140-160	2500-3000	6-12	150	$3-5 \times 10^{-5}$	190	0.8	Opaque	Attacked by strong acids & alkalis

Notes:
1. Data from references 11,18,23,24,26.
2. Determined by method ASTM D648-56 (23).
3. Specimen 1 cm^2 area abraded on slightly worn 600 grit SiC under load of 100 g. at rubbing speed of 10^4 cm/min. (23).
4. One-inch diameter mount on a wheel rotating at 250 rpm covered with synthetic swede cloth and charged with 4-8 μ diamond (23).

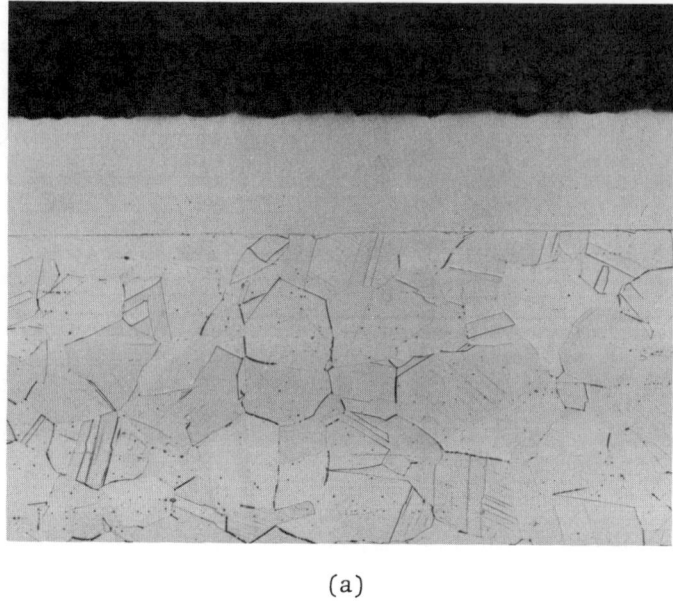

(a)

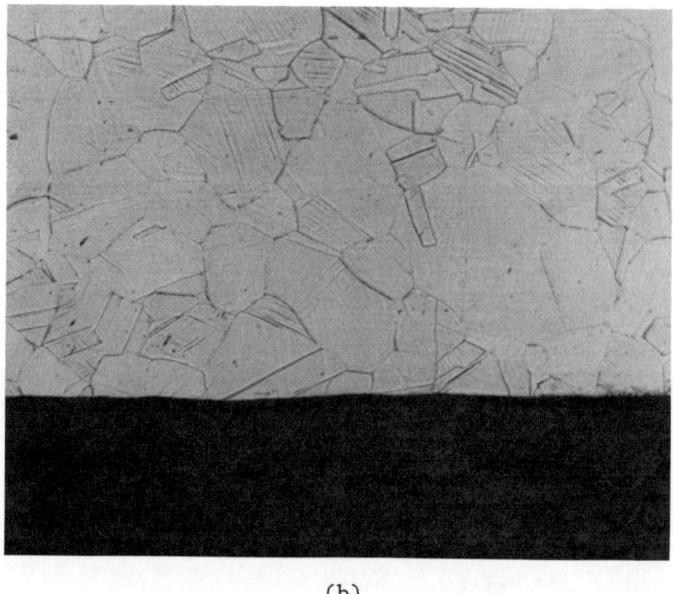

(b)

Figure 5. 200X Bright field illumination.
Longitudinal section of tube showing tight gripping of outside
diameter in (a) and shrinkage gap between plastic and inside
diameter in (b). (Reduced 20 percent for reproduction.)

plastic resins differ from the thermosetting in that they do not cure at the molding temperature. Rather, they become fluid at this temperature, and then must be cooled under pressure to produce the cured amount. Common mount defects, their causes, and remedies are illustrated in Figure 6 (7,23,24,26).

DEFECT	CAUSE	REMEDY
 COTTONBALL	RESIN DID NOT REACH PROPER TEMPERATURE INSUFFICIENT TIME AT MAXIMUM TEMPERATURE	INCREASE HOLDING TIME AT MAXIMUM TEMPERATURE
 CRAZING	RELIEF OF INTERNAL STRESSES UPON OR AFTER EJECTION FROM MOLD	COOL TO LOWER TEMPERATURE BEFORE EJECTING FROM MOLD TEMPER MOUNTS IN BOILING WATER

Figure 6. Mold defects in mounts made with thermoplastic resins (Refs. 7,26).

The properties of several thermoplastic resins are summarized in Table II (11,16,18,23,24,26). The heat and pressure required during the molding cycle are similar to that required by the thermosetting resins. With the addition of a mold cooling device, the mounting presses used for thermosetting resins may also be used for the thermoplastics. One advantage of most thermoplastic resins over the thermosetting types is that maximum pressure need not be applied until the resin is completely molten, thus avoiding damage to some specimens. Another advantage over the thermosetting resins is the transparent mount that can be obtained with the methyl methacrylate and formvar resins.

From Table II, we see that the thermoplastic resins are characterized by fairly low heat distortion temperatures. This results in softening and distortion of the mount if frictional heating during grinding and polishing is not properly controlled. Damage to the mount can also occur if insufficient heat filters are used when viewing small specimens under high intensity illumination, as Richardson has pointed out for PVC (24).

The thermoplastic resins possess fairly high coefficients of thermal expansion - generally somewhat higher than the thermosetting materials. As previously discussed, this high coefficient may be an advantage or a disadvantage, depending on sample geometry. Samuels points out that epoxies are the only mounting resins which show true adherence to the sample (23). Therefore, the thermoplastics, like the thermosetting resins, must rely on mechanical gripping which can be either enhanced or limited by differential ther-

TABLE II. TYPICAL PROPERTIES OF THERMOPLASTIC MOLDING RESINS (1)

Resin	Molding Conditions						Heat Distortion Temp.-°C (2)	Coeff. of Thermal Expansion in/in/°C	Abrasion Rate (3)	Polishing Rate (4)	Transparency	Chemical Resistance
	Heating			Cooling								
	Temp. °C	Press. psi	Time min	Temp. °C	Press. psi	Time min						
Methyl methacrylate	140-165	2500-4200	6	75-85	max	6-7	65	$5-9 \times 10^{-5}$		7.5	Water white to clear	Not resistant to strong acids & some solvents
Polystyrene	140-165	2500	5	85-100	max	6	65					
Polyvinyl formal	220	4000	-				75	$6-8 \times 10^{-5}$	20	1.1	Light brown, clear	Not resistant to strong acids
Polyvinyl chloride	120-160	100	nil	60	4000		60	$5-18 \times 10^{-5}$	45	1.3	Opaque	Resistant to most acids and alkalis

Notes:
1. Data from references 11,16,18,23,24,26
2. Determined by method ASTM D648-56 (23).
3. Specimen 1 cm² area abraded on slightly worn 600 grit SiC under load of 100 g. at rubbing speed of 10^4 cm/min. (23).
4. One-inch diameter mount on a wheel rotating at 250 rpm covered with synthetic swede cloth and charged with 4-8 μ diamond (23).

mal contraction. In extreme cases of mount shrinkage, whole sections of the mount may loosen and possibly fall out. For example, a longitudinal section of tubing may result in this type of mount defect. Richardson reports a simple remedy for this situation (24). By drilling one or more holes through the tube wall, additional bonding and support is provided for the portion of mount material on the tube I.D.

The abrasion and polishing rates for the thermoplastic resins are generally lower than they are for the thermosetting resins. Of the thermoplastics, formvar and PVC seem to have the best polishing characteristics (23,28).

The chemical resistance of the thermoplastic resins is good, except most are attacked by strong acids such as nitric and acetic, and some are at least partially soluble in organic solvents, such as acetone or chlorinated hydrocarbons. All show good resistance to the usual etching reagents, dilute acids, and alcohol.

Probably the major advantage of the thermoplastics is that some of them produce clear, transparent mounts. Their other properties, in general, are not significantly better than those of the thermosetting resins, and the thermoplastics do require a controlled cooling cycle which may be inconvenient. The thermoplastics do not completely solve the problems of edge retention and gaps between sample and mount.

Castable Mounts

The use of castable resins, or "cold mounting" plastics, has been growing in popularity for about 15-20 years. These resins offer certain advantages over the pressure molding resins, and possess properties which expand the mounting capabilities of the laboratory (6,7,21-23,30-40).

These resins are normally two-component systems, liquid-liquid, liquid-solid, or solid-solid (which must be melted for use). One component is the resin and the other component is variously known as the hardener, activator, or catalyst. Resins available for cold mounting applications include polyesters, polystyrenes, acrylics, and epoxides. (Polystyrenes and acrylics were also discussed under Pressure Mounts.) Various sizes and shapes of cast mounts are illustrated in Figure 7.

The castable resins are readied for use by thoroughly mixing the required amount of resin and hardener. Pot life of the mixture will vary from a few minutes to many hours, depending on resin type

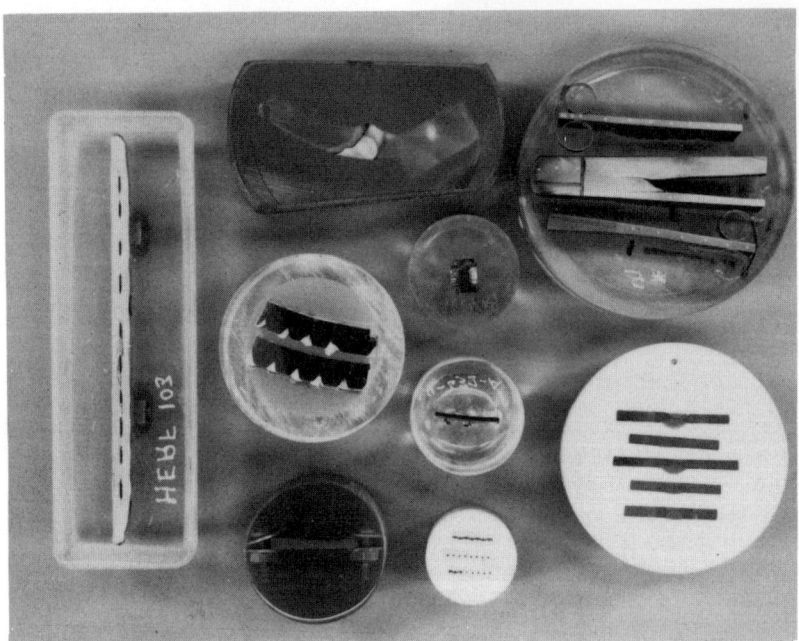

Figure 7. Some of the sizes and shapes of mounts possible with the castable resins.

and temperature of storage if the resin is not to be used immediately. Mounts are prepared by placing the sample into a re-usable mold, and then filling the mold with the resin. The mount is cured at room temperature or slightly above.

A variety of materials are used for molds, including glass, aluminum, brass, steel, teflon, and silicone rubber. Mold release agents such as vacuum grease or silicone oil are often required, depending on the mold material and the type of resin. Epoxies adhere to most materials, so epoxy mounts are probably most difficult to remove from the mold. However, silicone rubber molds have proved very convenient to use with the epoxies as well as the other castable resins (23,24,29,39). Although molds are normally made in the same range of standard sizes used for pressure mounts, the size and shape of the mold and resulting mount are limited only by the capabilities of the grinding and polishing equipment available. Mount defects common to the castable resins are illustrated in Figure 8.

The castable resins have a wide range of applicability in situations where pressure mounting is impossible, inconvenient, or ineffective. The mounting of very fragile or delicate samples is easily accomplished with the castable resins. Vacuum impregnation is often used to improve mount quality for these specimens (7,21-23,

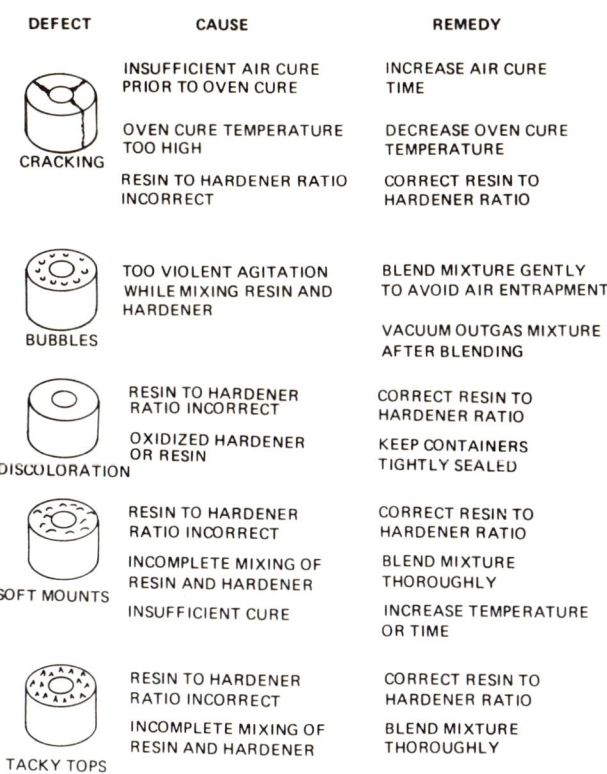

Figure 8. Mold defects in mounts made with castable resins (Ref. 7).

30-33). Several authors have reported on apparatus and techniques to accomplish vacuum impregnation, (23,24,31,33), but the simplest method is probably to use a small chamber such as an oven connected to a mechanical vacuum pump (21,22). Re-impregnation after successive grinding steps is sometimes necessary to insure specimen integrity or to eliminate voids on the mount face.

Relatively low curing stresses for the castable resins are an advantage in mounting delicate specimens. However, some specimens, because of their geometry, may be damaged by the curing stresses that are present. Crouse and Gray (38) investigated brazed T-joints which cracked after being mounted in castable resins. They were able to reproduce the cracks using several different resins, and attributed the defects to the curing stresses. To alleviate cracking in the joint, the specimens were nickel-plated prior to mounting. The joints did not crack, but the nickel plating was separated from the specimens, leaving an undesirable gap. The addition of pelletized alumina to the resins was effective in reducing the stress on the sample while maintaining integrity of the mount-to-specimen interface.

Castable resins have found wide use in the nuclear field, where the simple casting techniques are easily adaptable to hot cell or glovebox work (32-37). Chang, et al., (35) have reported a technique for minimizing the amount of resin which must be cured inside a glovebox. They make a pre-mold outside the glovebox which leaves a small conical or pyramidal cavity which is subsequently filled inside the glovebox with the sample and a small amount of uncured resin. This eliminates the use of dies inside the glovebox and minimizes contamination of the box atmosphere by gas released during curing.

The castable resins are also convenient to use when it is desired to alter the mount properties by addition of a filler material. Ground glass has been reported as an effective filler (20), and recent interest has centered on various grades of pelletized alumina (21,22). Fillers are very effective in improving abrasion and polishing characteristics, which are not especially good for the cold mounting plastics.

The epoxides exhibit many characteristics which make them excellent resins for use in castable mounts. They adhere well to most other materials, therefore, they need not rely on differential shrinkage to grip the specimen. Of the castable resins, they have the lowest shrinkage factor. Some epoxies cure rapidly, but others require a much longer curing time than the other castable resins. However, this is a definite advantage when vacuum impregnation is necessary.

Curing can be hastened by heating to about 60-70°C. Use of a radio frequency oven to accelerate curing has also been reported (40). Some epoxies can be stored under refrigeration for several weeks after the hardener and resin have been mixed. The excellent bonding characteristics of the epoxies makes them practical to use for double mounting or re-impregnation techniques. Some epoxies show excellent transparency. Chemical resistance of the epoxies is very good except in concentrated nitric or glacial acetic acid. The epoxies are sensitive to slight variations in the resin-to-hardener ratio. The resin and hardener can be purchased in pre-measured packets to minimize mixing problems. Some systems are toxic or may cause skin irritation, so adequate ventilation is required and skin contact should be avoided.

The primary advantage of the acrylics is their rapid cure rate. Consequently, they do possess a relatively high exotherm, which may be undesirable. The curing exotherm may be controlled somewhat by the mount thickness. Also because of their short curing time, they are not very applicable to vacuum impregnation. They show the highest shrinkage of the castable resins. Acrylic mounts may be clear to opaque. The acrylics are attacked by strong acids and are somewhat soluble in some organic solvents. They are probably less sensitive than the epoxies to variations in the mixture ratio. Breathing of the vapors or prolonged skin contact with the resins should be avoided.

The polyester resins exhibit less volume shrinkage than the acrylics, but they do rely on shrinkage to grip the specimen rather than bonding to the specimen. The polyesters cure fairly rapidly, but generally have longer curing times than the acrylics. They should therefore be more applicable to vacuum impregnation than the acrylics. They are not very sensitive to slight variations in the mixture. They show good chemical resistance to the normal metallographic reagents.

Probably the biggest advantage of the various castable resins is their versatility. By use of appropriate filler materials, the mount can be customized to the sample. Also, a wider variety of specimens, especially fragile or delicate items, can be mounted with castable resins than can be with the pressure-mounting techniques. On the other hand, the casting plastics must be measured fairly accurately, and generally require a longer curing time than the pressure-molding resins. They also exhibit relatively high shrinkage, which must be compensated for when mounting some types of specimens. The castable resins probably present a greater health hazard due to their vapors and possible skin irritant effects. However, these hazards are not great if reasonable care if used.

Special Techniques

In the foregoing sections, we have discussed the general procedures for mechanical, pressure, and castable mounts with only passing mention of some special problems which may be encountered. In this section, we will describe some of the techniques which have been reported for specific mounting requirements. These techniques will include mounting of small-diameter wires and tubes and thin sheet, edge retention, conductivity, and mount marking.

Mounting of Wire, Tubing, and Sheet

The mounting of small-diameter wire or tubing and thin sheet specimens can be rather challenging. Obtaining a truly accurate cross section, undistorted by tipping of the specimen, is the biggest problem. Good edge retention on these specimens is usually required also. Several authors have reported a variety of methods to achieve these ends (10,12,23,24,35,42-46), and their remarks are summarized here.

Double mounting techniques are very effective for wire, tubing, and sheet specimens. For example, holes or slots which are just larger than the specimen dimensions can be machined into a preformed blank of cured or uncured resin. The specimen is inserted into the preform, and depending on the type of resin which was originally used, is held in place by one of several means. If the preform is of a thermosetting resin, additional resin may be added and the curing cycle repeated. For thermoplastic resins, simply repeating the molding cycle without the addition of more resin will suffice. Epoxy resin can also be used to bond the sample into the preform. Figure 9a is an example of samples mounted in a bakelite preform, then cast in epoxy resin with a filler, and finally, a low-melting alloy was added to the back for conductivity.

Another double mounting technique starts by mounting the specimen horizontally in any desired resin. This first mount is then cut to reveal the cross-section of the sample. The sectioned mount is then re-mounted, with the specimen in the vertical position.

A mineral-filled epoxy with a putty-like consistency can also be used. The samples are simply pushed into the mixture in the desired orientation, and after this mount cures, it can be re-mounted into a standard size if necessary.

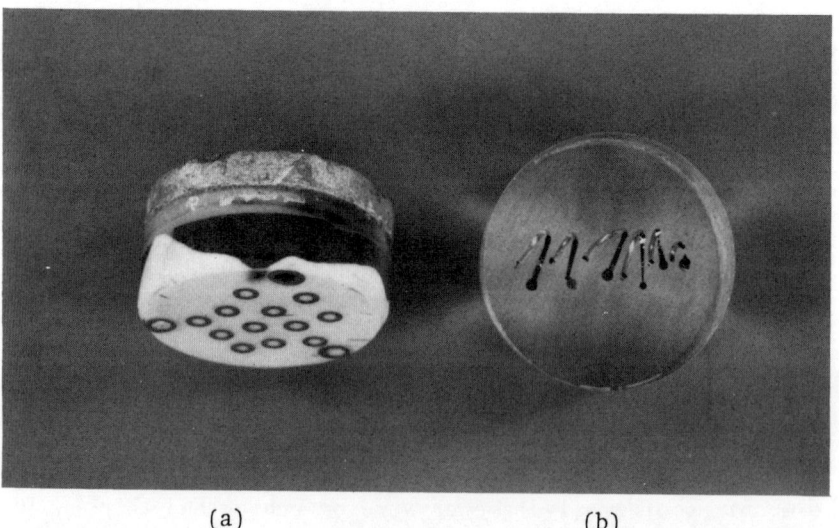

(a) (b)

Figure 9.a. A sample illustrating multiple mounting: First, bakelite preform drilled to accept samples. Second, cast into epoxy with alumina filler. Third, low melting alloy cast on back for conductivity.
b. Coiled wire specimen revealing longitudinal and transverse sections.

Pyrex or quartz glass has also been used successfully as an initial mounting medium for wires. In the case of tungsten wires, the glass capillary can be heated until it collapses onto the wire, insuring intimate contact. For other materials, a capillary is chosen with an inside diameter just larger than the wire diameter. After inserting the wire into the glass capillary, the composite is mounted in either a pressure-type or castable resin.

Wire specimens can sometimes be coiled carefully to reveal both longitudinal and transverse sections at the same time. This technique is illustrated in Figure 9b.

Sheet specimens can often be made self-supporting by bending them into an "S" or "U" shape, Figure 10a. A thin lead or copper strip can be used to make a self-supporting specimen, as in Figure 10b. Small specimens can also be supported with curved pieces of shim stock or small coils or spring clips. Several examples are shown in Figure 11. Stacks of sheet specimens are usually mounted with spacers between them. The stack may be bolted together before mounting, or the spacers may be cemented into place between the sheets. Usually it is advisable to arrange the spacers so they do not appear on the face of the mount after it has been ground and polished.

If the sheet specimens have friable surface layers, these layers can often be protected from damage by alternating the samples between sheets of thin plastic, such as 0.004- inch ethyl cellulose. After clamping the stack, it is mounted in acrylic resin. If an epoxy mount is preferred, polychlorotrifluoethylene film should be used.

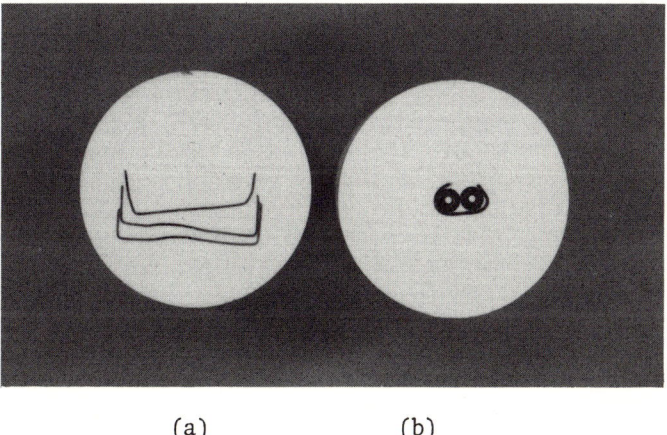

(a) (b)

Figure 10.a. Thin sheet specimens made self-supporting by bending.
      b. Copper strip used to make a self-supporting sample.

Figure 11. Examples of springs and clips used to support small specimens.

Mounting for Edge Retention

Retention of specimen edges and the preservation of surface features has long been an area of concern to the metallographer. The techniques we shall discuss here cover about 50 years of effort devoted to the examination of specimen edges (7,9,11,12,18-24,26, 32,36,41,47-52).

The quality of sample edge retention can be judged by the variation in height between the sample and the mount. This variation should not exceed the depth of field of the objective lens. So, for low-magnification viewing where depth of field is greater, more sample rounding can be tolerated than for high-magnification viewing. This may be an important consideration in determining which edge retention technique will be used. Samuels (23) has presented a comparison of the depths of field for typical objectives. His data are summarized in Table III.

In addition to specific methods which may be used to improve edge retention, there are factors which must be considered during specimen preparation which will affect sample rounding (23). During grinding, the abrasion rate of the plastic mount will almost always be greater than the rate for a metallic sample. The plastic will

TABLE III. DEPTHS OF FIELD OF TYPICAL MICROSCOPE OBJECTIVES (1)

Final Magnification	Objective Lense			Depth of Field, μm
	Magnification	Numerical Aperture	Type	
100	5.6	0.20	Dry	20
250	8.0	0.40	Dry	3
500	21.0	0.65	Dry	1
750	41.0	0.85	Dry	0.4
1000	58.0	0.95	Dry	0.1
1000	50.0	1.0	Oil	0.6
1500	75.0	1.4	Oil	0.2

1. Data from Samuels (23).

grind away faster leaving the metal edges exposed and rounded off. Finer abrasives will produce less edge rounding than the coarser grits, and fixed abrasive laps cause less rounding than the abrasive papers. Also, the abrasion rates of the various plastics differ and affect the degree of rounding. Samuels (23) rates formvar and PVC as best, followed by the allyl plastics, and epoxy and the phenolics as worst. The abrasion rate of the sample is another consideration, and generally the lower the abrasion rate of the sample, the more edge rounding may be expected.

After grinding, the specimen will normally stand in relief above the mount, unless it is a very soft metal specimen. The relative polishing rates of the sample and the plastic will have a bearing on whether the relief and edge rounding become greater or less. A napless polishing cloth will help to improve edge retention, but it may also cause heavier scratching of the sample. Unidirectional polishing by the "trailing edge" technique may improve edge retention, but this is not normally recommended. It is very important to avoid rocking of the sample during all stages of grinding and polishing.

With these considerations in mind, let us now discuss improvement of edge retention in more specific terms. One of the simplest methods of edge retention, and one that is fairly effective, is the use of mechanical devices such as those described earlier in this presentation. If clamping devices and spacers between specimens are selected to closely match the abrasion and polishing rates of the specimen, reasonable edge retention can be achieved. These methods can be used with or without plastic mounting resins. When plastic mounts are used, a common practice is to simply include a metal ring which surrounds the sample in the mount.

One of the oldest techniques, and still considered by many to be the best, is to plate the sample either by electrolytic or electroless deposition of metal. The plated sample can then be mounted in any convenient plastic resin.

The photomicrograph in Figure 5a shows an electrolytic nickel plate on stainless steel. A fairly thick plating can be applied electrolytically so that any edge rounding which does occur will be confined to the plating. The sample itself then remains flat.

The metals commonly electrodeposited include copper, iron, nickel, and zinc. Copper may be plated from a cyanide or sulfate bath; the cyanide bath is often used to form a flash coating, with heavier deposition being made from the sulfate bath. Deposition of iron can be accomplished from a ferrous chloride solution, but is reportedly difficult to accomplish. Electrolytic nickel from a nickel sulfate-nickel chloride bath is effective on ferrous alloys, nickel alloys, and metals of fairly high melting point. Zinc is plated onto zinc alloys from a cyanide solution. Compositions and operating conditions for several plating baths are presented in Appendix II.

Electroless plating can be accomplished with a variety of metals, such as copper, nickel, and silver. (See Appendix II). They have the advantage of being usable on nonconductive samples, and often they are used as an initial layer prior to electrodeposition. A commercially-available electroless nickel solution known as Edgemet is distributed by Buehler Ltd. Figure 12 shows examples of surface defects preserved by the use of this solution.

A problem which can be encountered with electroless platings is illustrated in Figure 13. In 13a, the outside diameter of a stainless steel tube plated with electroless nickel is shown. There is good contact between the mount, the nickel plating, and the stainless steel. The tube inside diameter is shown in 13b. Here,

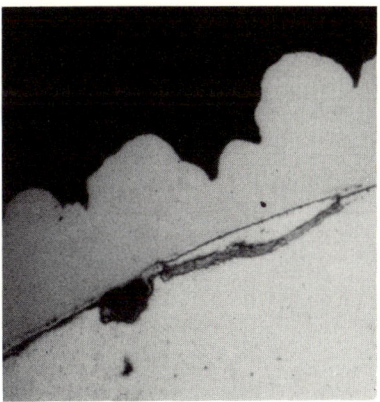

Figure 12. 1000X Surface defects on bronze wire preserved by Buehler's Edgemet solution (courtesy Buehler Ltd.). (Reduced 35 percent for reproduction.)

the shrinkage of the mounting plastic has pulled away the nickel plating, leaving a gap between the plating and the sample.

The biggest problem with electrodeposition is in obtaining a sufficiently clean specimen to insure adequate adhesion of the plating. Surface treatments used in industrial applications of electroplating are normally too harsh for metallographic samples and they may remove the features of interest. It may be possible to sufficiently clean the surface with detergent, solvents, or mild etchants if damage to the surface layers can be avoided.

Adhesion to the sample is also related to the internal stresses which are present in most electrodeposited metals. The residual stresses are surprisingly high, and may range from a few thousand psi compression to 100,000 psi tension, depending on the type of deposit. Kushner (49) has reported the approximate values of stress in various electrodeposited metals, and some of these data are presented in Table IV. It was also shown that residual stress is significantly affected by thickness of the plating and the type of plating bath that is used.

Some plating baths also have limited throwing power, and produce poor penetration of the coating into fine surface features. A thin electroless deposit can help alleviate this problem.

Electroless deposits do avoid some of the difficulties encountered with electrolytic deposits. They are reportedly stress-free, and they normally penetrate surface irregularities quite well. Also, any type of specimen may be plated, whether or not it is electrically conductive.

(a)

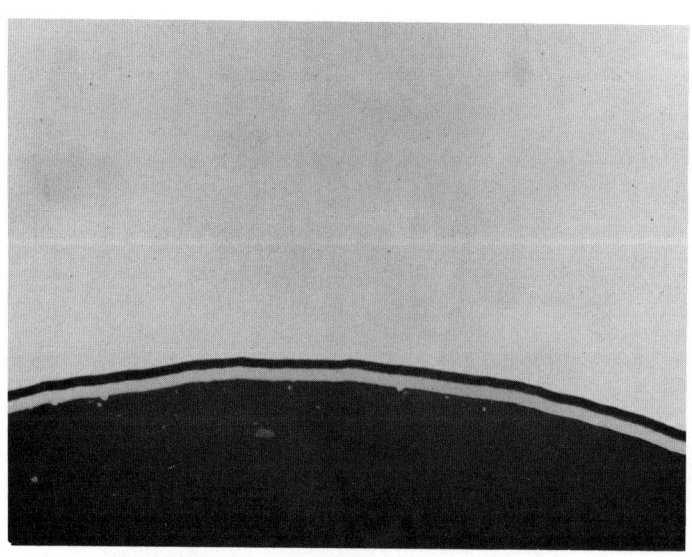

(b)

Figure 13. 200X Bright field illumination.
   a. Tubing O.D., good contact between mount, plating, and sample.
   b. Tubing I.D., shrinkage in mount has pulled plating away from sample.
   (Reduced 20 percent for reproduction.)

TABLE IV APPROXIMATE RESIDUAL STRESS IN ELECTRODEPOSITED METALS[1]

Metal	Stress in Electrodeposit [2] psi
Cd	-500 to -300
Zn	-1000 to -2000
Al	0
Ag	+2000
Au	+1500
Cu	+2000
Ni	+20,000
Fe	+40,000
Cr	+60,000
Rh	+100,000

(1) Data from Kushner (49)

(2) Approximate relative order of magnitude for thin films.

An additional advantage of metallic coatings is that they can help to improve the contrast between surface films and the mount material. Vapor deposition has also been used to apply coatings for improved contrast (48).

Another general method employed to improve edge retention is to make additions of a filler material to the mounting resin. The object is to alter the abrasion and polishing characteristics of the mount to more closely match the sample. An added benefit of the filler technique is that when some types of fillers are used, they effectively reduce the amount of shrinkage in the mount and help to eliminate the troublesome gap between the sample and the plastic.

Over the years, many types of filler materials have been used. Wood flour is usually added to bakelite, and mineral fibers are often added to diallyl phthalate. However, these additions are not especially effective in improving edge retention. Probably the most effective additive for thermosetting resins is short glass fibers, successfully used with diallyl phthalate as shown in Figure 14. Ground glass has been an effective additive to epoxy resins (20). The glass tube fused to tungsten mentioned earlier as an aid to mounting fine wires also served to improve edge retention of the wire (12).

Iron grit was advocated as a filler for edge retention (19), but there were difficulties with its use, and more effective fillers are now available. The grit had to be present in thin layers so the mounting plastic could fully penetrate it, and it could not be in contact with the specimen, or undesirable etching effects resulted when electrolytic methods were used. Also, the grit had a tendency to fall out of the mount if much grinding was required.

Probably the most popular filler now in use for edge retention is pelletized alumina, shown in Figure 15. This filler is available in three hardness grades and in two mesh sizes for matching to the sample hardness and shape. This material is now available in black as well as white. When the black alumina is used in conjunction with a mounting resin to which black die has been added, it is very effective in reducing reflected light from the mount surface. This is especially helpful in improving contrast of specimens being examined under polarized light. Figure 16 shows 0.005-inch beryllium

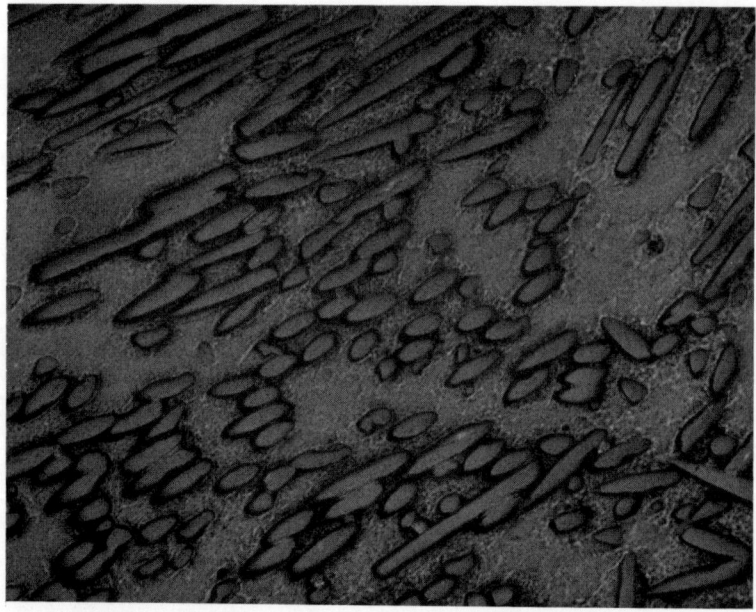

Figure 14. 200X Bright field illumination. Glass fibers added to diallyl phthalate for improved edge retention. (Reduced 10 percent for reproduction.)

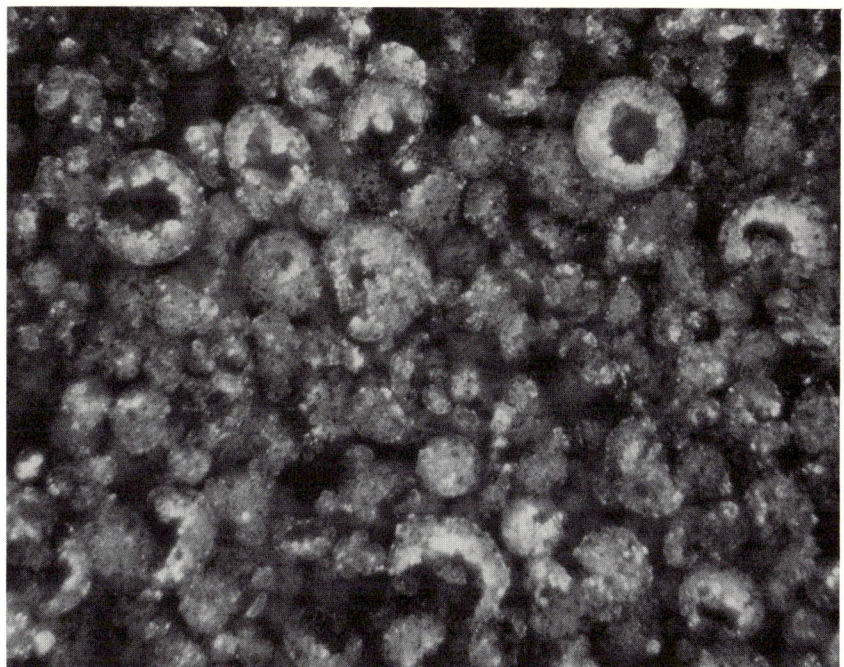

Figure 15. 200X Polarized light. Pelletized alumina filler for use in casting resins.

foil mounted with white alumina and with black alumina, and illustrates the greatly-improved contrast that results with the black filler. Optimum wetting of the alumina by the resin is best accomplished by vacuum impregnation.

The use of pelletized alumina does slow the grinding and polishing steps, and additional abrasive is required. For this reason, a double mounting technique is sometimes recommended in which the initial mount surrounds the area immediately adjacent to the sample with alumina. The second mounting step adds pure resin, so that the resulting final mount has a relatively small area of alumina exposed.

Mounting for Conductivity

The plastics used for metallographic mounts are normally electrical insulators, so special steps are necessary if electrolytic preparation methods are to be used. We will describe several methods of physically achieving electrical contact with the specimen and methods to make the mount itself electrically conductive (23, 24,27,53,54).

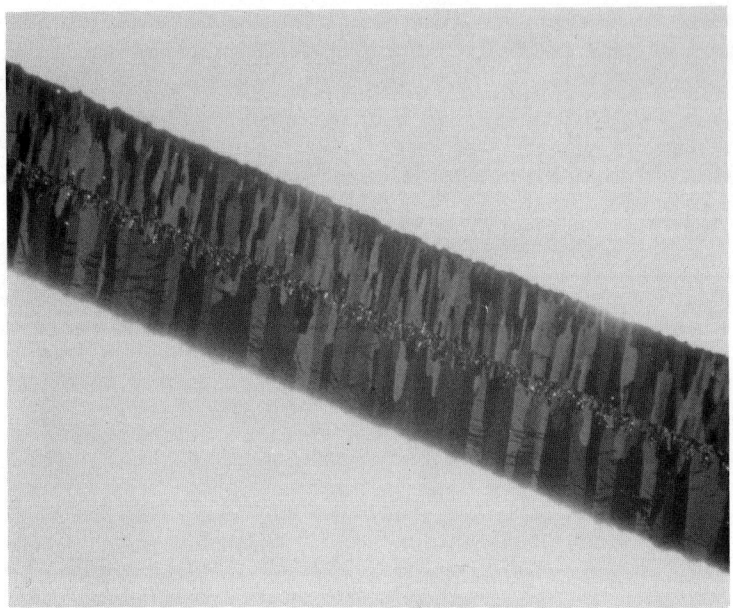

(a)

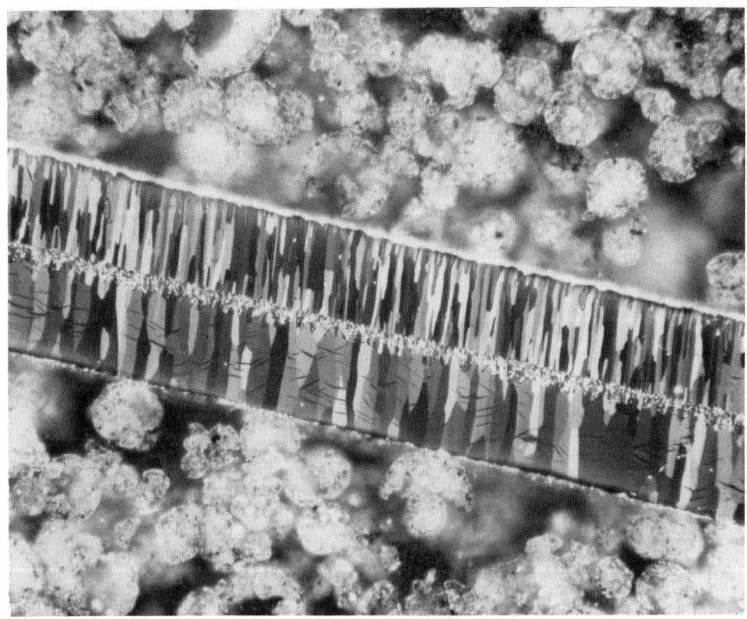

(b)

Figure 16. 250X Polarized light.
  a. Beryllium foil mounted in epoxy with white alumina filler.
  b. Beryllium foil mounted in black epoxy with black alumina filler.
  (Reduced 20 percent for reproduction.)

Contacting the sample surface with a probe results in uneven current distribution, so this simple approach is not always satisfactory. A better method is to drill a hole through the back of the mount and into the specimen. A screw is then self-tapped into the hole to act as the electrical contact. This method is not always satisfactory, however, since the sample may be damaged or electrical contact may not be sufficient. Also, it is possible to apply enough pressure when inserting the screw to actually pop the specimen partially out of the mount. If castable resins are to be used, leads may be attached to the specimen before mounting and left protruding from the back of the mount.

Another approach is to place metal balls or rods on the back surface of the specimen prior to mounting. After the mounting plastic is cured, the back of the mount can be machined to expose the balls or rods. A variation of this technique was described by Hoffman (53). His mounts were prepared in just the reverse order to those described above. A thick metal disc, which will eventually become the back of the mount, is placed into a mold, and the sample placed on top of the disc. Mounting resin is then poured into the mold to cover the specimen. After curing, the mount is ground to expose the specimen. This technique is supposed to work even if the specimen is a fine powder. A further modification reported by Richardson (24) includes the use of a conductive cement to insure positive contact of the specimen to the metal disc. In addition to providing the necessary electrical contact, the metal disc can serve as a clamping base for a magnetic chuck for hot cell use.

For some applications, the most satisfactory solution is to make the mount itself conductive. According to Samuels (23) a resistance of about 100 ohms from front to back of the mount is desirable. It is possible to achieve satisfactory conductivity by mixing approximately 10 volume percent metal with the molding plastic. However, this method is probably the least reliable. Good results can be obtained if the individual plastic particles are coated with a conductor. For example, PVC can be milled with carbon black to produce a conductive mounting material (23).

Another combination of conductor and mounting resin is the copper-filled diallyl phthalate powder which is commercially available. See Figure 17. This material can be used by itself or in conjunction with straight diallyl phthalate in a technique described by Ladroga (27). In this method, a small tube is used to confine the copper-filled resin to the area adjacent to the sample. Unfilled resin makes up the remainder of the mount. This technique is effective if a completely conductive mount face is not desired. For example, in a vacuum cathodic etcher, arcing can occur if the entire mount is conductive. This application of the technique was also described by Dobbins (54).

Figure 17. The use of copper-filled diallyl phthalate to achieve a conductive mount face.

Mount Marking

It may be appropriate to conclude this discussion of mounting methods with a few brief comments on the marking of mounts for identification. Hand scribers and vibrating-point tools are used extensively for this purpose. If the resulting markings are difficult to see, their visibility can be improved by inking over the scribe lines. Ink will adhere well to the roughened scribed areas, but it usually will not wet the smooth surfaces of most plastic mounts. Bartell (55) reported that 2 ml of ethyl alcohol added to 1 oz. of India ink acts as an effective wetting agent. Any portion of the mount can then be written on with the ink mixture.

A more permanent type of mount identification can be achieved if a clear mounting resin is used (56). In this case a small metal tag or a scrap of paper with the identification can be included in the mount with the sample. The identification is then permanently visible and protected with the sample.

A similar technique was reported by Clayton (57) for preserving the polished and etched surface of mounted specimens with a thin layer of lucite (or any other clear resin, presumably). The protected surface can still be examined at magnifications up to about 100X. This technique could be of value in cases where repeated examination is required, such as in the classroom.

CONCLUSION

If there is any one conclusion that can be drawn from the material presented here, it is that there is a mounting technique available for virtually any type of specimen. And the metallographer's first concern must be to select the technique which will best permit him to preserve the specimen and to prepare it for thorough examination. We have seen that each mounting technique and each mounting material has advantages and disadvantages, and no single material can fulfill every desirable requirement. However, by intelligent selection of materials and techniques, with proper diligence and care, and sometimes with a bit of imagination, this first step of metallographic preparation can foster success in the other steps that follow.

REFERENCES

1. Wyman, L.L., "A Mid-Century of Metallography-Retrospect and Aspect," Fifty Years of Progress in Metallographic Techniques, ASTM STP 430, American Society for Testing and Materials, 1968, p. 1-16.
2. Wever, F., "On the Background and Beginning of Metallography in Germany," The Sorby Centennial Symposium on the History of Metallurgy, Ed. by C.S. Smith, Gordon & Breach Scientific Publishers, New York, 1965, p. 163-169.
3. Smith, C.S., A History of Metallography, Univ. Chicago Press, Chicago, 1965.
4. Humphries, D.W., "Sorby: The Father of Microscopical Petrography," The Sorby Centennial Symposium on the History of Metallurgy, Ed. by C.S. Smith, Gordon & Breach Scientific Publishers, New York, 1965, P.17-41.
5. ABC's of Modern Plastics, Bakelite Company, Div. of Union Carbide Corp., 30 E. 42nd St., New York.
6. Kruger, O.L., Hughes, J.P., and Schmitz, F.J., Variable Curing Resins for Mounting Metallographic Samples, ANL-6712, Argonne National Laboratory, Sept. 1963.
7. Dvorak, J.R., Mounting Materials and Automation in Sample Preparation, ASM Metallography Symposium, Cleveland, Ohio, 2-3 April, 1970.
8. Schleicher, H.M. and Everhart, J.L., Metals and Alloys, 5, 1934, 59.
9. Ragatz, R.A., Mining and Metallurgy, 10, 1929, 372.
10. Ellis, O.W., Mining and Metallurgy, 9, 1928, 326.
11. Kehl, G.L., The Principles of Metallographic Laboratory Practice, McGraw Hill, New York, 1949, p. 49-58.
12. Wolff, U.E., and Fradetter, L.B., Metal Progress, 76, (2), 1959, 111.

13. Bender, J.H., A Technique for the Metallographic Preparation of Metallic Sodium, Proceedings, First Annual Technical Meeting, International Metallographic Society, 1968, p.33.
14. Brandenburg, C.F., "Present Methods of Metallographic Specimen Preparation for Retention and Identification of Inclusions in Steel," Symposium on Methods of Metallographic Specimen Preparation, ASTM STP 285, American Society for Testing and Materials, 1960, p. 29-36.
15. Krause, D.E. and Oesterle, J.F., Metal Progress, 24, (5), 1933, 33.
16. Wyman, L.L., Proceedings ASTM, Vol. 38, Part I, 1938, p. 511.
17. Kehl, G.L. and Church, J.S., Metal Progress, 50, (5), 1946, 1089.
18. ASTM Method E-3, ASTM Standards, Part 31, American Society for Testing and Materials, 1969.
19. Schaeffler, A.L., Metal Progress, 50, (4), 1946, 659.
20. Dahlin, R., Metal Progress, 82, (4), 1962, 146.
21. Calabra, A.E., "Maraset Epoxy Resin and Coors Pelletized Alumina as a Metallographic Specimen Potting Mixture", RFP-545, Rocky Flats Division, Dow Chemical U.S.A., Golden, Colorado, April 20, 1965.
22. Calabra, A.E., and Miley, D.V., "A Vacuum Potting Technique for Metallographic Specimens Using Black Epoxy Resin and Black Alumina Filler", RFP-1813, Rocky Flats Division, Dow Chemical U.S.A., Golden, Colorado, August 15, 1972.
23. Samuels, L.E., Metallographic Polishing by Mechanical Methods, American Elsevier Publishing Co., Inc., New York, 1971, p. 4-22.
24. Richardson, J., Optical Microscopy for the Materials Sciences, Marcel Dekker, Inc., New York, 1971.
25. Williams, R.S. and Homerberg, V.O., Principles of Metallography, McGraw Hill, New York, 1948.
26. Buehler Analyst, Buehler Ltd., 2120 Greenwood St., Evanston, Ill., 1971.
27. Ladroga, W.J. Jr., Metal Progress, 83, (2), 1963, 108.
28. Gibbon, G.L. and Furth, J.V., Metal Progress, 84, (3), 1963, 118.
29. Calabra, A.E., Metal Progress, 90, (1), 1966, 103.
30. Gray, R.J., "The Present Status of Metallography," Fifty Years of Progress in Metallographic Techniques, ASTM STP 430, American Society for Testing and Materials, 1968, p. 17-62.
31. Bertino, J.P., "Metallographic Preparation of Microspheres With Loose Concentric Shells", Proceedings, First Annual Technical Meeting, International Metallographic Society, 1968, p. 37.
32. Gruber, W.J., "Preparation of Irradiated Liquid Metal Fast Breeder Reactor (LMFBR) Prototypic Fuels for Metallography", Proceedings, Third Annual Technical Meeting, International Metallographic Society, 1970, p. 41.

33. Cochran, F.L. and Shoemaker, H.E., "In-Cell Vacuum Impregnation of Metallographic Specimens", Technical Papers of the Nineteenth Metallographic Group Meeting, ORNL-TM-1161, Oak Ridge National Laboratory, 1966, p. 173.
34. Gray, R.L., Long, E.L. Jr., and Richt, A.E., "Metallography of Radioactive Materials at Oak Ridge National Laboratory", Applications of Modern Metallographic Techniques, ASTM STP 480, American Society for Testing and Materials, 1970, p. 67-96.
35. Chang, J.V., French, P.M., Jacoby, W.R., and Shalek, P.D., "Examination of the Microstructures of Mixed Uranium-Plutonium Dioxide Fuel Pellets for Fast Flux Irradiation Studies", Proceedings, Third Annual Technical Meeting, International Metallographic Society, 1970, p. 41.
36. Gruber, W.J., Smith, F.M., and Watts, E.C., "Preparation of Irradiated Specimens to Retain Edges for Examination of Thin Surface Coatings," Advances in Metallography, RFP-658, Rocky Flats Division, Dow Chemical U.S.A., Golden, Colorado, October 20, 1966, p. 106.
37. Ritchie, E.E., "Evaluation of Dissolvable Epoxy Systems for Remote Metallographic Evaluations", Technical Papers of the Nineteenth Metallographic Group Meeting, ORNL-TM-1161, Oak Ridge National Laboratory, 1966, p. 103.
38. Crouse, R.S. and Gray, R.J., "An Effect of Curing Stresses in Epoxy Resins," Advances in Metallography, RFP-658, Dow Chemical U.S.A., Golden, Colorado, October 20, 1966, p. 118.
39. Hewette, P.M. II and Crouse, R.S., "Mounting of Metallographic Samples in Epoxy Resins Using Multiple Cavity Molds", Technical Papers of the Sixteenth Metallographic Group Meeting, NMI-4998, August 31, 1964, p. 1.
40. Crouse, R.S., "Accelerated Epoxy Mounting Using a Radio Frequency Furnace", Technical Papers of the Nineteenth Metallographic Group Meeting, ORNL-TM-1161, Oak Ridge National Laboratory, 1966, p. 99.
41. Rahn, D.J., Metal Progress, 76, (2), 1959, 109.
42. Sproule, G., Metal Progress, 46, (3), 1944, 484.
43. Broderick, S.J., Metal Progress, 46, (6), 1944, 1276.
44. zurHorst, C.G., Metal Progress, 47, (3), 1945, 509.
45. Duffner, R.L. and Norris, E.S., Metal Progress, 50, (4), 1946, 658-9.
46. Hochschild, V.J., Metals and Alloys, 20, 1944, 1614.
47. Cochran, F.L. and Wallace, W.P., "Electroless Nickel Plating of Graphite Specimens", GA-7057, General Atomics Division of General Dynamics, San Diego, California, May 2, 1966.
48. Godden, C.A., Metal Progress, 79, (1), 1961, 121.
49. Kushner, J.B., Metal Progress, 81, (2), 1962, 88.
50. Haynes, W.D., Tighe, N.J., and Kirkpatrick, H.B., Metal Progress, 81, (3), 1962, 112.

51. Owens, C.M., "A Metallographic Technique for Measuring Thin Films of Zirconium Oxide as Formed on Reactor Fuel Cladding", Technical Papers of the Nineteenth Metallographic Group Meeting, ORNL-TM-1161, Oak Ridge National Laboratory, p. 80.
52. Costas, L.P. and Hoxie, E.C., "Measurement of Thickness of Aluminum Oxide Film", Technical Papers of the Sixteenth Metallographic Group Meeting, NMI-4998, August 31, 1964, p. 26.
53. Hoffman, C.G., Metal Progress, 84, (3), 1963, 118.
54. Dobbins, A.G., "Metallographic Preparation of Holmium", Proceedings, First Annual Technical Meeting, International Metallographic Society, 1968.
55. Bartell, H.F., Metal Progress, 47, (5), 1945, 940.
56. Koppa, W., Metal Progress, 47, (2), 1945, 274.
57. Clayton, C.V., Metal Progress, 48, (1), 1945, 89.

APPENDIX I. SEVERAL LOW MELTING POINT ALLOYS FOR SPECIMEN MOUNTING

1. Woods Alloy: melting point 70°C

    Bi   50.0
    Pb   25.0
    Sn   12.5
    Cd   12.5

2. Lipowitz Alloy: melting point 70°C

    Bi   50.0
    Pb   27.0
    Sn   13.0
    Cd   10.0

3. Cadmium-Zinc Eutectic: melting point 265°C

    Cd   82.5
    Zn   17.5

4. Lead-Tin Eutectic: melting point 183°C

    Pb   38.1
    Sn   61.9

5. Cerrolow*117: melting point 47°C

    Bi   44.7
    Pb   22.6
    Sn   8.3
    Cd   5.3
    In   19.1

6. Cerrolow*136: melting point 58°C

    Bi   49.0
    Pb   18.0
    Sn   12.0
    In   21.0

7. Cerrobend*: melting point 70°C

    Bi   50.0
    Pb   26.7
    Sn   13.3
    Cd   10.0

*Trademark, Cerro Corp., New York

APPENDIX II. SOLUTIONS FOR ELECTROLYTIC AND ELECTROLESS DEPOSITION OF METALS

1. Copper, cyanide bath (Ref. 23. See also refs. 11 and 24)

    Solution: CuCN   20 g/ℓ
                     NaCN   30 g/ℓ
                     NaOH   1.5-3 g/ℓ

    Temperature: 45-60°C

    Voltage: 4-6 volts

    Current Density: .5-1 ma/cm^2

    Agitation: mild stirring

    Anode: copper

2. Copper, acid bath (Ref. 23. See also ref. 11)

    Solution: $CuSO_4 \cdot 5H_2O$ 170 g/ℓ
                     H_2SO_4 (conc.)   60 g/ℓ

    Temperature: 15-50°C

    Voltage: 1-4 volts

    Current Density: 10-20 ma/cm^2

    Agitation: mild stirring

    Anode: Copper

3. Iron (Ref. 23. See also ref. 11)

    Solution: $FeCl_2$ 288 g
                     NaCl   57 g
                     H_2O (dist.) 1 ℓ
                     Filtered for use

    Temperature: 70-100°C
                     (Must make up for evaporation loss with distilled water.)

    Current Density: .5-4 a/dm^2

    Agitation: Specimen (cathode) suspended and rotated @ 50 rpm.

    Anode: Ingot iron plate

APPENDIX II (cont'd)

4. Nickel (Ref. 23. See also ref. 24)

    Solution:
- $NiSO_4 \cdot 7H_2O$ 300g/ℓ
- $NiCl_2 \cdot 6H_2O$ 60g/ℓ
- Boric acid   40g/ℓ
- pH   4
- Add 1 part 30% H_2O_2 per 200 parts solution once a day

    Temperature: 40-70°C
    Voltage: 1-3 volts
    Current Density: 30 ma/cm^2
    Agitation: Vigorous stirring
    Anode: Anode nickel

5. Zinc (Ref. 23)

    Solution:
- $Zn(CN)_2$ 60g/ℓ
- NaCN   23g/ℓ
- NaOH   53g/ℓ

    Temperature: Room temperature
    Voltage: 1-4 volts
    Current Density: 10-15 ma/dm^2
    Anode: Zinc

6. Brashear Process for Silvering (Ref. 23. See also ref. 51)

    Stock Solutions:

    A. Silver nitrate 20g
       Water to 300 ml

    B. Potassium hydroxide 14g
       Water to 100 ml
       (Make fresh or store in polythene)

    C. Ammonium hydroxide, s.g. 0.880

    D. Dextrose 6.5g
       Water to 100 ml
       (Make fresh)

APPENDIX II (cont'd)

Mixing the
Silvering
Solution: To 3 volumes of A, add C until precipitate is just redissolved. Add A by drops until solution is straw-colored. Add 1 volume of B with stirring. Add C until solution clears. Add A until permanent precipitate forms. Filter and use same day. Dispose of unused solution the same day to avoid formation of explosive compound.

Silvering: Mix 3 volumes of silvering solution with 1 volume of D. Immerse specimen immediately. Silver for about 3 minutes, rinse in distilled water, apply second coat. Rinse and transfer immediately to desired electroplating bath.

7. Electroless Nickel (Ref. 47)

Solution: $NiCl_2 \cdot 6H_2O$ 30 g/ℓ

$NaH_2PO_2 \cdot H_2O$ 10 g/ℓ

$Na_3C_6H_2O_7 \cdot H_2O$ 100 g/ℓ

NH_4Cl_2 50 g/ℓ

NH_4OH to adjust pH to 8-10

Dissolve in order given in warm distilled water.

Plating: Bring solution to low boil, immerse specimens. Plating time about 1-1/2 hours.

ABRASIVE CUTTING IN METALLOGRAPHY

James A. Nelson and Robert M. Westrich

Buehler Ltd.

Evanston, Illinois 60204

INTRODUCTION

Abrasive cutting is the most widely used method of sectioning materials for microscopic examination and other material investigations. Conventional abrasive cutting using consumable wheels is the most popular method for routine metallographic sectioning because it is fast, accurate and economical. The quality of the cut surface obtained is usually superior to that obtained by other means and fewer subsequent steps may be required. Metal core diamond blades fulfill the demanding requirements of specialized applications such as ceramics, rocks, very hard metallics and printed circuit boards.

Together, these two methods of abrasive cutting offer a wide range of cutting characteristics which are capable of meeting the demands of virtually any material sectioning challenge. Only through a thorough understanding of the basic principles of abrasive cutting can effective use be made and the broad benefits derived from its application.

WHAT IS ABRASIVE CUTTING?

By definition, abrasive cutting is the sectioning of material, using a relatively thin rotating disc composed of abrasive particles supported by a suitable media. Ideally, each abrasive particle should function as an individual cutting tool. In a machine tool, as shown in Figure 1, the angle at which the tool contacts the work is fixed so that near ideal conditions are readily achieved (1). Abrasive wheels, however, are necessarily composed

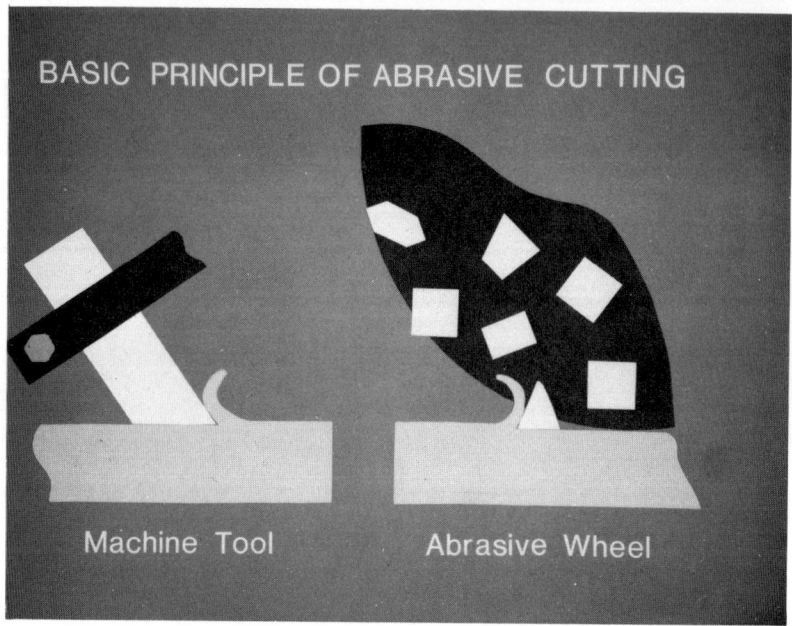

Figure 1. Basic principle of abrasive cutting

of randomly oriented abrasive particles which contact the work at various angles. Ideal cutting is, therefore, not achieved by every abrasive particle; the more favorably aligned ones will cut effectively, while the less favorably aligned ones will tend to produce grooves and frictional heat. The heat generated by the contact of abrasive particles in contact with the work would at first appear to be a significant problem. Actually, this very heat aids in keeping the abrasive wheel cutting rather than working the surface. The specimen is protected from excessive heating by the action of the coolant system of the cutter.

Metal matrix diamond blades differ sharply in construction from the consumable abrasive wheels. Figure 2 shows how the abrasive particles in diamond blades are confined to a relatively narrow rim on the periphery of the disc while abrasive wheels contain abrasive particles throughout. These particles are attached to the metal core with a resin, vitrious or metal bond, and may either be continuous or intermittent. Unlike the consumable abrasive wheels, the diamond blades do not break down and therefore require dressing when they become loaded with specimen material.

(a) Abrasive wheel

Diamond Blades

(b) Rimlock

(c) Resin bonded

(d) Continuous rim

(e) Metal bonded

Figure 2. Differences in construction of abrasive wheels and various diamond blade types.

FACTORS CONTROLLING SUCCESSFUL ABRASIVE CUTTING

Freedom from heat effects, flowed metal and distortion are prerequisites of acceptable abrasive cutting. To achieve these goals, three essential conditions must be met. They are illustrated in Figure 3. Failure to meet any of these conditions will result in specimens that are imperfectly cut, damaged, or in extreme cases ruined.

CONVENTIONAL ABRASIVE (CONSUMABLE) WHEELS

Correct Wheel Choice

All abrasive wheels are not the same. Wide variations of cutting performance are possible by controlling the following compositional parameters:

(1) <u>Abrasive Particles</u>: Silicon Carbide is preferred for cutting non-ferrous metals and non-metals. Alumina is recommended for ferrous metals. Coarse grained wheels generally cut heavier sections faster and cooler, but fine grains produce smoother cuts with less burring. Finer grained wheels are therefore recommended for cutting delicate materials such as thin-walled tubing.

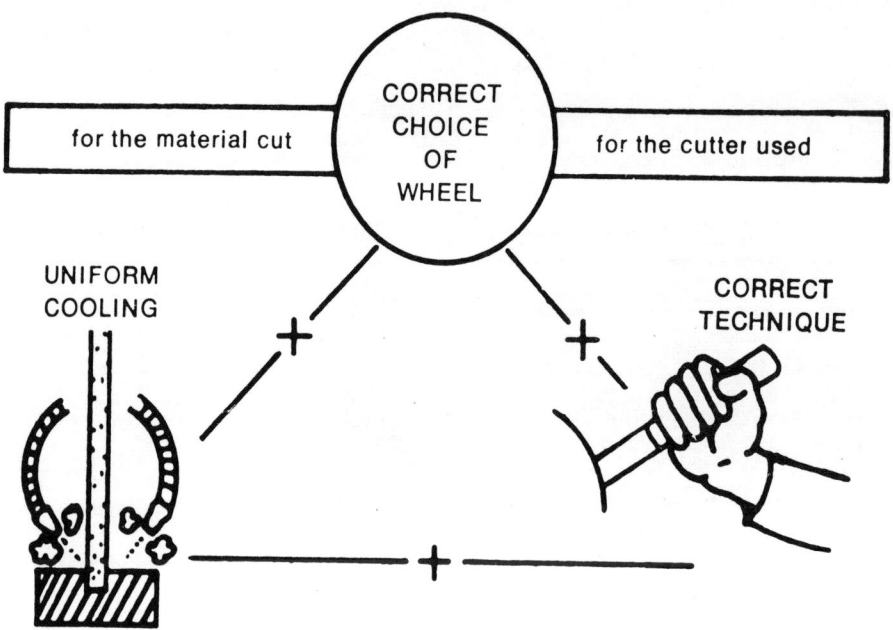

Figure 3. Essential parameters involved in successful abrasive cutting.

(2) <u>Bond</u>: Resin bonded wheels are generally used for dry cutting. They have very high cutting rates and are used in plant production cutting. Wet cutting wheels require a rubber or rubber-resin bond and are the type used in metallographic laboratories.

(3) <u>Porosity</u>: In rubber-resin wheels, the amount of bonding material and the percentage of free space controls the hardness or wheel grade. A more porous, less dense wheel (softer) breaks down faster because the abrasive particles are held more loosely. Softer wheels, generally preferred for cutting hard materials, are used because fresh sharp abrasive grains are more frequently exposed. Less porous, dense wheels are harder, break down slower, and are better for softer materials.

Abrasive wheels which show little or no wear are not performing satisfactorily. This is contrary to our natural inclinations which say that things which wear out at a noticeable rate must be defective. On the contrary, controlled wheel loss is an indication that the wheel bond is breaking down, exposing fresh abrasive grains for faster, more effective and cooler cutting. Wheels which fail to break down at a satisfactory rate will become glazed with specimen material with resultant poor cutting and excessive specimen heating. Any attempt by the operator to exert additional pressure will most likely cause overheating.

The acceptable rate of wheel loss is expressed by the M/W ratio:

$$LR = \frac{M}{W}$$

where LR = Wheel Life Ratio
M = Area of Material Cut
W = Area of Abrasive Wheel Consumed

(2)

In plant production cutting, resin bonded wheels are commonly used without a coolant. Rate of cutting is the main concern since this step probably precedes any heat treating. In this application an M/W ratio of 1.5:1 is considered acceptable. In other words, 1.5 times more material should be cut as wheel area consumed.

Metallographic laboratory cutting with specimen cooling requires the use of rubber or rubber-resin wheels and a ratio of 0.5:1 is considered an acceptable minimum. The wheel area consumed should be twice the material area cut because the chief consideration is to minimize burn, with economy being a secondary consideration.

These general property tendencies of abrasive wheels should not be considered as hard and fast rules. Rather, they should be viewed as guidelines in the selection and use of abrasive wheels. Individual material or alloy properties will ultimately determine the correct wheel choice.

ADEQUATE SPECIMEN COOLING

It has already been shown that frictional heat produced by the rapidly rotating abrasive wheel helps break down the wheel bond and promote free cutting. This same heat if allowed to build up within the specimen could significantly alter the structure. In steels, incorrect cutting due to inadequate cooling and incorrect technique could possibly result in the familiar "blue burn" or even heat checking of the cut surface. These cutting defects must be removed later by grinding to permit the true structure to be revealed. Abrasive wheels of different composition will differ in their ability to minimize burn. Correct wheel choice may therefore be very critical in certain applications, such as heat treated steels.

Adequate specimen cooling requires the sample to be subjected to a constant flow of coolant. Although water is a suitable coolant, the addition of coolant additive to provide some lubrication and inhibit rust is recommended. Submerged cooling is available on some cutters, like the one shown in Figure 4, and provides the maximum cooling for cutting sensitive heat treated specimens. The coolant also serves to flush away the waste products of cutting and to reduce the noxious odors which are inevitable in abrasive cutting with rubber wheel bonds.

Although the purpose of the coolant is specimen cooling, it also contacts the wheel and has a significant influence on the way it wears. Some effects of wheel profile on burr and cutting accuracy are shown in Figure 5 (3). Not only should specimen cooling have an adequate flow rate but it should also be uniform. Uniform cooling promotes a rather flat wheel profile or the even more desirable concave profile which produces a minimum of burring. A rounded wheel profile indicates the wheel is too hard, a condition which produces heavy burrs.

Non-uniform cooling causes the abrasive wheel to wear in an uneven manner, producing the characteristic "chisel effect". When cutting heavier sections, this could lead to curved cuts resulting in wheel seizure, stalling and finally, broken wheels.

Specimen cooling therefore should be characterized by adequate flow to prevent burning and uniform flow to promote even wheel-wear.

Figure 4. Abrasive cutter designed to perform submerged cutting.

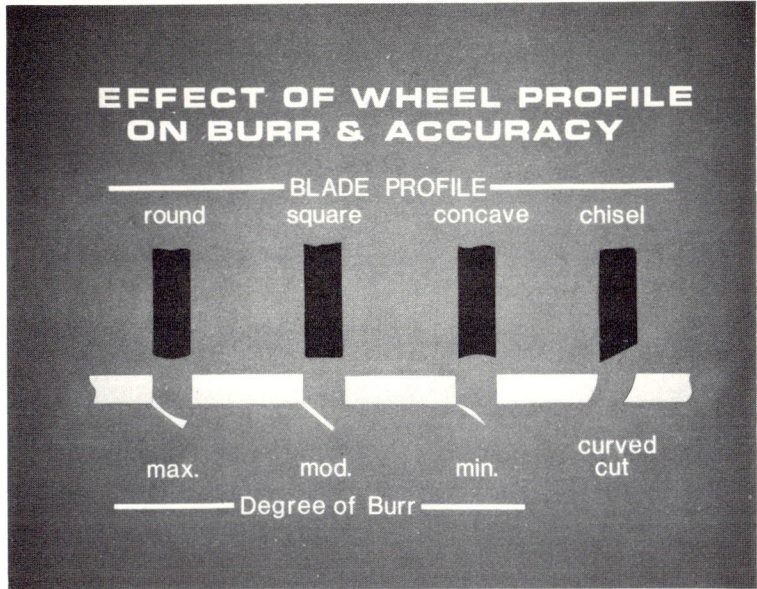

Figure 5. Some effects of wheel profile on burr and cutting accuracy.

CORRECT TECHNIQUE

The correct abrasive wheel choice and the most careful specimen-cooling are of little value unless the operator applies sound technique. Specimens must first be rigidly fixtured because loose specimens could result in broken wheels, damaged specimens and in extreme cases, injury to personnel.

Cutting pressure should be applied judiciously and if there is considerable resistance to cutting, a different wheel composition should be tried. Sometimes this may be avoided by applying a light oscillating stroke which helps promote wheel breakdown. If the cutter stalls frequently, even with very light pressure, it has insufficient horsepower for the job. Table 1 offers suggested action to take when the abrasive wheel operator is faced with a particular problem.

Some examples of successful abrasive cutting of irregularly shaped samples are shown in Figure 6. Another challenging application (not shown) was the cutting of steel reinforced rubber tires, where very hard and very soft materials had to be cut simultaneously.

Safe practice is always a part of correct technique. Safety guards and electrical interlocks should always be respected and no

TABLE 1. ABRASIVE CUTTING TROUBLE SHOOTING CHART

PROBLEM	POSSIBLE CAUSE	SUGGESTED REMEDY
Burning (Bluish discoloration)	Overheated Specimen	1. Increase coolant rate 2. Lighter cutting pressure 3. Choose softer Wheel
Rapid wheel wear	Wheel bond breaking down to rapidly	1. Choose harder wheel 2. Lighter cutting pressure
Frequent wheel breakage	1. UNeven coolant distribution 2. Loose specimen fixturing	1. Uniform coolant distribution 2. Rigid fixturing of specimen
Resistance to cutting	Slow wheel breakdown	1. Choose softer wheel 2. Reduce coolant flow 3. Use oscillating stroke
Cutter stalls	Too light cutter for the work	1. Use heavier cutter 2. Limit sample size

attempt should be made to avoid their use. Safety glasses should be worn when cutting, even if the apparatus is well guarded. Maintenance should never be done until the cutter is completely disconnected from the source of power.

By choosing the correct wheel, applying adequate uniform specimen cooling and applying correct cutting technique, good cutting using consumable abrasive wheels may be expected.

FACTORS CONTROLLING SUCCESSFUL ABRASIVE CUTTING -

METAL MATRIX DIAMOND BLADES

The exceptional high hardness and resistance to fracturing of diamond makes it a natural choice as an abrasive for cutting hard materials. Diamond abrasive blades are made of diamond bort that has been crushed, graded and chemically cleaned. The carefully

ABRASIVE CUTTING IN METALLOGRAPHY 49

Figure 6. Irregularly shaped samples cut by conventional abrasive cutting techniques.

graded particles are attached to the metal core using resin, vitrious or metal bonding in either a continuous rim or intermittently, as in rimlock blades (as shown in Figure 2). As in abrasive wheels, the choice of a suitable diamond blade is governed by the material application and the equipment available to make the cute.

RIMLOCK WHEELS

Metal bonded rimlock wheels consist of metal discs with hundreds of small notches uniformly cut into the periphery. Each notch contains many diamond particles which are held in place with a metal bond. The sides of the wheel rim have a serrated texture and are considerably thicker than the core itself, a construction that does not lend itself to delicate cutting. When cutting the more ductile materials, the need for more frequent dressing of the blades should be expected.

Rimlock blades are recommended for the bulk cutting of rocks and ceramics where considerable material loss may be tolerated. Kerosene or mineral spirits are used as the coolant lubricant and a constant cutting pressure or feed must be maintained to avoid damage to the rim. Rimlock blades should be used in special cutters such as the one shown in Figure 7.

CONTINUOUS RIM RESIN BONDED BLADES

Continuous rim resin bonded wheels consist of diamond particles attached to the rim of a metal core by resin bonding. These blades are suitable for cutting very hard metallics such as tungsten carbide and non-metals such as high alumina ceramics, dense fired refractories, metal-ceramic composites and others.

Resin bonded diamond blades may be used in conventional metallographic cutters with the normal water base coolant or in a special cutter for non-metals such as the one shown in Figure 8. Considerable care should be observed in the use of resin bonded wheels

Figure 7. Cutter designed for heavy cutting of nonmetals.

ABRASIVE CUTTING IN METALLOGRAPHY

Figure 8. A machine suitable for cutting nonmetals using a continuous rim diamond blade.

to prolong their life. Cutting pressure should be light and constant; an oscillating stroke should not be used.

WAFERING BLADES

Wafering blades are very thin, small diameter metal core blades which are used to make low distortion thin sections or wafers for electron microscopy and many other specialized applications. These blades are usually constructed of diamond, powdered metals and fillers that are pressed, sintered and bonded to the metal core. Wafering blades are available in both high and low diamond concentrations; the lower concentrations are better for harder materials, particularly the non-metals, while the higher concentrations are better for the softer materials (4).

Wafering blades should be used on equipment like the low speed saw shown in Figure 9 which offers control of the wheel speed and cutting pressure. By selecting the proper concentration blade, adjusting the blade speed and selecting the correct weight, the optimum conditions for the material may be achieved. Precision wheel placement, constant effective coolant application and automatic shut-off are all desirable features of a low speed, high precision

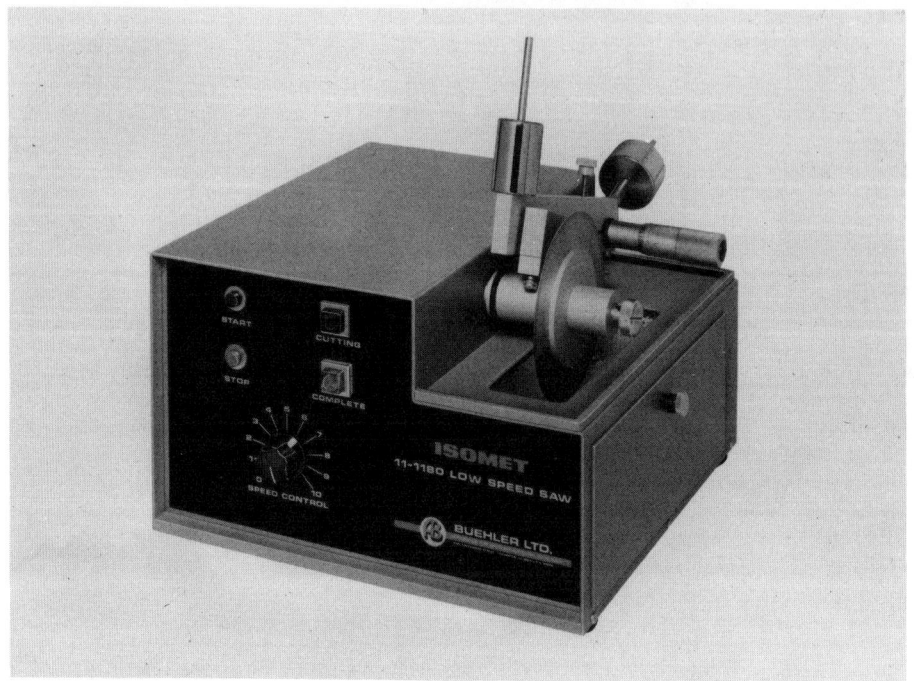

Figure 9. Low speed precision saw

wafering saw. Some of the many and varied applications of low speed saw cutting are shown in Figure 10.

SUMMARY

Table 2 summarizes the general areas of abrasive sectioning applications. It may be seen from this table that there are various routes to take in solving material sectioning problems. Abrasive wheels or diamond blades should be selected that are suited for the particular material to be cut. The wheel or blade rotation speeds should fall within the range indicated to achieve realistic cutting rates.

It should also be noted that maximum operating speeds printed on the specific blade or wheel should never be exceeded in the interest of safety.

Abrasive sectioning, as we have shown, is applicable to a broad spectrum of material sectioning problems. The successful application of this technique is dependent on an intelligent understanding of the material to be sectioned, which should result in a correct choice of wheel or blade. It also requires equipment that provides rigid fixturing and adequate, uniform specimen cooling.

TABLE 2: SUMMARY OF RECOMMENDED PARAMETERS
REQUIRED TO ACHIEVE SUCCESSFUL ABRASIVE CUTTING

APPLICATION	OPTIMUM S.F.M.	RECOMMENDED WHEEL OR BLADE
Low Distortion, Precision Cutting	0 - 300	Metal Bonded Diamond Blade
Soft Non-Metals - Fine Cuts	2000 - 3000	Continuous Rim Diamond Blade
Hard Non-Metals - Rough, Fast Cuts	3000 - 4500	Continuous Rim or Rimlock Diamond Blade
Metals - Submerged Cutting	4000 - 5000	Rubber or Resin-Rubber Abrasive Wheels
Metals - Wet Cutting	6000 - 8000	Rubber or Resin-Rubber Abrasive Wheels
Dry Cutting	9000 - 11,000	Resin Bonded Wheels

CAUTION: Maximum operating speeds printed on each blade should never be exceeded to assure safe operation.

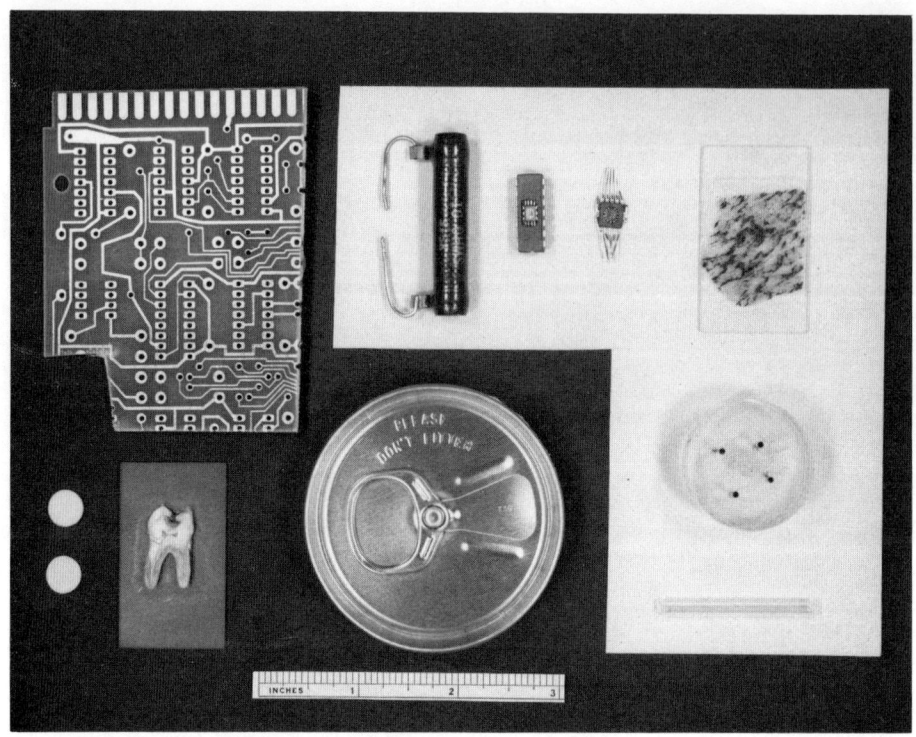

Figure 10. Some varied applications of low speed precision saw cutting.

Finally, correct technique based on knowledge, experience and common sense will complete the three-fold requirements for successful abrasive cutting.

While other highly specialized, expensive and more demanding techniques may do a specific sectioning job better, no other approach offers more advantages and has achieved a wider acceptance than abrasive sectioning.

REFERENCES

1. Biechele, Howard I., "Abrasive Cutting", reprint from Grinding Finishing, Hichcock Publications, 1966.
2. Clemens, Ray S., private communication.
3. Biechele, Howard, Howard., Ibid.
4. Ahmed, W. and Woodbury, J.L., "Isomet Wafering Blade Evaluation", Buehler Ltd., R & D Laboratory Memo, 1972.

THE USE OF WIRE SAWS FOR METALLOGRAPHIC SECTIONING

H. B. McLaughlin

Laser Technology, Inc.

North Hollywood, California 91604

INTRODUCTION

There are many methods available to the metallographer or ceramographer to cut samples. Each has its place and each excells in some phase of cutting. Some methods cut exceedingly fast, while others cut relatively slow. The damage done to the sample is usually a function of the speed of cutting. When materials are cut at high speed or under high force with an abrasive saw blade, considerable heat is generated at the cut. On some samples this is of little consequence. However, on many materials, such as semiconductors, this damage cannot be tolerated. Hairline cracks and craze caused by forceful cutting must be polished out before the material may be examined with the microscope.

The traditional methods of removing the sample from the bulk material use hack saws, band saws and abrasive cutoff wheels. These methods are generally rapid, but, as mentioned earlier, they often produce much damage to the surface of the cut which requires time and material to remove.

Other problems with these methods are encountered. For example, the diameter of an abrasive or diamond disk saw required for a cut is always twice the size of the specimen plus the diameters of the flanges holding the disk. As the weight of the disk goes up, so does the inertia of the wheel. This type of saw usually runs at high speed and generates considerable heat. Even band saws equipped with diamond blades have a problem; they involve a high capital investment.

Internal cutting disk saws are a special type which are made of thin material stretched over a hollow wheel. Diamonds are attached to the edge of a hole in the middle of the wheel. This method has the advantage of using a much thinner disk because it is pulled up taut from the outside diameter. However, the size of the specimen which can be sliced is limited to the size of the hole in the disk.

Spark erosion saws employ an electrical discharge between the electrode and the specimen to be cut. This method works only when the material being cut is a conductor.

We will now turn our attention to the wire saw. Although the wire saw is one of the oldest methods of cutting stone and its history dates back to early Biblical times, we are concerned with it in its present form.

When semiconductors were first invented, all the materials which were used were of a rather fragile nature. As the industry progressed and new crystals were brought to market they not only became more sophisticated in performance but generally they became softer and more fragile. The entire industry began looking for a method of cutting these crystals without causing damage to the surface being cut. Probably no other industry is so conscious of surface damage. When the surface of a semiconductor crystal is damaged the electrical properties may be changed significantly. In fact, the cutting method used was found to affect the performance of the crystal to such an extent that it fell outside the specification tolerances and had to be scrapped.

Every avenue of severing materials was explored to seek a method which would not damage significantly the material. All the various methods of cutting mentioned above were tried, but with many of them the yield of acceptable parts approached zero, so a whole new concept of cutting was needed.

THE WIRE SAW

The wire saw was tried for cutting crystals for the semiconductor industry and seemed to give a cut which was acceptable. The ancient, rather crude, rock saw was brought indoors, scaled down and built as a precision instrument. In some cases it was completely automated. Some of the instruments in use today can run for hours unattended while cutting undamaged slices from even the most fragile crystals. It cuts to tolerances and with finishes that are almost unbelievable.

The wire saw takes several forms with respect to the means of removing material. For example, a simple solvent such as water [1] or a chemical solvent such as an acid (2) may be used to effect a cut depending on the specific material. The most important form of the wire saw, however, is the abrasive saw.

Various methods have been devised for drawing the wire across the specimen. The endless wire saw consists of a loop of wire which is fastened together at the end. One such saw is shown in Figure 1. This wire is driven in one direction. The oscillating wire saw passes a wire back and forth across the sample usually with a short stroke. An extension of this technique employs a long length (up to 100 feet) of wire which is fed from a capstan across the work and back onto the capstan. The direction of the capstan is reversed at each end of the stroke. The capstan is further shuttled back and forth to maintain the alignment of the wire with respect to the pulleys. Such a wire saw is shown in Figure 2.

Abrasives

Any crystalline material can be used as an abrasive in wire sawing providing the abrasive is harder than the specimen to be cut.

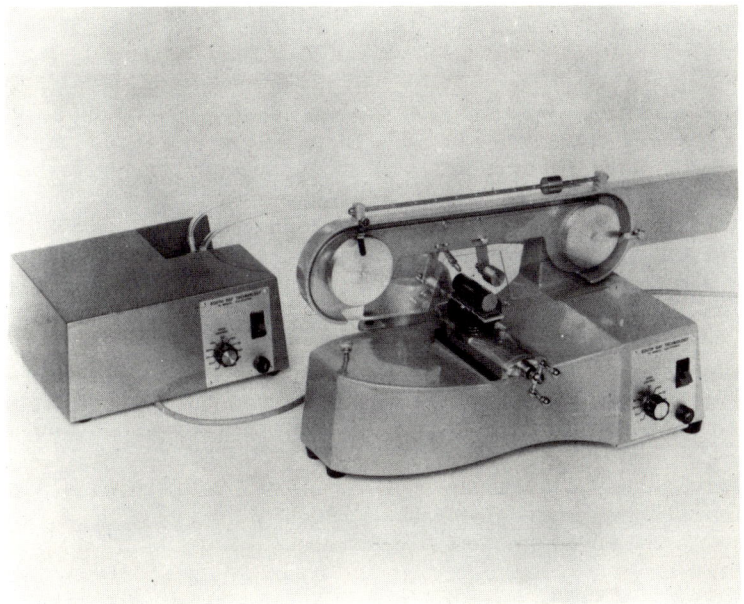

Figure 1. Wire saw employing an endless loop. Courtesy of South Bay Technology.

Figure 2. Wire saw employing shuttling capstan. One loop of wire comes off and over the idlers; it is this segment of wire that performs the cutting. The wire is unrolled at one end and rolled up at the other end simultaneously. The capstan shuttled back and forth keeping the wire in alignment with the pulleys. Courtesy of Laser Technology Inc.

The life of the abrasive is a factor to be considered. An abrasive which does not "break down" in use is considered best. Although natural abrasives such as emery, garnet, etc. have been used extensively in the past, newer man-made abrasives are harder and have better wearing qualities. However, the best overall abrasive found to date is diamond; this combines great hardness and unsurpassed durability. These desirable features of diamonds must be weighed against their high cost.

Abrasives to be used on wire saws should be accurately graded in order to produce a uniform finish to the cut surface of the specimen. There are two methods of applying abrasives to the wire:

1. Loose abrasive is mixed with a liquid vehicle as a slurry to be applied at the kerf behind the wire.
2. The abrasive is bonded to the wire.

In the first method, part of the abrasive stays with the specimen and erodes the wire. Furthermore much of the abrasive is wasted; this precludes the use of diamond in a slurry. In the second method, all the abrasive moves with the wire to cut the specimen. Thus only a fixed quantity of abrasive is employed and diamond then becomes economically feasible.

Lubricants

On nearly all types of cutting a lubricant is used. Some are used to prevent the cutter from wedging in the cut. Others are used to prevent heat buildup which can damage the material. In wire sawing with diamond impregnated wire, water is used. This is not used to lubricate the cut, nor is it used to prevent heat buildup. The amount of heat generated is negligible and lubrication of the wire is unnecessary. Water is used to wash out the sawdust which would accumulate above the wire thus preventing the easy exit of the wire when the cut is complete. For certain applications, such as cutting "moon rocks", air can be used to remove the cuttings and nozzles from vacuum cleaners can be used to catch the dust at each side of the cut.

Force Between the Wire and the Specimen

As force is increased between the wire and the specimen the bow in the wire increases even though the wire is under maximum tension. Little is gained in cutting time by increasing the force. When the force is increased excessively, the bow becomes so great that the wire has a tendency to wander. This wandering of the wire increases the kerf. Since when wandering occurs more material is being cut away, cutting time increases. This also sacrifices wire life. Thus, a high force with the resulting wider kerf is a poor alternative to the lighter force with a straighter wire and a more accurate cut. The lighter force also produces a better finish. If the cut is to be flat at the bottom the saw should be allowed to dwell for a short time with no force. A good figure for force between the wire and the specimen is 50 to 150 grams when a high tensile 0.008-inch diameter wire is used. The minimum kerf produced by the wire saw was important in the preparation and study of some of the recently returned "moon rocks". "Moon rocks" were without a doubt the highest priced and the scarcest material on earth. They had to be cut before studies could be made. Every gram of material had to be accounted for. Wire saws and diamond impregnated wire were used as this minimized the loss.

Tension of the Wire

The tension on the wire is the most important factor for good sawing. When a wire is limp, it has a tendency to wander. When it is under high tension it becomes taut and straight. The higher the tension, the straighter the wire becomes. The straighter the wire, the better quality of the cut. When a wire is under proper tension it gives out a high-pitched musical note when plucked.

The material used to make the wire determines the upper limit of the tension under which it can be run. Since tungsten has the highest tensile strength of materials available in wire form, this would be the obvious choice. However, tungsten work-hardens and breaks very quickly when run on a wire saw. This limits the life of the wire.

A heat treatable steel alloy has been developed which has a tensile strength of over 500,000 psi. This material is second only to tungsten. This material is so hard that it cannot be economically charged with diamonds, so it is plated with copper. The core holds the tension, the copper holds the diamonds. This combination results in a wire which can be run under about twice the tension of ordinary wire of the same size.

The wire saw is capable of cutting a thin slice from the middle of the bulk material where the cut goes only part way through. This method is used by museums to take samples from artifacts for analysis. This is done by making parallel cuts using a thin wire. The slice can be sawn across at the bottom of the cut thus removing it without damage; this leaves a very small notch.

When a firm, hard, tough, specimen is to be cut and where the surface damage is of little or no problem the fastest and most economical method of cutting usually is best. For example, a diamond impregnated wire 0.015-inch diameter and charged with 60 micron diamonds would be chosen. The tension on the wire would be in the order of 6000 to 8000 grams. The machine would be run up to 2000 feet per minute. The force between the wire and the specimen would be between 200 and 500 grams. If a sample is limited in amount, fragile, high priced, and/or delicate it must be handled gently. For such materials a wire 0.003-inch diameter impregnated with 8 micron diamonds would be selected. The force between the wire and the crystal would be between 10 and 35 grams. The tension on the wire would be between 500 and 750 grams and the wire would travel 60 to 100 feet per minute. Experience with this abrasive wire saw on other materials showed that it was an ideal instrument for cutting certain metallographic specimens.

SAMPLE PREPARATION FOR METALLOGRAPHY

Removing Sample from Bulk Material

Wire saws are available in a variety of designs. Some move the specimen into the wire, some move the wire into the specimen, some run horizontal, some run vertical. A saw in which the wire runs vertical is of great advantage if a specimen is to be removed from bulk material. In this case the material is attached to an X-Y table and the material moved into the saw.

Easy Preparation of Macrographic Specimens

A wire when running under high tension can oftentimes make cuts of sufficient quality for macrographic specimens without further polishing.

Cutting Potted Assemblies Open for Inspection

Analysis studies of electronics failures require that many potted assemblies be examined. Once potted, however, it is difficult to take the item apart to find the trouble. A new method to take these devices apart is to saw the assembly into slices with a wire saw. Slices as thin as 0.010-inch are practical. Many times this method can pin-point the source of trouble.

Cutting Thin-Walled Tubing

Nearly all mechanical means of cutting thin-walled tubing causes damage. If the tubing is metal, the edges bend over leaving burrs which have to be removed. Generally the cost of removing the burrs is greater than the cost of the tubing. If the tubing is quartz or glass, most methods of cutting cause considerable break-out damage. A good method of overcoming this problem is to bundle the tubing before cutting. This can be done by inserting a quantity of tubes into a heat-shrinkable tubing jacket. When heat is applied to the jacket the tubes are nested tightly together for cutting. The tubing can be sliced to any desired length. In this way slices can be handled easily, photographed, and ready for subsequent metallographic preparation.

Tungsten and Boron Filaments

Much work is being done presently on laminated filaments for structural members made of boron-coated tungsten wire. The ratio of strength to weight in these materials is extremely high. The wires are placed in "lays". Each lay consists of a number of wires all laid out parallel to each other. Each succeeding lay is placed at an angle with regard to the previous one. A great problem exists in cutting these materials as the cutting angle with the wires vary all the way from right angles to parallel. Wire saws have proven to be an outstanding method of cutting this type of structural material. Further information in this area is given by Bertone (3).

Cutting Honeycomb

Honeycomb structural material is made in various metals and non metals. It is available in paper (Figure 3), nylon (Figure 4), aluminum, titanium (Figure 5), stainless steel, ceramic (Figure 6), and inconel (Figure 7 & 8), to name only a few. The removal of burrs after cutting by most methods is extremely expensive. Most methods which will cut this material without leaving burrs are extremely complicated. In one such method, the cells are filled with wax, fastened to a substrate and cut on a vertical mill with a fly cutter using very light cuts. The whole assembly is then transferred to a surface grinder where several passes are made to remove the burrs. The part is turned over and all the operations are repeated on the other side. There are many methods equally as complicated. A wire saw will cut honeycomb at any angle with respect to the cells and leave a minimum of burrs. To illustrate the versatility of the wire saw, it will be within the realm of possibility in the near future to cut airfoils from honeycomb to finished dimensions, in both thickness and curvature by using wire saws.

Cutting Thin Slabs

Cutting with wires can be favorably compared to lapping. The wire with abrasive gently erodes its way through the specimen. Nearly all methods of cutting, except wire sawing, require clearance so the blade does not wedge in the cut. In wire sawing the wire touches the kerf on each side at a point of tangency. No place on the wire is it larger than the diameter. There is no tendency for the wire to wedge and jerk the part. This, coupled with the fact that the weight of the wire on the work is small, leads to little inertia to break a slice loose before the cut is complete.

Sample Preparation of Thin Sections for Optical Microscopy

When a sample must be sufficiently thin to be examined by transmitted light it is too thin to cut directly. Polishing is required to produce the extremely thin cross section. A good method of doing

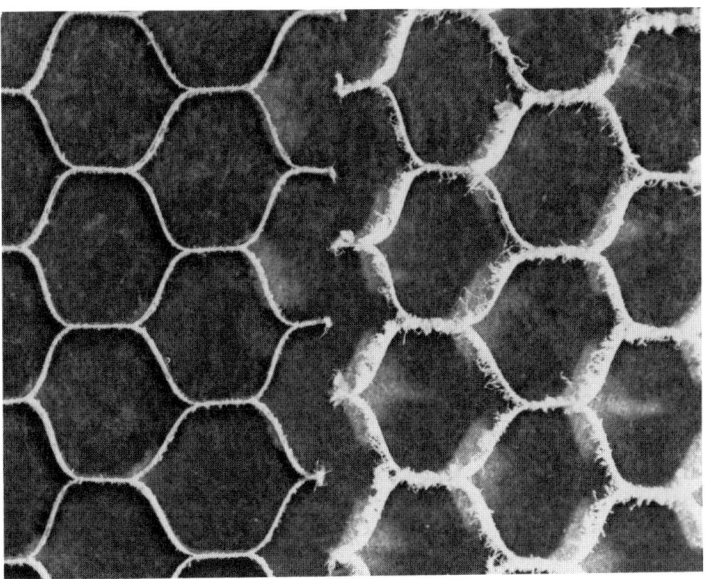

Figure 3. Photograph showing comparison between cuts on phenolic impregnated paper honeycomb produced by a wire saw (left) and a band saw (right). Notice the absence of burrs on the wire-sawn piece. Original magnification 2X. (Reduced 15 percent for reproduction.)

Figure 4. Photograph of a Nomex honeycomb showing a smooth cut with a complete absence of burrs. Original magnification 5X. (Reduced 15 percent for reproduction.)

Figure 5. Cross section of titanium honeycomb showing the absence of burrs. The cut was made at 30° to the cells. Original magnification 4X. (Reduced 15 percent for reproduction.)

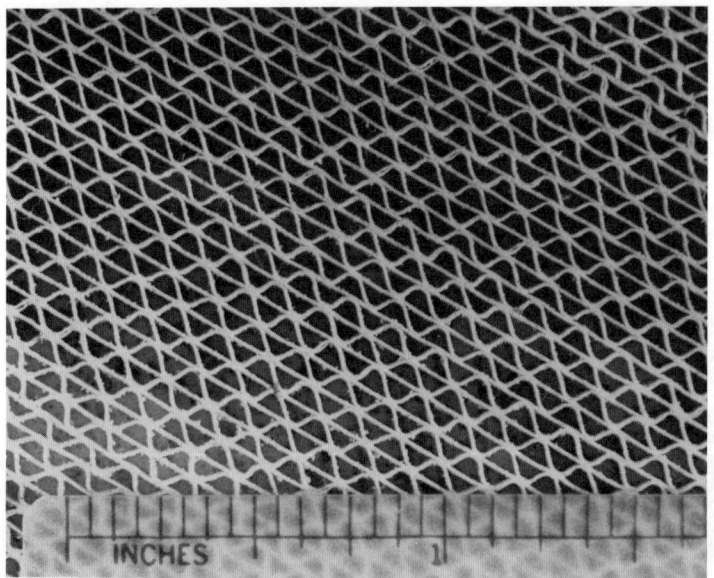

Figure 6. Cross section of ceramic honeycomb showing absence of chipping that is characteristic of wire sawing. Original magnification 2X. (Reduced 15 percent for reproduction.)

THE USE OF WIRE SAWS FOR METALLOGRAPHIC SECTIONING 65

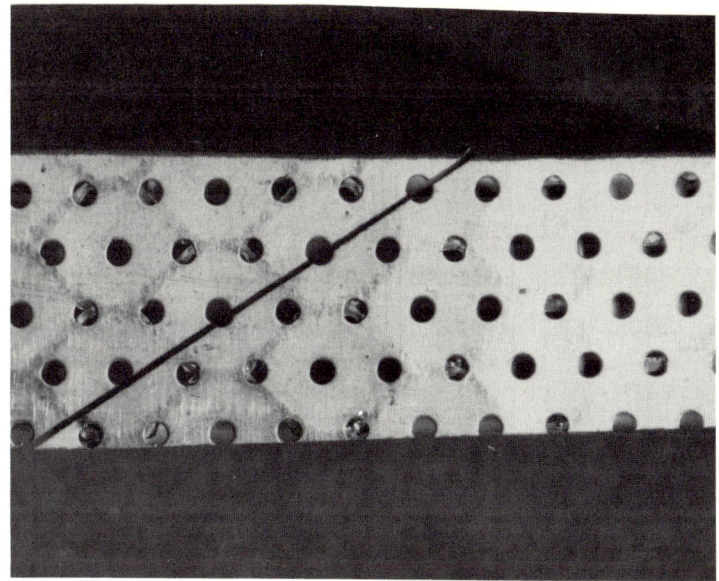

Figure 7. Photomacrograph of a section of welded inconel honeycomb showing the straightness of the cut. Note the absence of any distortion or burrs along the cut. Original magnification 2X. (Reduced 15 percent for reproduction.)

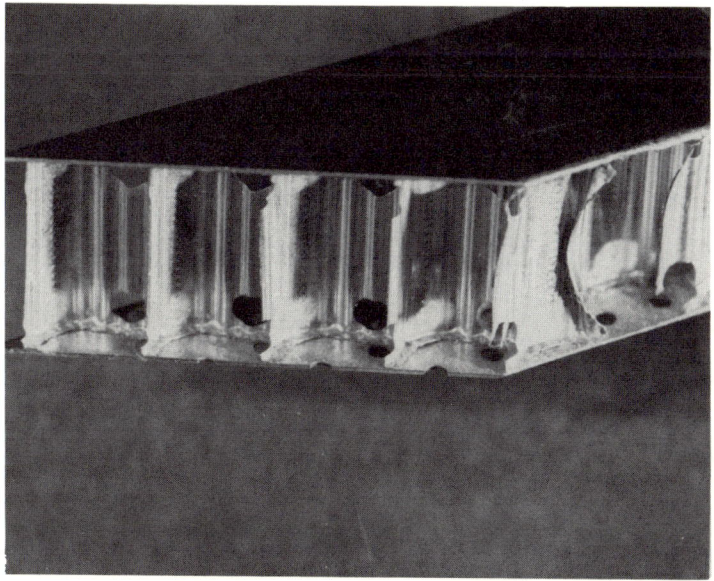

Figure 8. Side view of honeycomb shown in Figure 7. The cap strips and honeycomb are cut with equal ease. (Reduced 15 percent for reproduction.)

this is to polish one side of the slice optically flat and cement it to a flat microscope slide. The microscope slide is then waxed to a crystal-carrying wheel and the specimen is polished extremely thin and parallel to the first side. This method is practiced in sample preparation of oil well cores and other geological materials. Figure 9 shows a thin section of a rock prepared in this manner. Another method, which saves material, is to cut a slice from the specimen, polish the cut surface optically flat, cement the bulk material to the microscope slide, and then hold the specimen in the wire saw by the glass slide, cutting as close to the glass slide as possible. Cuts can be made by this method to well under 0.003-inches. Crittendon (4) has mentioned that an article being published in England claiming slices of 15 micron thickness are being cut by the British Museum using a wire saw. In this way the sample is supported during cutting by the microscope slide thus eliminating the likelihood of breakage.

SAMPLES FOR ELECTRON PROBE MICROANALYZER AND ION PROBE MASS ANALYZER

Both the electron probe microanalyzer and the ion probe mass analyzer are extremely sensitive instruments. Impurities on the surface of samples analyzed by them present quite a problem. Impurities come from various sources; some come from airborne dust and surface dirt, others come from sample preparation.

Figure 9. Thin section of rock sample.

When a specimen is being cut, the cutter causes smearing. This smearing action shows up as the displacement of chemical phases to areas of the sample where they do not normally exist. The amount of smear produced at the cut is usually a function of the method of cutting and the speed at which the specimen was cut. A wire saw, when run under proper tension and at the correct speed, causes minimum smear.

Any type of cutter used to cut a specimen from materials is subject to wear. Many of the materials to be cut are of an abrasive nature. When the cutter runs against this material the surface of the cutter is eroded away. Small particles or slivers from the cutter may become imbedded in the specimen. This contamination can usually be polished or etched off. However, the diamond impregnated wire tends to minimize this contamination.

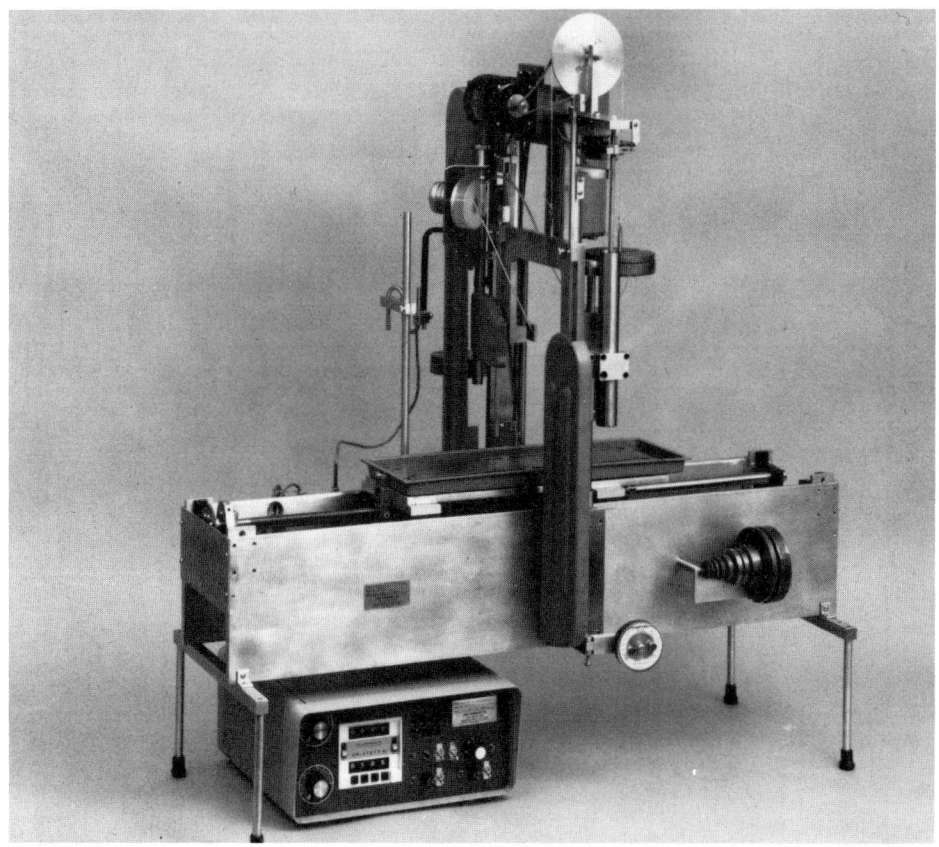

Figure 10. Automated capstan type wire saw capable of producing serial sections of a material automatically. Courtesy of Laser Technology Inc.

SAMPLES FOR X-RAY DIFFRACTION ANALYSIS

Often it is necessary to have an oriented specimen of material. The wire saw equipped with a 3-axis stage will permit easy selection of the desired cutting direction. The cutting direction is extremely critical in semiconductor industry.

THE 10 PARAMETERS REQUIRED FOR SUCCESSFULLY CUTTING WITH WIRE SAWS

1. When using slurry as an abrasive, it should be applied at a uniform rate.
2. All pulleys over which the wire passes must be in the same plane.
3. There must be no lost motion between the pulley shafts and bearings.
4. There must be no end play in pulley shafts.
5. The groove in the pulleys must be concentric with the shafts.
6. The work holder on which the specimen is placed must be held at right angles to the plane of the wire at all times.
7. The wire must be drawn up taut and kept under uniform tension throughout the cut.
8. The force between the wire and the work must not vary throughout the cut.
9. When the cut is started, it should not be stopped until it is completed.
10. The speed of the wire should be set at the beginning of the cut and remain unchanged until the cut is completed.

AUTOMATED WIRE SAWS

Recent advances in wire saw technology permits the automatic cutting of a block of material in slabs of any desired thickness with a minimum loss of material in saw kerf. Such a saw is shown in Figure 10. This saw may have some application in metallography for the production of serial samples.

REFERENCES

1. Fallier, Jr., C. N. Rev. Sci. Instruments, 32, (1961) p. 32.
2. Hunt, M. D., Spittle, J. A. and Smith, R. W., Rev. Sci. Instruments, 44 (1967) p. 230.
3. Bertone, T. J., "The Metallographic Sample Preparation of Fiber-Reinforced Composites", Metallographic Specimen Preparation, Edited by J. L. McCall & W. M. Mueller, Plenum Press, New York, (1974) pp. 251-274.
4. Crittendon, Private Communication.

ELECTRIC DISCHARGE MACHINING

Joseph Barrett

Imanco-Metals Research

Monsey, New York 10952

INTRODUCTION

Compared to other fabrication techniques, spark machining is a relatively unknown process. However, for certain applications it is the only method that can be used. Conversely, if other methods can achieve satisfactory results, then spark machining is probably not the best method. Hence, there are applications for spark machining, often referred to as EDM (Electric-Discharge Machining) that are unique to this process.

This paper is intended to describe the use of EDM in the preparation of thin specimens for TEM applications.* In most situations, the samples obtained by spark machining will be sufficiently thin and damage free to permit final thinning by electro-polishing techniques without major degradation in flatness.

Spark machining was first described by Joseph Priestly in 1768. Essentially it is a process that uses sparks in a controlled manner to remove material from a conducting workpiece in a liquid dielectric. This dielectric is usually kerosene or transformer oil, but in special cases, other liquids such as liquified gases have been used. The Russians, during World War II, perfected the process from its original purpose of making fine metallic powders into a viable machining process for various materials. The incentive was the lack of diamond powders needed for machining hard metals.

---
* Listed at the end of the paper are several general references on EDM which the reader can refer to for more specific information on the technique.

Spark machining is a fabrication method in its own right. A spark-gap is generated between the tool and the sample, and the material is removed from the sample in the form of small, microscopic craters. Depending on the relative hardness between the sample and the tool, and to some degree the polarity of the discharge and type of generator, material can be removed quite effectively and accurately. No contact between the tool and sample is required, and it is this feature which permits material to be machined with very little damage. Furthermore, material can be removed from the sample in the same shape as the tool. This means that one can sink non-circular holes through a sample or even enter at grazing incidence. If desired, one could make a tool in the same shape as a butterfly wing and machine a similar negative shape from the sample.

For commercial applications, the largest use of spark-machining is in the shaping of intricate dies from hard material such as tungsten carbide. Other uses include the machining of small holes as fine as 5 microns in foils only 0.001" thick. Other applications include the removal of broken taps, the generation of narrow slots, the ability to machine in inaccessible positions, the ability to machine fragile and soft material with very little damage, and the attainment of excellent surface finish without burrs. Because of the attachments that can be used, all normal machining operations can be performed including drilling, slicing, turning, shaping and planing.

The initial preparation of metallographic samples for optical and transmission electron microscopy can be performed on EDM machines. Resulting samples can have a surface finish of 5 microinches C.L.A., exhibit excellent edge definition and be less than 0.005" thick. These samples can be in either single or poly-crystalline form, and can be machined virtually without strain.

A number of commercially available spark machines are on the market today. Most are for engineering applications where low damage and minimal strain is of secondary importance. A few, such as the "Servomet", are designed specifically for laboratory applications where low damage, precision machining of critical metal samples is required. These samples could cover the range in hardness from indium single crystals to tungsten carbide rods, and the end product could be in almost any shape dictated by the tool that is used.

Most spark machines consist of a worktable, tool holder with servo system, bath for dielectric fluid and an electronic package which controls the spark. Coordinate slides for translation of the tool holder in both X- and Y- directions is also required. The slides should be rigid enough to prevent flexing, yet light enough to permit rapid up and down response to the servomechanism. Three types of spark generators are available, but the oldest and most useful for laboratory applications is the relaxation circuit.

This derives its energy from capacitor discharge through a resistor. A spark is produced by rectifying the A.C. mains and applying the resulting D.C. to a relaxation circuit. When the capacitor reaches the dielectric breakdown voltage, a spark is generated across the work gap. The condenser is then recharged and the cycle repeated. Since the breakdown voltage is porportional to the work gap, the gap must be controlled within very close tolerances for efficient operation. The voltage is used to control a servo-system which maintains the work gap at its optimum level. A range of 100-200 volts is a typical working voltage on a laboratory instrument with spark gaps between 0.0005" to 0.015". On laboratory instruments, erosion rates for maximum cutting energy will range between 0.7 cc/hour on tungsten carbide to 21 cc/hour on lead.

The material produced by the disintegration of the tool and workpiece, and by the decomposition of the dielectric is called swarf. The usual dielectric is kerosene. Sparking is done while the sample and tool are totally immersed in the dielectric. The dielectric must be kept clean to achieve the full accuracy limitation of the instrument, and this is routinely accomplished by using a pump and filter attachment. For extremely fine surface finishes, extra spark ranges can be selected, with a corresponding decrease in erosion rates.

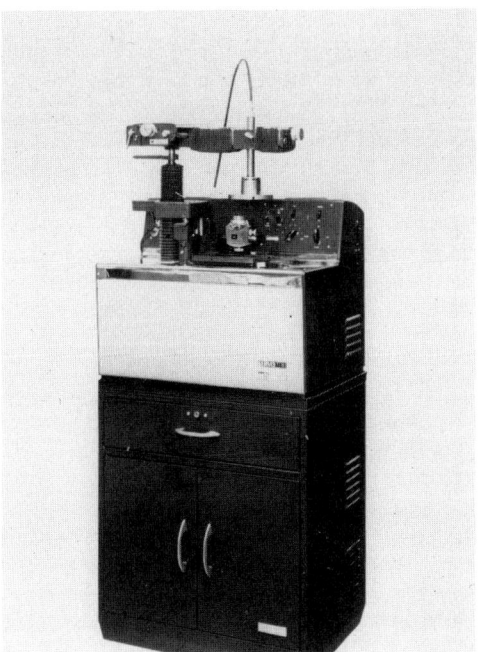

Figure 1. Servomet EDM machine showing cross-slides, spark-planar and goniometer. (Set-up position).

For extremely accurate work, vibrations must be avoided. Some are eliminated by careful design of the instrument, but the tool must also be selected to prevent "fluttering" during the cutting stage, and the sample must be rigidly mounted. Almost any conductor may be used as the tool, but brass is the most common. The tool will also erode during the spark machining operation and allowances for this must be made.

As a general rule, the resistivity of the sample must not exceed 100 ohm-cm. Marginal semi-conductor materials can sometimes be lowered to this range by using hot oil. For working with radioactive samples, the tank and servo-mechanism can be placed in a hot-cell and controlled remotely from the console.

With the possible exception of chemical etching, no viable process exists today for machining metallic samples with near zero damage. Spark machining will generate some damage to the sample, and this damage can extend to several millimeters or more in depth if precautions are not taken. Two criteria exist for assessing depth of damage. One is the "Depth of Detectable" damage. This is the depth at which the structure is altered as measured by the most sensitive process available. The "Depth of Significant Damage" is the depth to which damage can be tolerated for the application intended.

Four zones can be defined in the spark-affected surface layer. The most strongly affected layer is the "melted zone." This zone can extend from fractions of a micron on laboratory instruments that use relaxation circuitry to hundreds of microns on the long pulse, high-power rotary generator machines used for heavy machining applications. In the EDM process, sparks melt a shallow crater of metal in the melted zone. Most of this is ejected at the end of the spark. Some residual liquid material remains and freezes epitaxially onto the solid below. This leaves the melted layer in tension and the layer beneath in compression. Deep melted layers can cause cracking, and X-ray diffraction photographs are generally not available from this layer.

The second layer is the "chemically-affected zone". In this layer, the chemical composition has changed. This may be due to reaction with the dielectric and the tool, diffusion of impurities, and reaction with the dielectric. The diffusion zone is generally very small due to the time involved.

The third layer is the "microstrained zone" which is subjected to large compressive forces, both during the heating cycle and later during the shrinkage of the rapidly frozen molten layer. This zone can be detected by optical microscopy and is characterized by the presence of twins, slip, phase changes and sometimes microcracks.

ELECTRIC DISCHARGE MACHINING   73

The fourth layer is the "micro-strained zone". Damage in the layer can only be detected by counting dislocations. Slip, twinning or cracking does not occur.

In order to illustrate the flexibility of the spark machining process, it is assumed that oriented single crystal discs of aluminum are to be cut from a 1-inch diameter by 3-inch long randomly oriented single crystal rod. The discs will be planed to a thickness of approximately 0.005 inches and then machined to the required diameter. A further requirement is that the flat surfaces must be parallel to within a few microns, have a surface finish of 5 microinches C.L.A. and exhibit very little detectable damage.

The crystal is first mounted on a two-circle goniometer designed specifically for the spark machine. The goniometer table is insulated at the factory from the rest of the goniometer to prevent arcing of its internal parts. The crystal is attached to the table using a low melting point wax, and the orientation is determined on an X-ray machine using back-reflection Laue techniques. Corrections are made on the goniometer while on the X-ray machine to achieve the proper orientation, and the assembly is then transferred to the goniometer track on the spark machine. Since the wax is non-conducting, a conducting wire must be connected to the crystal. Later, conducting epoxys are used for planing, but wax is used initially since it is quickly removed, yet strong enough for holding larger specimens.

Figure 2. Spark planar with aluminum crystal mounted on goniometer.

A spark planar is now used to expose the desired crystallographic plane. Essentially, the planar is a rotating disc, usually made of brass, that erodes a flat surface on the sample. Rotation of the planar is required to minimize grooving and ensure that the ejected residue is swept clear. The spark machining is done at least 1 inch below the surface of the dielectric. For fast erosion rates, higher spark-energy ranges are used. As the area of the planed surface approaches the desired size, finer spark ranges are used. Precision laboratory machines should have six or more spark ranges, with the finest range giving the best surface finish and lowest damage. Once a surface finish appropriate to the selected spark range is attained, usually only a few minutes is required on each subsequent range to remove the characteristic damage from the previous range. When completed, the finished surface should have a flatness of 5 micro-inches C.L.A. This surface will now become the reference plane.

The rod is now demounted using a suitable solvent, and remounted using a conducting adhesive. The adhesive is usually Duco cement mixed with fine graphite powder. The specimen is repositioned on the goniometer such that a cut parallel to the reference face can be made using a wire slicer. The slicer consists of a sliding wire which generates a spark between the wire and the crystal. It is similar to an abrasive wire saw, but no contact between the sample and wire is made. The resulting wafer should be sufficiently thick, preferably 0.040 inches minimum, so that the sample does not curl during slicing and also to permit additional stock removal during the thinning operation.

The wafer now has one flat spark-planed surface and one wire cut surface and is approximately 0.040 inches thick. The wafer must now be reduced in thickness by planing to 0.005 inches while achieving the required flatness and parallelism.

A quick-setting table is now used, and to be certain that the mounting table is parallel to the planing wheel, the mounting table surface is planed. Without disturbing the table, the spark-planed face is mounted to the table with conducting epoxy resin. A slight even pressure is applied to the sample during the curing stage to ensure an even layer of adhesive. After curing, the micrometer depth-stop is set on the spark machine to allow approximately 0.035 inches of stock removal. The first 0.010 inches of stock removal can be performed using an intermediate spark range, and successively finer spark ranges are used until the final thickness is achieved. When switching to each finer range, the planar wheel should be translated to a fresh, unsparked area of the wheel to maintain flatness. Total planing time is approximately 2 hours.

Figure 3. Spark planar and quick-setting table.

Without disturbing the sample, the planing wheel is now removed, and replaced with a trepanning tool. The trepanning tool works in a similar manner to a cookie cutter, and is basically a brass tube such that the I.D. is approximately 0.002 inches greater than the diameter of the required discs.

Trepanning is carried out on fine spark ranges, and the tool is permitted to slowly traverse completely through both the sample and conducting epoxy and slightly into the mounting table. Upon completion of the cycle, the trepanning tool is raised and repositioned by means of the cross slides for the next disc. When completed, the discs can be removed from the table by soaking in dimethyl formamide. The resulting discs should have a surface finish of 5 micro-inches C.L.A.; their faces should be parallel to within a few microns, and the depth of significant damage should not extend more than 25 microns from either side. This layer can later be removed during the final electro-polishing process.

REFERENCES

1. M. Cole and I.A. Bucklow, Metals Research, Internal Report No. S11.
2. H. K. Lloyd and R. H. Warren, J. Iron & Steel Inst., (March 1965), pp. 238-247.
3. L. E. Samuels, J. Inst. Metals, $\underline{91}$ (1962-63), pp. 14-18.
4. P. Beardmore and D. Hull, J. Inst. Metals, $\underline{94}$ (1966), pp. 14-18.
5. M. Cole, I.A. Buckow and C.W.B. Grigson, Brit. J. Appl. Phys., $\underline{12}$ (1961), pp. 296-297.
6. G. Tibbets and F.M. Propst, Rev. Sci. Inst., $\underline{34}$ (1963), p. 1268.
7. I.A. Bucklow and L. E. Drain, F. Sci. Instruments, $\underline{41}$ (1964), p. 614.
8. G. V. Smith, Proceedings of Conference on Electrical Methods of Machining and Forming, London (Inst. Elect. Eng.) (1967), p. 48.
9. Servomet Handbook, 1966: Melbourn, Royston, Herts (Metals Research Ltd.).
10. M. Cole, I.A. Buckow, and C.W.B. Grigson, Brit. J. Appl. Physics, $\underline{12}$ (1961), p. 296.
11. Briers, Dawe, Dewey and Brammar, "A Technique for the Rapid Preparation of Thin Foils for Electron Microscopy from Bulk Materials", Aeon Laboratories, Egham, Surrey.
12. C. Luke and K. E. Puttick, Brit. J. Appl. Phys. , $\underline{12}$ (1971) pp. 967-973.
13. I.A. Buckow and L.E. Drain, J. Sci. Inst., $\underline{41}$ (1964) pp. 614-617.

DIAMOND COMPOUNDS FOR POLISHING METALLOGRAPHIC SPECIMENS

R. D. Buchheit and J. L. McCall

Battelle-Columbus Laboratories

Columbus, Ohio 43201

INTRODUCTION

Diamond compounds are perhaps the most common abrasive material in use today for intermediate and final polishing of metallographic specimens. And, even though a significant quantity of the micron-size diamonds that are produced each year are consumed in the preparation of metallographic samples, only very limited technical information has been made available to the metallographer to enable him to determine what characteristics a diamond compound should have for his particular application. Very few scientific studies of diamond polishing in metallography have been reported. Studies of the influence of polishing speed and polishing time (1) and a study comparing the polishing behavior of "normal" diamond to a compound containing diamonds made from a powder produced by an explosive compacting method (2) have been reported. Some work regarding the use of diamond compounds for lapping (3,4) has been reported also, but little else can be found in the literature. Unfortunately, it has not been possible, or economical, for even the largest metallographic laboratories, who use only a few-thousand-dollars worth of diamond abrasives each year, to perform any significant scientific evaluations of the available diamond compounds to find out which compound is best suited for his purpose. Therefore, a metallographer must choose the diamond material he is going to use on the basis of some intuitive belief, perhaps based on many years of experience, that one material is best for him. Whether it is the optimum material perhaps is not known, since feedback from metallographer to diamond supplier is practically nonexistent. It is our belief that many of the factors that influence a metallographer to purchase a particular type of diamond compound are based

initially on the color of the carrier, the type of dispenser, the
price per gram, the literature supplied by the manufacturer, and,
perhaps, the sales technique used by the diamond seller. Then the
metallographer determines through usage of his selection how well
the diamond compound performs for his application.

In an effort to provide some technical information to begin to
establish a more sound basis on which metallographers could choose
the correct diamond compound for their use, and, perhaps, to obtain
some information which would communicate to the diamond suppliers
what types of compounds would be best for metallographic purposes,
controlled laboratory experiments were made of several different
types of diamond compounds. These experiments were designed so
that one or more of the variables which are known to influence a
diamond compound's performance were altered and evaluated under
controlled conditions. The results of this work were not intended
to provide a complete basis, but it is hoped that some of those
characteristics which are important to diamond polishing have been
identified, and hopefully increased communication between metallographer and diamond compound manufacturer will result. If this
were to occur, both should benefit. In fact, the primary purpose
in discussing the experimental program we have been conducting is
to describe an approach that we believe can be beneficial for the
evaluation of metallographic diamond compounds. The results presented in this paper are intended to be representative rather than
provide a basis upon which conclusions can be made. This will be
done elsewhere (5).

Before the experimental program and results are described, it
might be useful to discuss in general terms what diamond compounds
are, and what characteristics of them might influence their performance as metallographic polishing agents.

Uzanas (6) has described a diamond compound as "a mixture of
diamond powder in an organic base medium or vehicle which serves
as a means of carrying the diamond to the workpiece". Generally,
there are three types of diamonds which are used in metallographic
compounds; natural, manufactured, and reclaimed. Natural are
previously unused diamonds which have been crushed. Manufactured
comprise several types which are made by one of several recently
developed synthesis methods. Reclaimed are diamonds recovered from
wastes formed during previous use.

In addition to the type of diamond, various characteristics of
the diamonds which affect their performance include their size, size
distribution, and their shape. These characteristics are defined
by a grading system developed by manufacturers, distributors, and
users in cooperation with the Office of Commodity Standards of the
National Bureau of Standards (7). The particle size, which is

defined as equal to the diameter of a circle having the same area as the profile of the particle, generally is determined for sub-sieve size diamonds by some type of Stokesian separation. Shapes generally are described in regard to how sharp the corners of the particles are. Some particles are defined as slivers or shales (8). A sliver is a particle having a needle or rodlike shape whose major-to-minor axis ratio is greater than 3:1. A shale is a particle having a thin plate-like shape.

The type of carrier plays an important role in the performance of a diamond compound; however, the particular characteristics required of a carrier are not well known. Historically, diamond was mixed with olive oil (9). Now the carrier can vary from common vegetable oils to very sophisticated formulations and may include thickening agents, emulsifying and dispersing agents, antioxidants, corrosion inhibitors, and colorants. Generally, the color of the compound is a code to the particle size which makes identification simple and minimizes the possibility of using the wrong size abrasive.

The amount of the diamond in the carrier varies greatly from manufacturer to manufacturer. Generally, it is believed that the more diamond there is in a compound the more one is getting for his money. However, this is not the case, and, as will be shown, there can be either too little or too much diamond for optimum performance.

It is understood, however, that the compound is only part of the polishing operation in metallography. Variables which can influence the polishing characteristics of a particular setup include the polishing machine, the type of cloth used, how the compound is dispersed on the cloth, how much of the compound is used, the type and amount of extender used, the polishing wheel speed, the sample holder (whether mechanical or manual), the way the sample is moved on the polishing wheel, the pressure of the sample on the wheel, and perhaps many more too numerous to mention. Finally, the type of material being polished must be an important variable to the use of diamond abrasives. The optimum polishing method for one material very likely may not be optimum for another, and, indeed, may not even include diamond as the polishing abrasive.

EXPERIMENTAL POLISHING PROGRAM

Obviously, a single experiment cannot evaluate all of the above-mentioned variables, even in a cursory way. Nevertheless, the studies that have been performed were designed in such a way that hopefully all but one of the possible variables were controlled at a time.

In this program, 39 different diamond compounds were selected for examination. The diamonds ranged in size from 1/4 to 15 microns. The concentrations of the diamonds ranged from light to heavy. Some of the compounds were commercial materials obtained from several different suppliers and some were compounds especially made for the program. Included were compounds containing manufactured diamonds, some with natural diamonds, and others with reclaimed diamonds. The commercially obtained compounds included some reported to be soluble in water, some soluble in oil, and some soluble in both.

Although most materials were obtained directly from manufacturers along with some information regarding their characteristics, none of this information was accepted and all the compounds were characterized in the program.

Carrier Solubility

A rather qualitative test was used to evaluate the relative solubility of each of the compounds in water and in kerosene. A commercial grade of filtered kerosene and laboratory-distilled water were used for the solubility tests. Small 10-mg amounts of each compound were placed in 10 drops of either water or kerosene in spot-plates. The compounds were stirred using a microspatula and, after allowing the solutions to settle for 60 seconds, a photograph of the spot-plate was taken. Figure 1 shows representative results from six of the compounds. The spots labeled with numbers followed by W were dispersed in water. The same compound dispersed in kerosene is labeled with the same number followed by the letter K. All of the compounds shown were soluble in water, but as can be seen, their solubility in kerosene varied considerably. The differences in solubility were recorded by assigning a rating of very soluble, soluble, insoluble, or very insoluble to each compound for both the water and the kerosene. All 39 were found to be either very soluble or soluble in water, whereas only four of the 39 were soluble, and 11 partially soluble, in kerosene.

Diamond Concentration

The concentration of diamonds in the compounds was determined by separating and weighing the diamonds from a measured weight of compound. An initial weight of about 1 gram of compound was placed in a test tube. The diamonds were separated from the carrier by dissolving the carrier in distilled water and washing the residue with distilled water and alcohol with intermediate centrifuging and decanting. The centrifuging was done at 16,000 rpm for 10 minutes which should have removed all diamonds greater than 0.1 micron. Following the final washing, the remaining material was

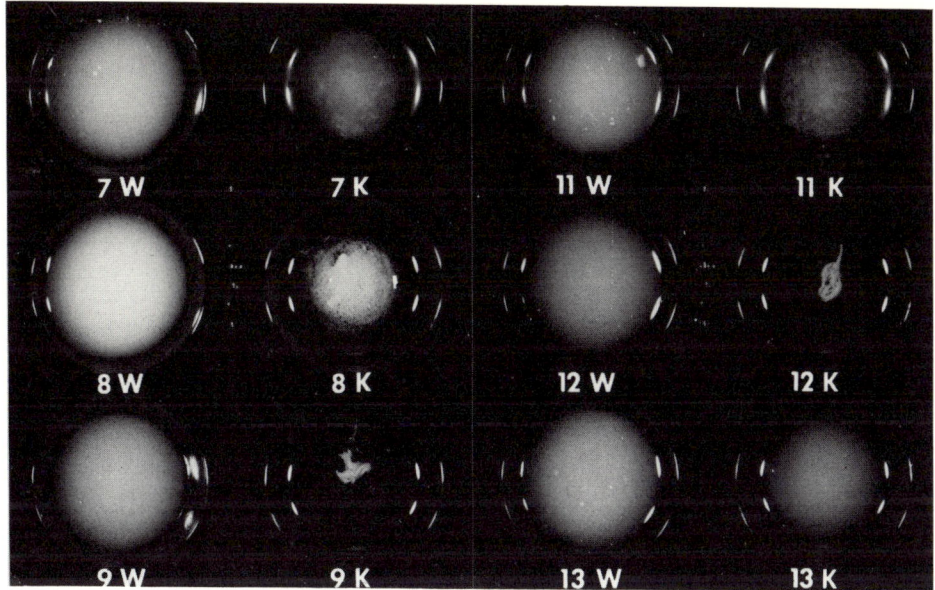

Figure 1. Appearance of representative diamond compound solubility tests.

dried at 180 F for 1 hour. Weighing of the dried samples showed that the concentration of diamonds in them ranged from less than 5 weight percent to greater than 24 weight percent. X-ray diffraction analysis of the dried samples verified them to be diamond. For those compounds where the concentrations were reported by the manufacturers, the concentrations measured in this way agreed well with the values reported.

Diamond Particle Size and Size Distribution

The size and size distribution of the diamonds in each compound were determined by making microscopic measurements on the diamonds separated from the carriers. For the 6-15 micron diamonds, the measurements were made on scanning electron microscope photographs by methods first reported by Schneider (8). The smaller diamonds were measured on transmission electron micrographs. For this program, the maximum projected length of each particle was measured and recorded rather than the equivalent diameter normally used. This provided higher values than the NBS method (7); the values obtained were believed to have more significance with regard to a material's polishing performance. In many of the compounds a few large particles were found which, in most instances, were slivers with one long dimension. Figure 2 contains a representative distribution curve for (a) one with a rather closely sized compound and (b) one with a rather wide size range. Both compounds were commercial materials reported to be 6-micron. The average particle size was mathematically computed from the data.

Diamond Particle Shape

A belief that has been held by many, including the authors, is that sharp angular diamonds make better polishing abrasives for metallography than do rounded ones. (This belief is expressed in several of the manufacturers' literature also.) In fact, it has been suggested that diamond grading should include a shape criterion, perhaps this being even more significant than size (9). The results of our studies have shown that this belief is not entirely correct and that shape is only one of many characteristics that affect a diamond's performance. For the shape determinations, scanning electron micrograph stereo-pairs were taken of the separated diamonds and examined. Qualitative descriptions of the shapes were assigned including blocky, sharp-angular, rounded, and those containing slivers or shales. Figure 3 contains scanning electron micrographs of (a) representative sharp-angular particles and (b) rounded particles.

Polishing Performance Experiments

Several different experiments could undoubtedly be designed to evaluate and compare diamond compounds. The experiment established for the present evaluation was a polishing experiment designed to hopefully permit the change of only one variable at a time. For this, all the factors which possibly could vary had to be fixed. This meant that the evaluations were limited in that they imposed very restricted conditions under which the performance of all of the diamond compounds were compared. Other sets of conditions likely may change the response of a particular compound and, for this reason, additional studies are encouraged.

To control the polishing procedure and eliminate the operator's bias, an automatic polishing device was used (a). This permitted a constant polishing wheel speed (set at 103 rpm) and a constant specimen motion relative to the wheel (23 rpm in a 4-inch circle counter to the wheel). A total weight on the specimen surface during all of the experiments was held constant at 1985 grams. As the specimen lost weight during grinding and polishing, additional weight was added to the back of the specimen to maintain the total weight constant. The test specimen being polished was a National

---

(a) Whirlamet Automatic Polishing Device, Buehler Ltd., Evanston, Ill.

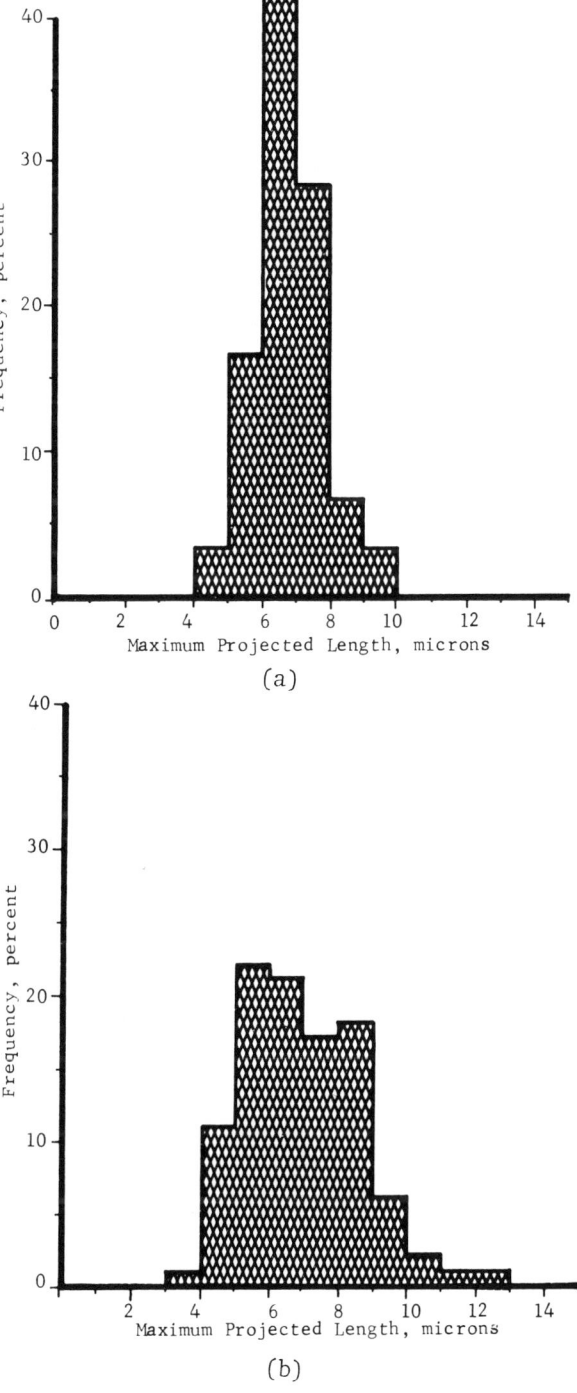

Figure 2. Diamond particle distribution curves for representative 6-micron diamond compounds.

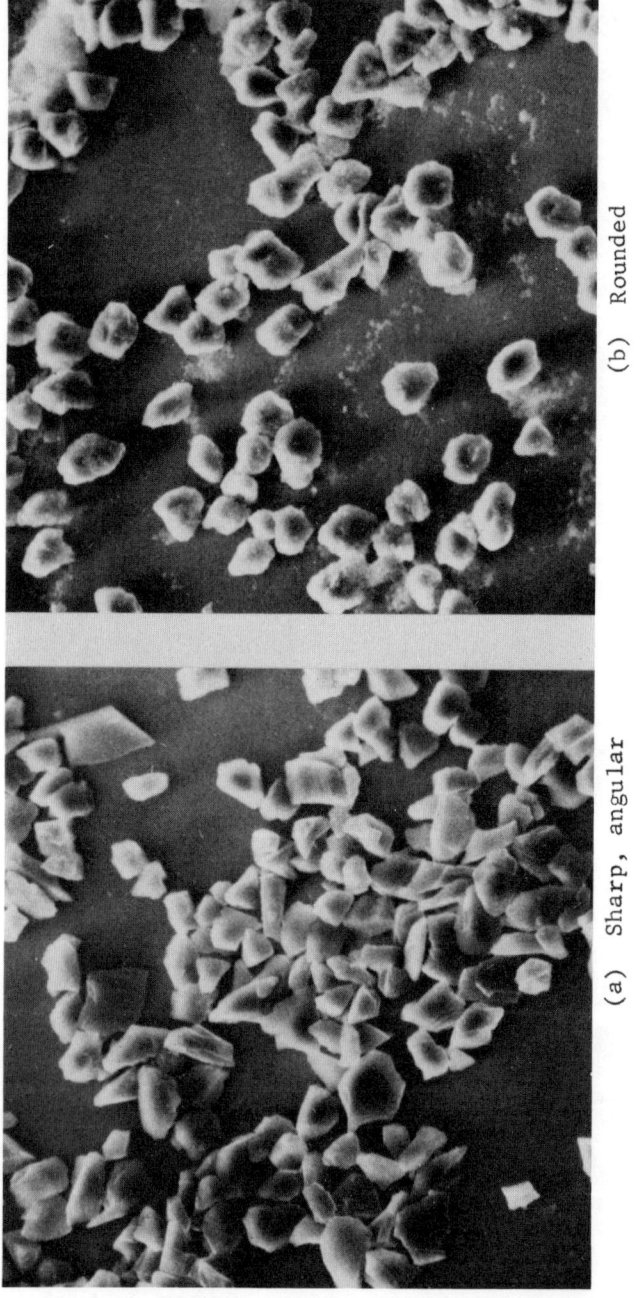

(a) Sharp, angular (b) Rounded

Figure 3. SEM micrographs of representative diamond particles.

Bureau of Standards Rockwell "C" 40 steel test block which was machined to 1-inch diameter. The cloth was a commercial nylon cloth (a). The cloths were charged with controlled amounts of diamond which had been placed in 5-gram syringes. The amounts charged were determined by weighing the syringe before and after extruding the diamond. Approximately 0.19 gram of compound was used on each cloth. The compound was applied in several radial swabs in the area of the cloth being contacted by the specimen. Twenty-five drops of a universal extender were applied at the beginning of each polishing experiment, after 5 minutes of polishing and after 30 minutes of polishing. No additional extender was added during the remainder of an experiment.

The initial condition of the NBS specimen surface was a 320-grit silicon carbide grind. To provide a uniform condition, all specimens were hand-ground on a 320-grit abrasive paper for 10 strokes using "medium" pressure. All strokes were made in the same direction and on unused portions of the abrasive paper. To assure that identical starting conditions were achieved, the surfaces were evaluated on a Taly-Surf Model 4 to determine the centerline average, CLA, roughness in microinches. The Taly-Surf stylus traverse was made across the diameter of the specimen in a direction perpendicular to the 320 grinding marks. Surface roughness resulting from each 320-grit grinding operation was controlled to about 17 ± 3.0 CLA.

After the evaluation of the initial surface, the specimen was polished in the automatic polishing device for a period of 5 minutes. It was then removed, the surface cleaned in alcohol and dried, and the surface roughness was determined on the Taly-Surf. A qualitative evaluation of the surface was also made and photomacrographs of the surface taken. The specimen was then returned to the polishing device and polished for an additional 10 minutes. It was then removed and reground on the 320-grit silicon carbide paper. After measuring the surface roughness from the 320-grit grind, the specimen was returned to the polishing device for 5 additional minutes. At this point the surface roughness was again measured with the Taly-Surf and its appearance noted and recorded. This procedure of alternate 5- and 10-minute polishing steps and Taly-Surf roughness measurements was repeated for a total of 185 minutes which gave a total of 14 evaluation points for each polishing experiment (i.e., at 0 min, 5 min, 20 min, and then at every 15-min interval up to 185 min).

Figure 4 contains a series of Taly-Surf recordings from one of the polishing experiments. The degradation of the diamond's polishing ability is evident from these recordings.

(a) AB Nylon Polishing Cloth, Buehler Ltd., Evanston, Illinois.

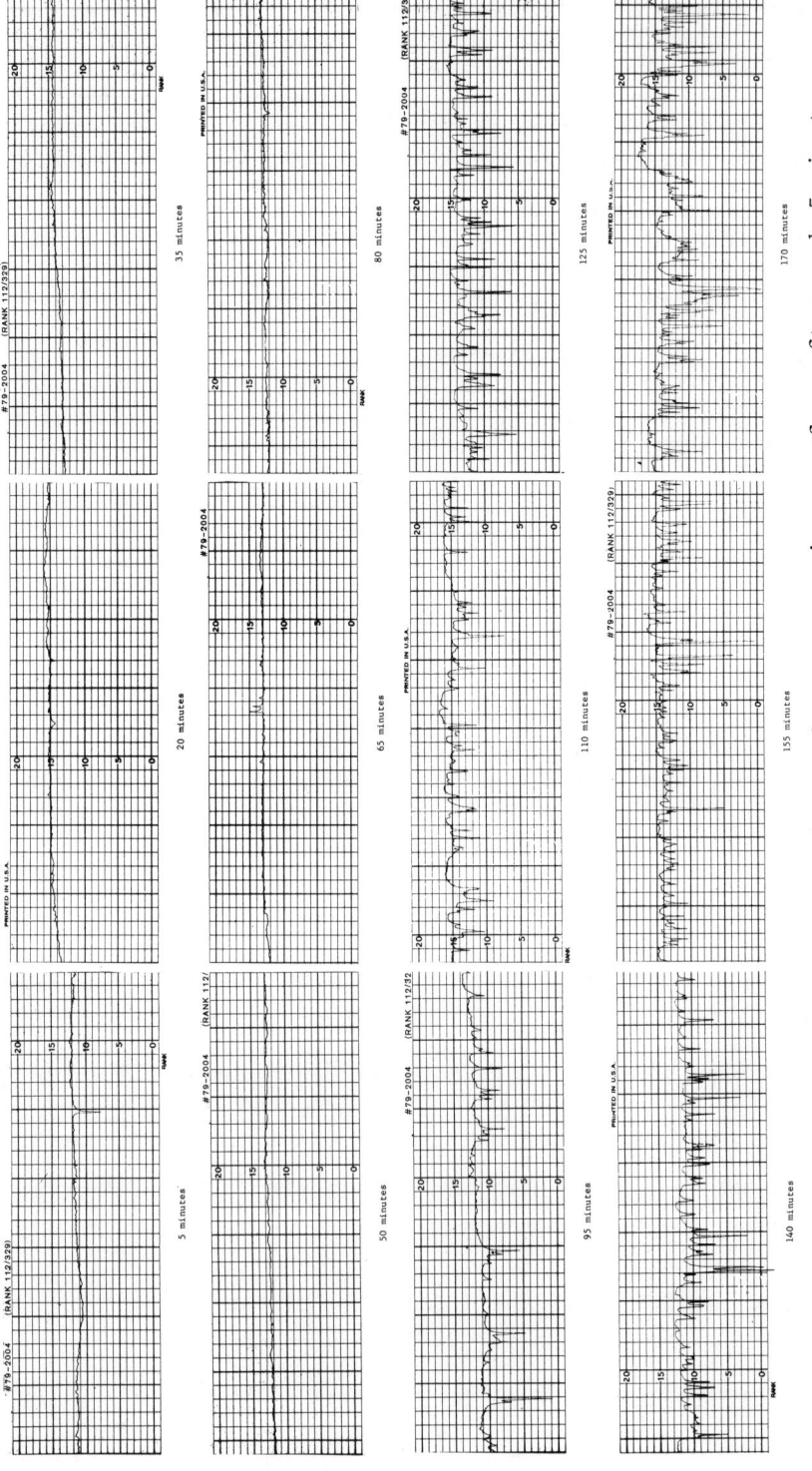

Figure 4. Representative Taly-Surf profiles of a test specimen surface after each 5-minute polishing interval (vertical magnification = 20,000X, horizontal magnification = 100X).

Figure 5 contains representative curves of surface roughness expressed as CLA in microinches versus polishing time for several of the 6-micron diamond compounds. Although it is not our intent here to report comparisons of one characteristic of a diamond to another, the data in Table 1 describe some of the measured characteristics of the compounds shown in Figure 5. As can be seen, some of the compounds were still polishing after 185 minutes of use practically as well as they were at the beginning of the experiment, e.g., 2-MN and 4-SN, whereas the performance of others began to degrade rapidly soon after the experiment was initiated, e.g., 19-RM and 15-MN. One material, 6-RN, failed to do much polishing even during the first 5-minute period. This was a compound containing rounded diamonds, but another compound with similar-shaped diamonds, 14-RM, polished quite satisfactorily.

In addition to the Taly-Surf evaluations of the surface of the sample, the appearance of the surface was used as a method of evaluating the performance of the diamond compound. This was done by noting the time at which the 320-grit grinding scratches were no longer removed during the first 5 minutes of each 15-minute polishing interval. A series of photomicrographs of the appearance of the sample from a representative polishing experiment is shown in Figure 6. This series illustrates that, after an accumulated total of 90 minutes diamond usage, the 320-grit grinding scratches were not removed during the first 5-minutes of each 15-minute polishing interval thereafter.

Effect of Diamond Concentration in Compounds on Their Polishing Performance

One specific characteristic of diamonds which was investigated in the program was the effect of variations of the amount of diamond particles in the compounds on their polishing performance. Six compounds were especially made for this with everything being identical except the concentration of diamond. Figure 7 is a plot of the area under polishing curves (of the type shown in Figure 5) versus diamond concentration. Arbitrary units are used on this plot because of the proprietary nature of the data.

Examination of Diamonds After Use

Upon completion of the polishing experiments, the diamonds from a portion of several of the cloths were collected and reexamined by scanning electron microscopy to see what changes had occurred in them during use. This was done by cutting several approximately 1-inch square swatches of the cloth from the polishing area and suspending these in a small beaker of benzene. The swatches were ultrasonically agitated for approximately 10 minutes to free the

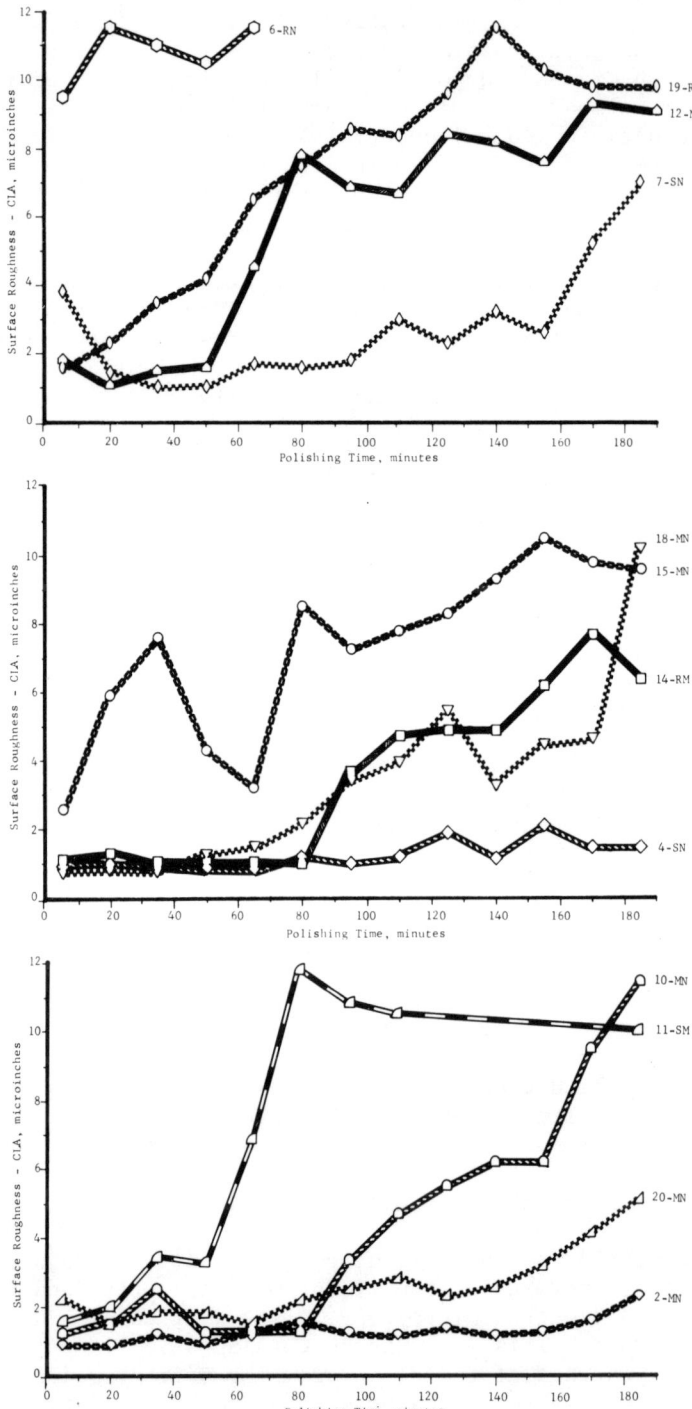

Figure 5. Surface roughness versus polishing time for several 6-micron diamond compounds.

TABLE 1. CHARACTERISTICS OF DIAMOND COMPOUNDS USED IN POLISHING EXPERIMENTS

Identification	Diamond Type	Diamond Concentration in Compound, percent	Average Particle Size, microns	Size Range, microns	Relative Solubility Distilled Water	Relative Solubility Kerosene	Shape Description
2-MN	Natural	24.0	6.7	3.2-12.7	Yes	No	Medium-sharp edges
4-SN	Natural	8.5	6.7	2.6-12.9	Yes	Yes	Sharp edges
6-RN	Natural	4.9	7.2	2.7-17.5	Yes	No	Rounded, some shales
7-SN	Natural	11.8	7.3	3.2-14.2	Yes	Yes	Sharp edges, some shales and slivers
10-MN	Natural	18.0	7.6	4.2-11.4	Yes	Partial	Medium-sharp edges
11-SM	Manufactured	9.9	7.2	3.2-12.4	Yes	Partial	Sharp edges, some shales and slivers
12-MN	Natural	24.2	7.1	2.3-11.9	Yes	No	Medium-sharp edges, some slivers
14-RM	Manufactured	18.1	6.3	4.4-10.0	Yes	Partial	Very rounded
15-MN	Natural	13.2	6.6	4.0-12.1	Yes	No	Medium-sharp edges
18-MN	Natural	18.7	6.7	4.5-9.6	Yes	No	Medium-sharp edges to rounded
19-RM	Manufactured	23.9	6.8	3.7-11.4	Yes	Partial	Very rounded
20-MN	Natural	12.2	6.8	3.8-12.2	Yes	No	Medium-sharp edges, some shales and slivers

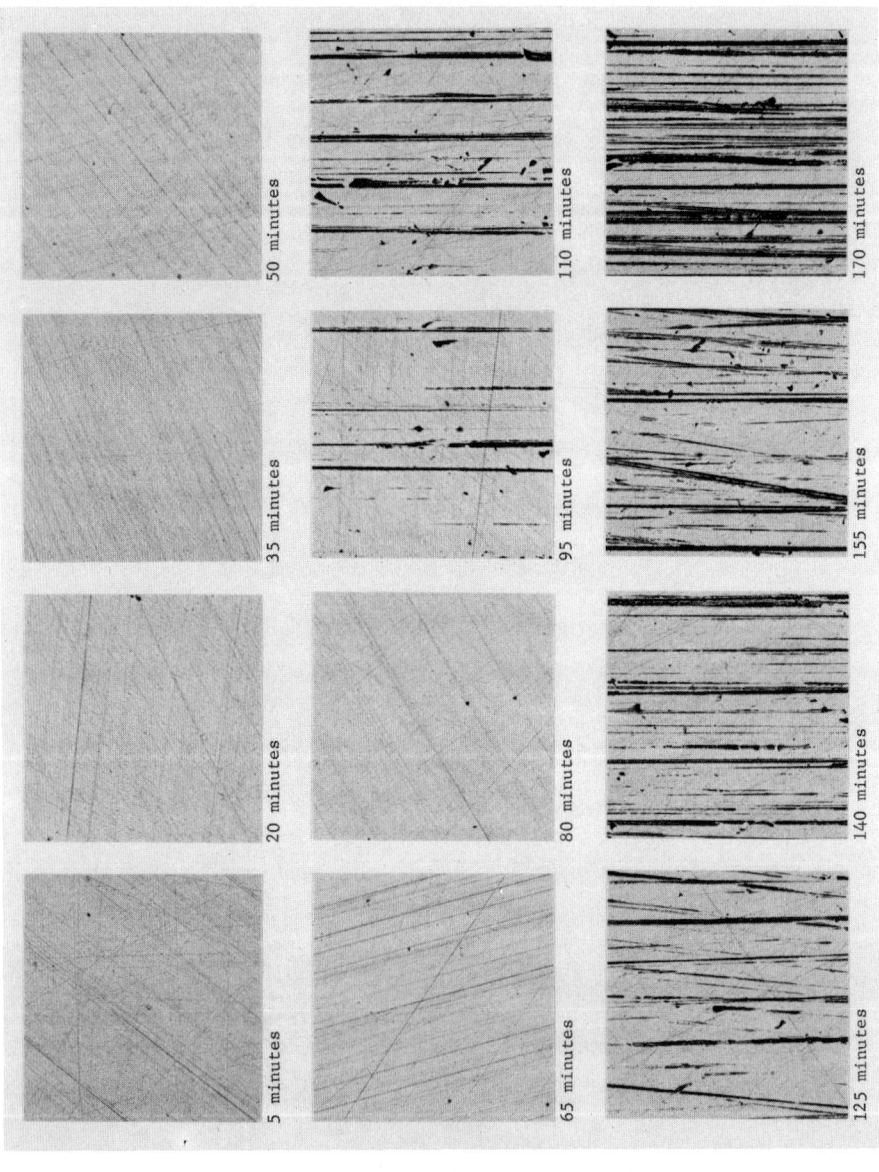

Figure 6. Representative test specimen surface appearance after each 5-minute polishing interval - 100X. (Reduced 35 percent for reproduction.)

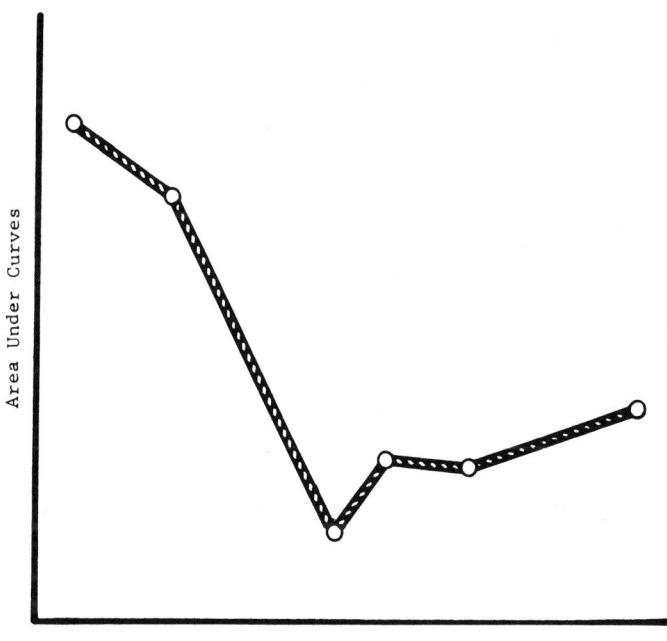

Figure 7. Effect of diamond concentration in compound on polishing performance.

diamonds from the cloths. The solid residue was separated from the benzene by centrifuging. The residue collected was found to be highly magnetic, indicating it contained a considerable amount of steel removed from the polished test specimen. To remove the steel, the residue was treated in hydrochloric acid. Following centrifuging and washing, the residue was examined and found to contain numerous fibers from the nylon polishing cloth. These were removed by treating the residue in sulfuric acid. Finally, after several washings in alcohol and centrifuging, followed by drying, a residue of chiefly diamonds was provided which could be examined by scanning electron microscopy.

Figure 8 contains examples of diamond particles from three of the identical compounds before use and after 185 minutes of use. The appearance, including size and shape, is, for all practical purposes, identical. The reason for the degradation of the polishing characteristics without a change in the diamonds is not fully understood, and additional research is planned to investigate this. Our present belief is that changes in the carrier may provide the clue.

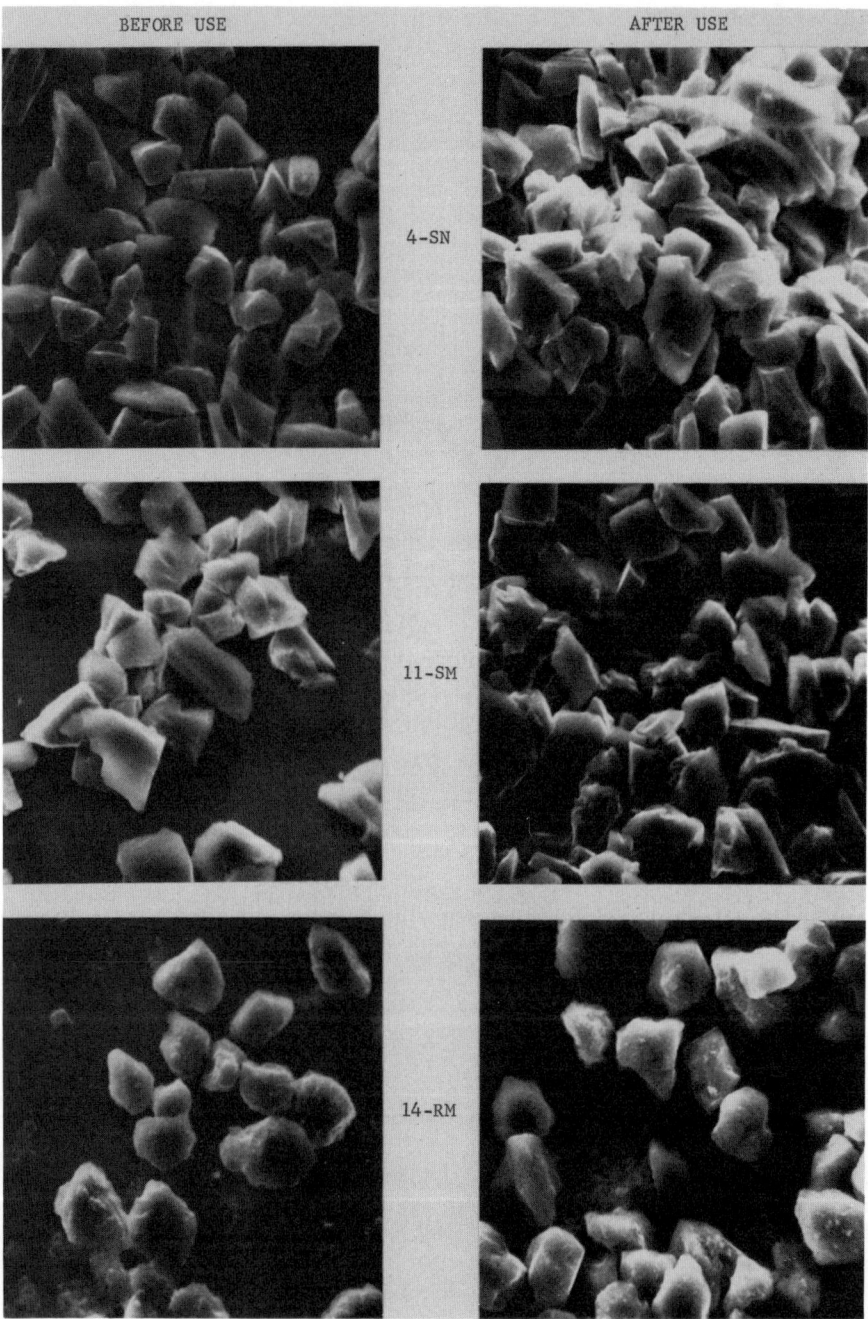

Figure 8. Appearance of diamond particles before and after use in polishing experiments.

SUMMARY

It has been shown that many characteristics of a diamond compound influence its performance during the polishing of metallographic samples. The best material for metallographic use will be achieved only after the influence of all these many variables are understood and optimized.

We would like to conclude this paper by describing an experience of our laboratory. A few years ago we were enticed into trying a new diamond compound for our use. Our metallographers evaluated material only on the basis of the quality of surface that could be achieved and reported that it behaved as well as what we had been using. Since the new material had a price almost 30 percent lower than what we had been using previously we quickly made the change. However, after 12 months of using the new material, we found our diamond consumption had almost doubled, when the number of specimens prepared remained essentially unchanged. Although we had none of this material to evaluate by the method described in this paper, we feel confident that the material would follow curves similar to those that give a good polish after a short period of use but then degrade rapidly with continued use.

ACKNOWLEDGMENTS

The experimental and diamond characterization work reported in this paper was performed by John Poole and George Wheeler. The authors are indebted to them for their exceptionally fine work and for the extra effort exerted by them during its performance. Also to be acknowledged is the early work on this program by Russel Staub who played a large part in designing the polishing experiment. Finally, this work would not have been possible without the financial support of Buehler Ltd. and to them we will ever be thankful.

REFERENCES

1. G. Ondracek, B. Leder, C. Politis, "Quantitative Metallography of Metal-Ceramic Composites", Praktische Metallographie, 5 (1968) p. 71-84.
2. K. Spiler, "The Influence of Different Diamond Polishing Pastes on the Polished Surface Quality of Cermets", Praktische Metallographie
3. R. E. Reid and S. L. Kanaba, "Performance, Characteristics of Natural and Manufactured Diamonds in Compound Form", Proceedings: The Industrial Diamond Revolution, November, 1967, Columbus, Ohio. Published by Advanced Technology Committee, Industrial Diamond Association of America, Inc.

4. Y. Yarnitsky, "New Approach to Parameters in Diamond Polishing", Science and Technology of Industrial Diamonds, Edited by J. Burls, Industrial Diamond Information Bureau, London, 1967, pp. 175-185.
5. R.D. Buchheit and J. L. McCall, to be published.
6. R. A. Uzanas, "A Guide to Diamond Compounds", Annual Diamond Directory, 1972-1973, Issue of Cutting Tool Engineering.
7. "Grading of Diamond Powder in Sub-Sieve Sizes", Commercial Standard CS261-63, U.S. Government Printing Office, Washington, D.C.
8. E.J. Schneider, "Diamond Powders in Micron Sizes", Science and Technology of Industrial Diamonds, Edited by J. Burls, Industrial Diamond Information Bureau, London, 1967, pp. 161-174.
9. S. Tolansky, The Strategic Diamond, Contemporary Science Paperbacks, Oliver and Boyd Ltd., Edinburgh, 1968.

LINDE ALUMINA ABRASIVES FOR METALLOGRAPHIC POLISHING

F. R. Charvat and P. C. Warren
Union Carbide Corporation, Crystal Products Department
San Diego, California
    and
E. D. Albrecht
Buehler Limited
Evanston, Illinois

INTRODUCTION

Most abrasives in use depend upon one or more forms of classification to yield a particle size and particle size distribution found empirically to make them suitable for a given lapping or polishing process need. The purpose of this paper is to describe in some detail the special manner by which LINDE* Alumina Abrasives are prepared by chemical means without need for subsequent classification. This makes them especially suitable for metallographic sample preparation.

The forms and types of alumina and the methods of preparation of the various aluminas are numerous; they are the subject of an extensive body of literature and technology. The more common methods for synthesizing alumina into a useful abrasive form include the fusion of bauxite and the use of Bayer-process alumina. These are supplemented by processes which calcine alpha alumina monohydrate and alpha alumina trihydrate to transition forms of alumina. Where melting processes are used, various techniques are employed to force rapid crystallization to yield small crystallites which can be either classified into mesh sizes or subjected to further attrition methods to obtain micron-sized alumina. In either case, some further form of classification is required and these processes generally are not capable of providing low or sub-micron classified particles.

Alumina is the most widely used synthetic abrasive because of its high hardness, high melting point, and low chemical reactivity

---
* Trademark, Union Carbide Corporation

with respect to the material being worked. As a result, alumina abrasives are incorporated into bonded products, such as wheels, belts, and papers, or used in loose abrasive form in lapping and finishing operations.

For LINDE Abrasives, the properties of alumina are supplemented by an extremely uniform control of particle size within a narrow range at the sub-micron level. Because of its ability to successfully finish a broad range of materials, including ceramics, single crystal oxide materials, ferrites, ferrous and non-ferrous metals, composites, plastics, and glass, it has been extremely useful as a classic metallographic finishing media. Readily compatible with aqueous and non-aqueous vehicles, cloth and synthetic material laps, it provides a straightforward low-cost polishing procedure.

METHOD OF PREPARING LINDE ALUMINA

LINDE Alumina Abrasive is based on the use of alum as a raw material with subsequent calcination of the alum directly to the final abrasive product. In this general respect, it can be considered an extension of metallographic polishing alumina preparation in the classic method described in many metallographic texts.

It is useful at this point to review the history of the development of LINDE Alumina Abrasives to show why they exist as a special class with respect to other aluminas. At the start of World War II, Linde Company, then a Unit of Union Carbide, contracted with the U.S. Government to establish a production capability for ruby and sapphire single crystal material as required for the manufacture of jewel bearings. The flame fusion or Verneuil crystal growth technique originally developed in Europe was selected and adapted by the Linde Company. This growth technique requires that the raw material for crystal growth be a finely-divided alumina of the highest attainable purity. No commercially-available alumina materials then and now fully satisfy the special needs for crystal growth which includes a need for extremely low levels of calcium, magnesium, iron, and silicon impurities.

As a result, it was necessary to design and install an alum recrystallization process capable of reducing total cation impurities to less than 50 ppm. Of the several processes possible, the one selected and in use today is that process which reacts aluminum sulfate with ammonium sulfate to form the double salt, ammonium aluminum alum. This process follows the path shown in Figure 1.

While it is possible to employ double and triple recrystallization steps to further enhance purity, as has sometimes been found necessary for crystal growth of raw material, only one step is usually employed. The size of the alum crystals produced and their

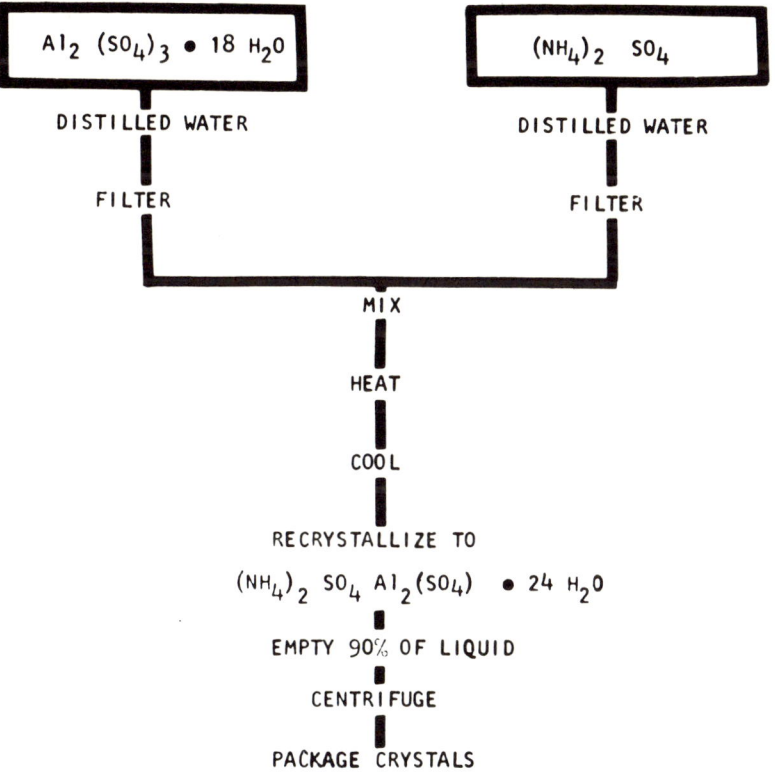

Figure 1. Preparation of ammonium alum.

uniformity is a measure of the control of the process. Relatively large crystals are indicative of good purity control.

The alum is then subjected to a closely-controlled calcination process where dehydration and decomposition of the alum result in a transition alumina, the gamma phase exhibiting cubic crystal structure. The dehydration and decomposition process is depicted in Figure 2.

The dehydration-decomposition sequence results in several drastic changes in volume and changes in state as shown in Figure 3. For instance, at 100°C, the alum crystals melt to a viscous, corrosive liquid which increases in viscosity as the liquid boils and as the water of hydration is driven off. At 400°C, a tough porous cake forms which has a volume three times that of the original alum volume. Above 400°C, the ammonium sulfate components commences to be driven off resulting in an aluminum residual product which is reduced in volume and is now in the condition of a friable

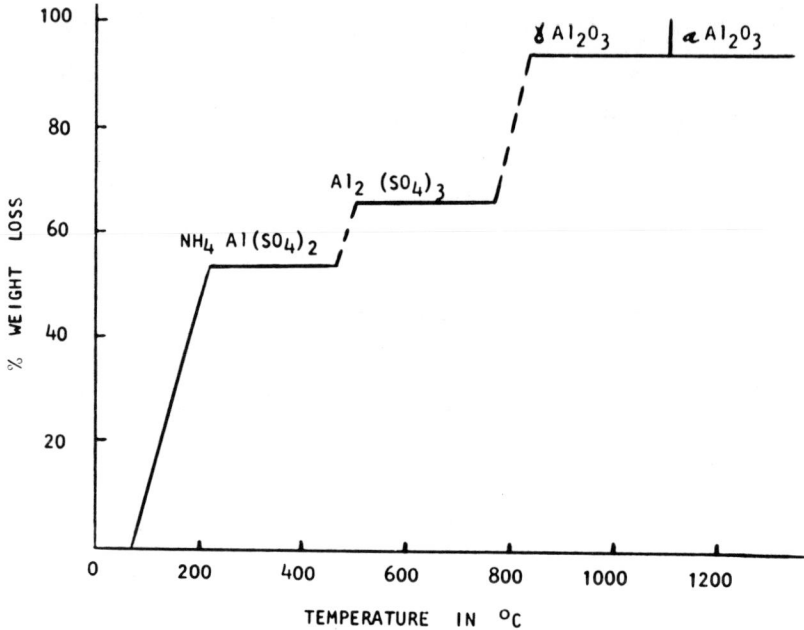

Figure 2. Effect of temperature on the decomposition of $(NH_4)_2Al_2(SO_4)_4 \cdot 24\, H_2O$

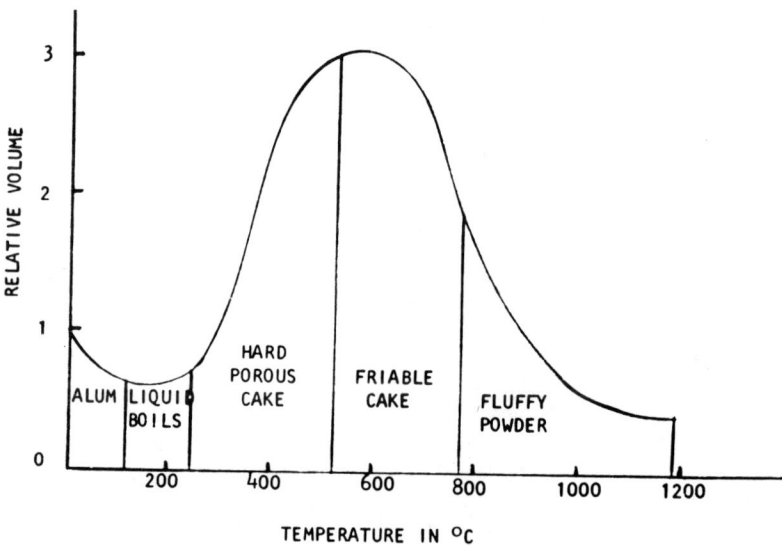

Figure 3. Effect of temperature on the relative volume and state of ammonium alum during calcination to Gamma and Alpha alumina.

cake. Further increase in the temperature through the region of 800° to 1000°C results in the formation of gamma alumina which is a light fluffy powder. The gamma alumina produced at this stage has a surface area of about 120 square meters per gram with particles having a size from 50 to 200 angstroms. The powder produced at the end of this calcination is called "boule" powder and is the basic raw material for crystal growth.

Boule powder is not considered to have sufficient abrading power to be useful as an abrasive at this stage. In order for any of the LINDE Abrasives, LINDE B, LINDE A, or LINDE C, to be processed into their final form, it is necessary at this stage to recharge calcination containers with boule powder for final calcination.

In the case of LINDE B, calcining at only a modest increase in temperature for the specific time cycle causes sufficient growth of the gamma alumina particles to a larger size and the transformation of about 10 to 15 percent by weight of the alumina to the alpha phase with particles ranging to 0.1 microns. The net decrease in surface area to 80 square meters per gram results in a particle size estimated to be 0.05 microns average size. The powder is of abrasive quality and suitable for final finishing uses. The name LINDE B traces to the term boule powder and explains for many the use of the B designation for the finest of the powders.

LINDE A, or the 0.3 micron alumina, is essentially the inverse of LINDE B with about 90% by weight of the gamma starting material converted to the alpha phase with a slightly higher temperature calcination cycle. The remaining 10% remains in the gamma phase. The surface area of LINDE A typically runs from 7 to 20 square meters per gram with the average equivalent spherical particle size centered at 0.3 microns.

Since LINDE A is the most widely-used of the three LINDE Abrasives, and because it is the most difficult to control, some detail on it is in order. Because of what is thought to be very rapid crystallization in the alpha phase, the particles of alpha are generally anhedral in character. Figure 4 is an electron micrograph of the LINDE A type. The particle shape most generally thought to be representative of the powder is that of two particles joined by a neck equivalent to the diameter of the particle and having the strength of the ultimate particle. The aspect ratio is about 2:1. The fact that the anhedral character of the particle offers no sharp cutting edges does not help the fit of LINDE A into the polishing mechanism theories that dwell on cutting edges as a requirement of the abraiding particle.

At this point it is useful to point out that the kinetics of the transformation from gamma to alpha alumina are much more detailed than can be reasonably presented here. Figure 5, however, is a

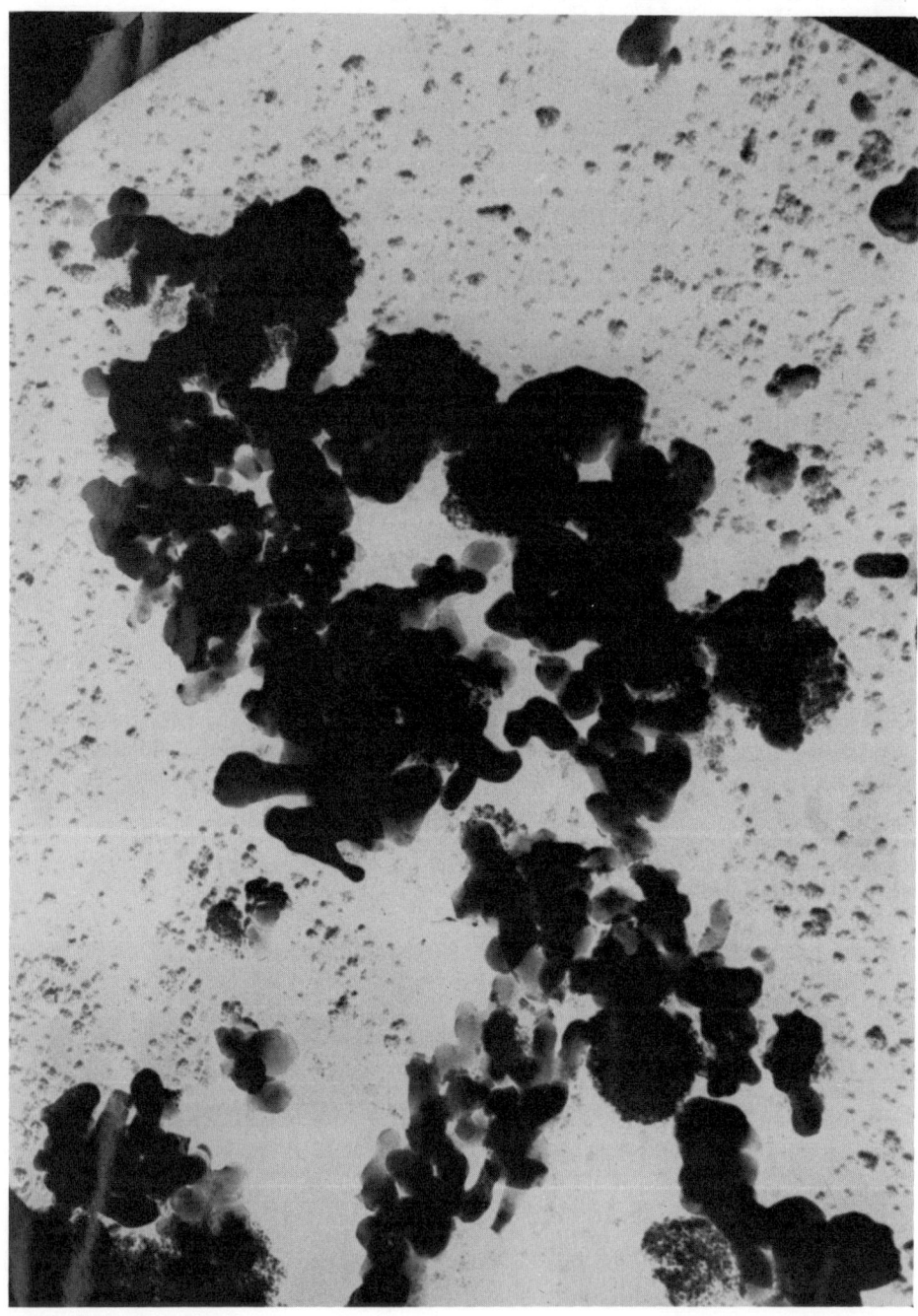

Figure 4. Electron micrograph of LINDE type 0.3A alumina at 45,000X. (Reduced 20 percent for reproduction.)

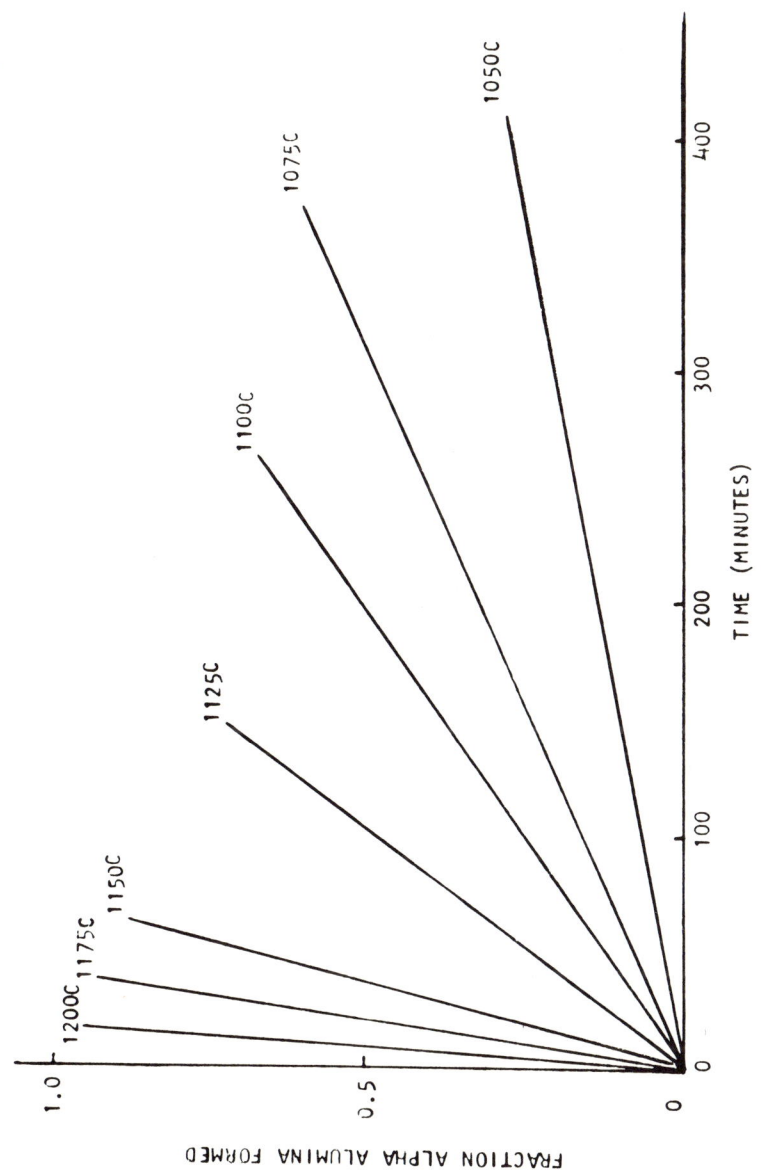

Figure 5. Representative kinetics of the Gamma to Alpha alumina phase transformation.

general representation of the various approaches in calcining one might take in considering the time-temperature regime. It is important to point out that purity control has a great deal to do with how reproducibly this transformation can be effected at a specific process control temperature. It is also important to point out that the powder itself is an extremely good insulator and care must be taken to insure that the powder mass is neither too thick nor too thin to avoid over or under calcination.

Summing some of the foregoing, the volume and state changes starting with alum, the need to maintain purity, the known fact that alterations of the physical state of the decomposition products will affect ultimate alpha particle formation, and the need for close thermal control within the powder mass itself requires the use of relatively small containerization and batch processing. A number of continuous processes have been tried at scale for replacement of the present process without achieving the success of reproducibly producing an abrasive product equivalent to LINDE A in its traditional form. As a result, production costs are relatively high in comparison with all other aluminas.

It is worthwhile to mention that LINDE A in special controlled form is useful as a raw material for high-density, fine-grain ceramics. These processes are sensitive to powder aggregates and agglomerates and, therefore, LINDE A has been studied extensively to determine its particle size, distribution, and the strength of agglomerates and aggregates. Most of these studies reveal particle size distributions other than a tight distribution around 0.3 microns but deal at great length with the various particle size distribution techniques and the methods by which dispersion of powder is made. In general, it is thought that some large and loosely-bound agglomerates exist between 10 and 44 microns while somewhat more strongly bonded aggregates exist between 0.3 microns and 10 microns. For lapping and polishing operations, however, it is thought that the mechanical action breaks up most of these bonds with relative ease. Counting either manually or with computer scanning of SEM micrographs has confirmed the 0.3 micron average equivalent spherical diameter particle for LINDE A.

Completing the LINDE Aluminas is LINDE C which requires an increase in temperature such that containerization integrity starts to reach an economic maximum. Here, complete conversion to the alpha form is attained with further growth of the particle size to 1 micron. Surface area is about 2.5 square meters per gram and process control is important to avoid sintering of particles into aggregates of large size.

Returning to manufacturing operations, once calcination is concluded to the specific type required, the powder is screened and then assembled as a lot, usually 1000 pounds. This lot is then blended and screened for further control of uniformity; then the powder is packaged.

QUALITY CONTROL PROCEDURES

Before detailing quality control procedures, it is worthwhile to point out that the quality of boule powder after the first calcination run is subject to continual checking for purity as seen in crystal growth yield. Secondly, LINDE A, under the designation of LINDE Ceramic Alumina is closely monitored for chemical impurities and in tests run by customers verifying the sinterability of the powder prior to production commitment of powder lots. Therefore, LINDE Alumina Abrasives receive a constant control influence from their co-product's end uses.

Abrasive powder quality control checks include the following:

1) Sedimentation test which is empirically coupled to polishing performance data.
2) Surface area and chemical tests to determine degree of phase change in LINDE A and particle size control in all powders.
3) Moisture loss on heating and loss on ignition to check completeness of calcination process.
4) pH tests in D.I. water to test degree of calcination.

The use of polishing powder test setups has been used for some time in the past and is currently being evaluated for revision. Historically, these tests have determined stock removal on tool steel and Lucite and a scratch-test analysis performed on single crystal rutile which is slightly softer than alumina.

In the early 60's a designation of Semiconductor Grade and Metallographic Grade was made using these polishing tests to segregate powder into the uses compatible with scratch and stock removal characteristics. However, in the past five years, these designations have been eliminated since the uniformity of the powder produced has been such that it is no longer necessary to make a distinction in the type of powder. It is recognized that in certain uses one specific lot of powder will perform better than another while the lot having poor characteristics for one use may be optimum for another use. In all cases, rejections of powder from a very broad customer base for all causes is essentially negligible.

TECHNIQUES FOR USING ALUMINA POWDERS IN METALLOGRAPHIC POLISHING

To produce a well-prepared polished metallographic specimen, it is necessary to properly select a combination of abrasive, cloth, extender, polisher speed and pressure. The spectrum of acceptable combinations is nearly as broad as the variety of materials being polished. The following discussion is limited to some of the more widely accepted procedures for using LINDE metallographic aluminas.

Prior to the introduction of diamond polishing abrasives, LINDE A or C and B Alumina were widely used as the standard rough and final polishing steps in metallographic sample preparation. With the introduction of diamond polishing compounds, the coarser aluminas, A and C, were sometimes replaced by one or more diamond polishing steps. Diamond polishing compounds teamed with the popular LINDE B produced significant improvements in polishing quality and material removal rates. Regardless of this diamond intrusion, LINDE A and C continue to play important roles in the final polishing of certain refractories, ceramic and composite metal alloys. Figure 6 shows that both A and C have a slightly higher hardness resulting from the alpha crystal form. This property, in concert with the proper particle size, can have a significant effect on polishing characteristics in specific applications.

Application to the Polishing Cloth

There are three popular means of applying aluminum oxide polishing powders to the polishing cloth. One is to shake dry powder from a jar with a perforated cap onto a premoistened cloth. Generally, the wheel is rotated during this application. Only when a

PROPERTY	MICROPOLISH LINDE A	MICROPOLISH LINDE B	MICROPOLISH LINDE C
Chemical Formula	Al_2O_3	Al_2O_3	Al_2O_3
Form	Alpha	Gamma	Alpha
Crystal System	Hexagonal	Cubic	Hexagonal
Hardness, Mohs'	9	8	9
Particle Size, microns	0.3	0.05	1.0

Figure 6. Physical properties of metallographic polishing aluminas.

variable speed polisher is used can the technician insure application without waste.

The second method is to mix a slurry by the addition of distilled water to a partially filled flask of aluminum oxide powder, or vice versa. This suspension is applied using a flask with a spout top so that the application may be carefully controlled. As the alumina tends to settle out of suspension, it is necessary to shake the contents before application to insure a uniform slurry. When one is preparing highly corrosive samples, it may be necessary to substitute light oils or inert chemicals for water as the vehicle or suspending agent.

The third method of application is the newer premeasured and mixed suspensions which keep the alumina permanently held in suspension. While these slurries are initially slightly more expensive, they have a number of advantages such as:

1) Convenience - No bottle shaking to resuspend.
2) Cleanliness - Dry powder is not scattered about the laboratory.
3) Safety - No particulate matter to inhale while mixing.
4) Economy - Less wasted powder.
5) Control - More accurate application is possible.

Alumina suspensions may generally be diluted to permit the metallographer an opportunity to select precisely the consistency of suspension he prefers.

Consistency of Slurry

The correct balance of abrasive and liquid is an important consideration. A slurry that is too wet will tend to produce pits by washing out non-metallic inclusions. Rounding of the specimen edges due to prolonged polishing time may also occur. Conditions that are too dry may cause ineffective polishing and disturbed metal at the specimen surface. The slurry consistency should be thin enough to allow the abrasive in such a way that maximum efficiency of material removal is achieved. Scratch-free polishing is accomplished when the abrasive penetrates the cloth to a depth that only the cutting edges of the abrasive contact the specimen. There is no accurate way to define the correct consistency required, as each individual material application must be given consideration.

Technique for Polishing

Hand polishing on a rotating lap requires considerable manual dexterity and experience. Figure 7 shows prepared suspension being applied to the polishing wheel. It should be noted, however, that

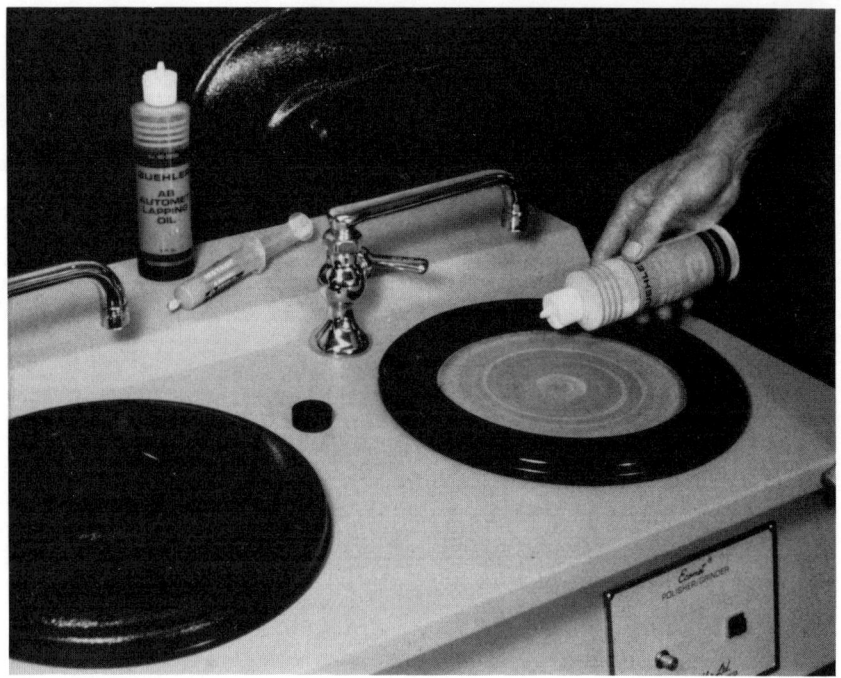

Figure 7. Polishing wheel set up showing proper application of aluminum oxide slurry to polishing cloth.

any specimen polishing using slurries requires that the sample be rotated or moved counter to the wheel rotation. This is necessary to avoid directional effects. Use of semiautomatic devices will simplify polishing with alumina while creating a more uniformly polished surface.

Attack polishing is a technique developed for the simultaneous polishing and etching to remove flowed material from a variety of difficult to prepare materials, especially those with a tendency to smear. When attack polishing, the metallographer adds a dilute etchant to the alumina powder or to the previously prepared slurry. This is applied to a cloth-covered stainless steel or other type of relatively inert polishing wheel. Generally, the normal etchant used to reveal the material's microstructural details is suitable for attack polishing. The chemical inertness of the LINDE aluminas makes them particularly useful in attack polishing applications.

Aluminum oxide powders are also adaptable to other specialized techniques such as skid polishing and vibratory polishing where large quantities of abrasive are used in a rather thick slurry or suspension.

SUMMARY

The close tolerance process controls applied to the manufacture of LINDE aluminum oxide abrasives are responsible for their continued high quality. In metallography, it is extremely important that the particle size remain the same from lot to lot and that the distribution of particle sizes is well controlled within extremely narrow limits. The metallographer must be able to depend upon the product being the same each time it is purchased and he cannot tolerate the presence of oversized particles or an abundance of undersized particles in his polishing powders. No matter if the metallographer is repeatedly preparing tens of similar samples day after day or if his work includes a wide variety of materials, he cannot accept the question of whether today's abrasive powder will perform exactly as yesterday's. It is continued high quality and dependability which has made LINDE polishing aluminas the metallographer's standard over the years.

THE EVALUATION OF FINAL POLISHING ABRASIVES

William C. Coons

Lockheed Missiles and Space Co., Inc.

Palo Alto, California 94304

INTRODUCTION

What is the ultimate criteria for the determination of the most satisfactory abrasive or preparation thereof for the final polishing of all metallographic samples?

For many years most of the producers and distributors of metallographic polishing abrasives and abrasive preparations have been promoting their products by quoting a particle size such as 3.0, 1.0, 0.3, 0.1, 0.06 and 0.05 microns. From these designations of particle size it seems logical that if one were to use the finest abrasive, listed as 0.05 microns, that the best finish, i.e., scratch free surfaces, even on the softest material samples, will be produced by its utilization. The author, with thirty-eight years experience in the polishing of an extremely wide variety of materials, knows that this is not a valid criteria for acceptance of a polishing abrasive.

A year and a half ago, Gregory and Schuyler (1) verified this knowledge by studying various aluminas by observation in the scanning electron microscope. They reported that all of the 0.05 and 0.06 micron aluminas examined contained agglomerate particles measuring up to 60 microns except one material, the Adolph Meller Agglomerate Free 0.06 CR Micron Alumina. However, they reported that even this designated agglomerate free 0.06 micron abrasive did indeed contain agglomerates 4 to 8 microns in size.

It is mystifying as to how anyone can determine a difference of 0.01 micron between two of these materials and arrive at 0.05 and 0.06 micron designations. Perhaps this is merely a manufacturer's method of differentiating his product from the more commonly known designation of 0.05 micron. In this case they might just as well have classified the abrasive as 0.04 micron.

Utilization of the scanning electron microscope as a tool to characterize abrasives could be considered a criteria for acceptance of dry powder abrasives. However, liquid vehicle abrasive preparations are not characterized easily in this fashion and many laboratories are not equipped with a SEM for such evaluation purposes.

In answer to the opening question, the only criteria for judging the best polishing abrasive or abrasive preparation is whether all material samples, regardless of hardness or abrasion resistance, can be final polished in a reasonable period of time and produce surfaces which are void of scratches, surface damage, smear or artifacts and can be etched by immersion only (no swabbing) and yield the finest possible delineation of microstructure. This has been, and always will be, the author's criteria for the assessment of the merit of a metallographic polishing abrasive regardless of any manufacturer's or vendor's claims of particle size.

The following describes the method employed in this study to evaluate abrasives and presents the results obtained.

ABRASIVE MATERIALS TESTED

Table I lists the name, type and sources of the abrasives which were evaluated and Table II lists the addresses of the suppliers.

In the case of the liquid preparations, they were used as bottled or diluted slightly with distilled water. The powder abrasives were mixed with distilled water with the amounts of abrasive approximating the liquid vehicle abrasives.

It should be emphasized that the primary intent of this paper is to present a method of evaluating abrasives and does not constitute an evaluation of all available products. Furthermore, all of the abrasives tested were procured within the last six months and represent currently available materials.

EVALUATIONS PROCEDURES

If a sample of a soft material can be polished and have a scratch-free surface, then no scratches should be anticipated on harder materials. Using this basis of reasoning, samples of small cast buttons of aluminum having Rockwell B hardnesses in the range of 40-45 were selected for the abrasive evaluation tests.

TABLE I. ABRASIVES EVALUATED IN POLISHING EXPERIMENTS

Trade Name	Material	Advertised Particle Size	Form Supplied	Major Suppliers *	Cat. No.
Micro Polish B (Linde B)	Gamma Al_2O_3	0.05 micron	Powder	Buehler Ltd.-Union Carbide	40-6301
Meller - Regular	99.98% Pure Al_2O_3	0.06 micron	Powder	Adolph Meller Company	None
Meller - Agglomerate Free	99.98% Pure Al_2O_3	0.06 micron	Powder	Adolph Meller Company	None
Glennel	Gamma Al_2O_3	0.05 micron	Powder	The Glennel Corporation	None
"C-RO"	CeO-Alpha Al_2O_3-Cr_2O_3	None	Liquid **	The Klarnell Company Burrell Corporation Leco Corporation	None 535-85 80-830
"Finish-Pol."	CeO-Alpha Al_2O_3-Fe_2O_3	0.05 micron	Liquid **	The Klarnell Company Leco Corporation Buehler Ltd.("Miromet"]μ)	None 810-831 40-6355

* Complete addresses given in Table II.

** Abrasive in distilled water vehicle.

TABLE II. ADDRESSES OF MAJOR SUPPLIERS OF TESTED ABRASIVES

Supplier Name	Address
1. Buehler Ltd.	- 2120 Greenwood St., Evanston, Ill. 60204
2. Adolph Meller Co.	- P.O. Box 6001, Providence, Rhode Island
3. The Glennel Corp.	- 908 Chickadee Lane, Westchester, Pa. 19380
4. The Klarnell Co.	- 8463 Medford St., Ventura, Calif. 93003 <u>or</u> 10421 Lansdale Ave., Cupertino, Calif. 95014
5. Leco Corporation	- 3000 Lakeview Ave., St. Joseph, Mich. 49085
6. Burrell Corporation	- Fifth Ave., Pittsburgh, Pa.

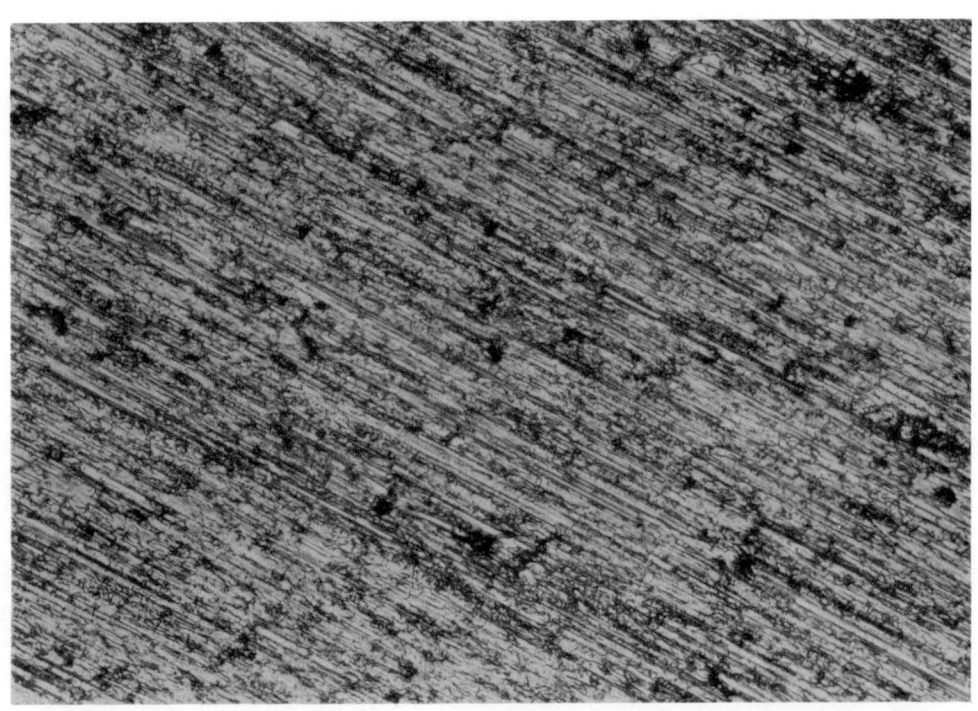

Figure 1. Appearance of the surface of an aluminum sample as-ground on 600-grit SiC paper. 500X. (Reduced 20 percent for reproduction.)

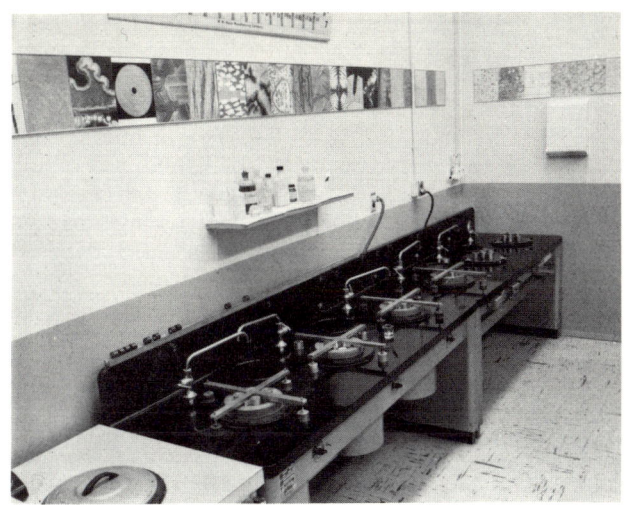

Figure 2. Lockheed Palo Alto research laboratory's polishing set-up.

The samples were mounted in red bakelite and the surfaces were ground through 600-grit SiC papers. This produced a surface as illustrated in Figure 1.

For many years a two-stage polishing system has been employed at the Research Center of the Lockheed Missiles & Space Company, Inc. By this system over 99% of the samples of material encountered can be prepared in essentially the same manner. Briefly, samples are rough polished with 6 or 1 micron diamond abrasives and final polished using a fine metallic oxide abrasive preparation. Both of these polishing steps are accomplished by automated means with devices similar to the "Fini/Pol" Polishing Attachment (a).

Figure 2 is a photograph of the polishing tables and the automated equipment in use at the Lockheed Palo Alto Research Laboratory showing the diamond polishing setup at the far right and the "Fini/Pol" like devices on the four wheels in the foreground.

Details of the manner in which the abrasive evaluation polishing tests were conducted is as follows:

Step 1. A ground sample was rough polished for 30 minutes on nylon cloth using 1 micron diamond paste as the abrasive and a weight of 240 grams. Figure 3 shows the appearance of a surface thus prepared. It should be noted that in these tests the weighted samples were not rotated on the diamond wheel and therefore the resulting scratches are unidirectional rather than random.

(a) Leco Corporation, 3000 Lakeview Ave., St. Joseph, Michigan 49085.

Step 2. Final polishing was performed on "Microcloth" using a portion of an abrasive or abrasive preparation sufficient to thoroughly cover the wheel. A rough diamond polished sample was placed in a 140 gram weight, put into the polishing device and was polished for 1 hour. In order not to scratch the surface of a finished polished sample by a cotton swab removal of abrasive from the surface, the sample was cleaned ultrasonically in methyl alcohol, then rinsed with alcohol and warm air blast dried.

Step 3. The resulting polished surface was examined and photographed using bright field and polarized light.

Step 4. The sample was then etched by immersion for 1 minute in a very mild etchant. A stock solution composed of 5 ml HNO_3, 2 ml HF, 3 ml lactic acid, 50 ml glycerine, 250 ml distilled water was prepared. Etching was performed in a solution consisting of one part of the stock solution and 10 parts of distilled water. The sample was photographed using bright field illumination.

This entire procedure was repeated for the testing of each abrasive listed in Table 1. Between each test the polishing device was thoroughly cleaned and a fresh cloth was applied to the polishing bowl.

"Microcloth" or its equivalent was selected as the best cloth for final polishing. This decision was based on experience and the fact that "Microcloth" was the only cloth that did not appreciably scratch a well polished sample when it was spun on the cloth for 5 to 10 minutes using only distilled water.

To illustrate the fact that the polishing cloth is a very important ingredient in a polishing system, a side experiment was performed wherein a sample was prepared as heretofore described using "Finish/Pol" abrasive on "Microcloth" and "Finish/Pol" abrasive on "Miramet" cloth.

RESULTS OF EVALUATION TESTS

Figure 4 illustrates the polishing results on a sample of 2014 aluminum and is representative of the finish produced by the Micropolish B 0.05 micron and the Meller regular 0.06 micron alumina abrasives. It will be noted that the scratches are random because of the spinning action imparted to a weighted sample during polishing in the "Fini/Pol" like device.

THE EVALUATION OF FINAL POLISHING ABRASIVES

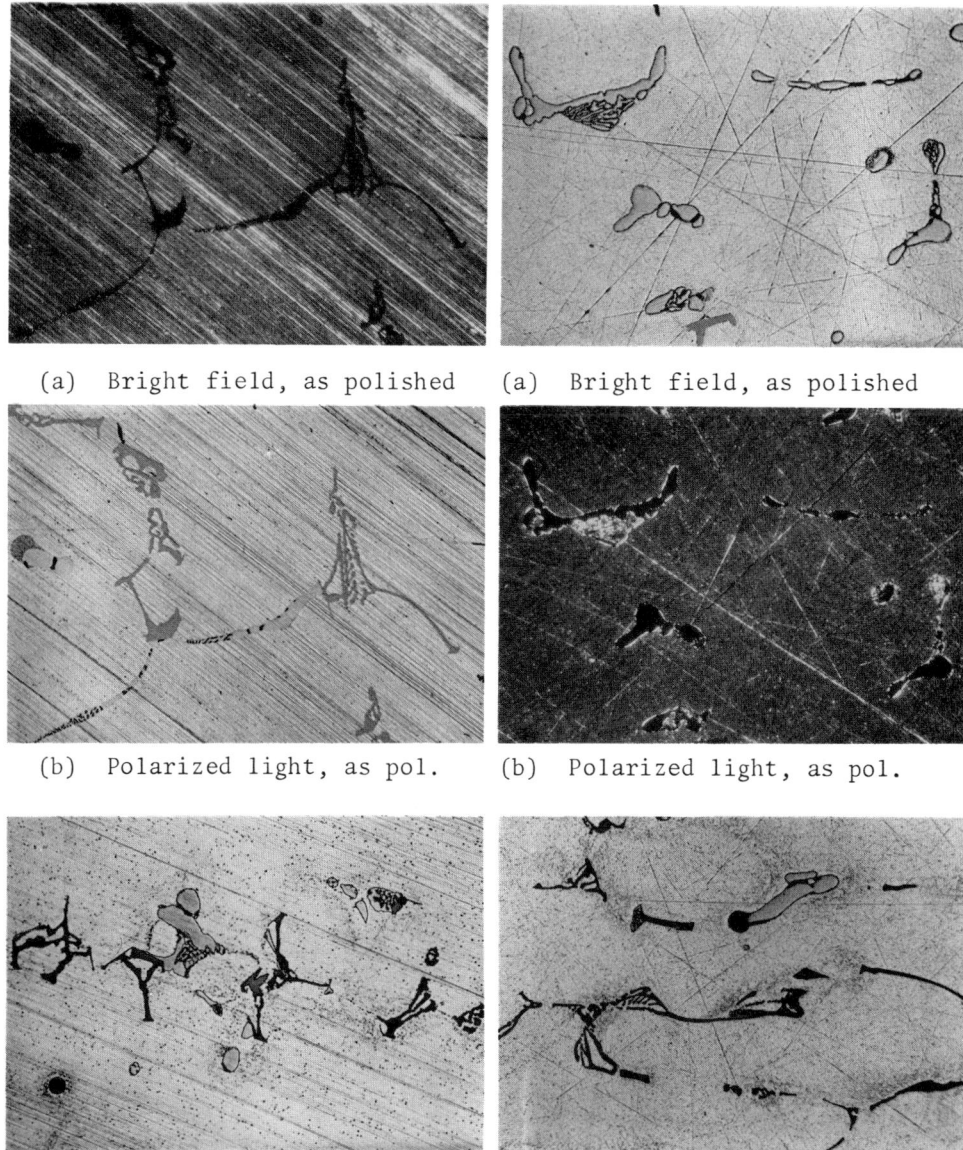

(a) Bright field, as polished (a) Bright field, as polished

(b) Polarized light, as pol. (b) Polarized light, as pol.

(c) Bright field, etched (c) Bright field, etched

Figure 3. Appearance of aluminum alloy specimen polished with 1 micron diamond on nylon cloth. 500X.

Figure 4. 2014 aluminum alloy sample final polished with 0.05 amd 0.06 micron aluminas. 500X.

(Reduced 60 percent for reproduction.)

Scratching was excessive and the mild immersion etching did not reveal clearly any finite details of microstructure indicating that surface smear existed on the polished surface.

The results obtained with the Meller agglomerate-free 0.06 micron and the Glennell 0.05 micron were slightly better and similar and are represented by Figure 5. However, etching disclosed that a clean smear free surface was still not obtained.

The best polish was obtained with the "C-RO" and "Finish/Pol" mixed oxide abrasive preparations. As shown in Figure 6, only a few faint scratches were visible at 500X.

Etching revealed a very fine precipitate in the regions surrounding the massive intermetallic compound particles and the aluminum matrix appeared to be much cleaner and exhibited very little smear. The precipitate observed in the sample as polished with "C-RO" and "Finish/Pol" was not disclosed after etching when the sample had been polished with the 0.05 and 0.06 micron alumina abrasives.

Figure 7 is a photomicrograph of a sample of as-cast 2021 aluminum final polished on "Microcloth" using "Finish/Pol" abrasive. When the same sample was repolished on "Miramet" cloth using "Finish/Pol" abrasive, the surface appeared as in Figure 8. These results attest to the important role of a polishing cloth and illustrate how a perfectly satisfactory abrasive may be condemned when it is the cloth that is the ingredient which promotes scratches.

DISCUSSION

The degree to which any abrasive will scratch the surface of a sample depends upon (1) the uniformity and size of the abrasive particles, (2) the characteristic of the surface on which polishing is being performed and (3) the pressure which is applied during polishing. When a sample of 2014 was inserted in a light 40 gram plastic weight and was final polished for 2 hours in the "Fini/Pol" device using "Finish/Pol" abrasive on a "Microcloth", a perfect polish was obtained in an extremely easy fashion, as illustrated in Figures 9 and 10.

The ultimate criteria for the evaluation of a metallographic polishing abrasive has thus been established. However, the matter of particle size classification has not. It is questioned whether an abrasive having a truly uniform maximum particle size of 0.05 micron would polish a sample in a reasonable period of time. If it did, it certainly would not leave scratches which could be observed on the optical microscope.

THE EVALUATION OF FINAL POLISHING ABRASIVES

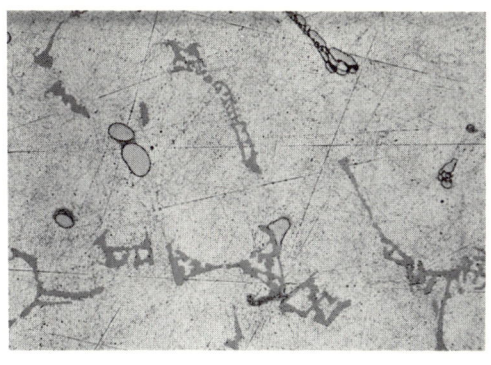

(a) Bright field, as polished   (a) Bright field, as polished

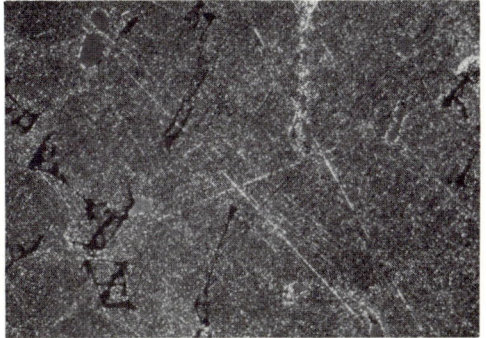 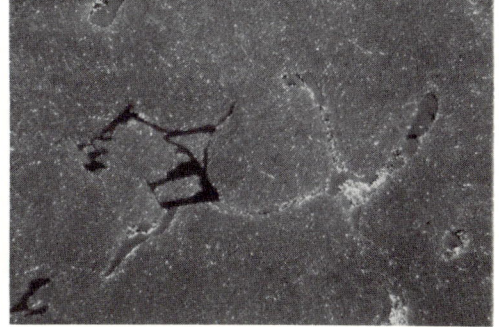

(b) Polarized light, as pol.   (b) Polarized light, as pol.

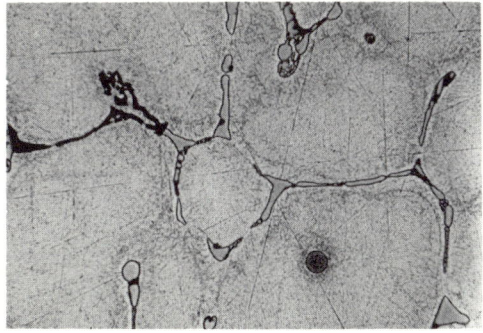 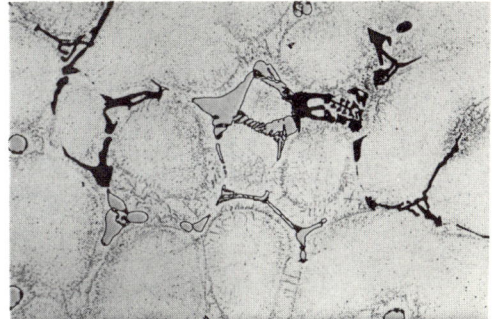

(c) Bright field, etched   (c) Bright field, etched

Figure 5. 2014 aluminum alloy sample final polished with Meller agglomerate-free 0.06 micron or Glennell 0.05 micron aluminas. 500X. (Reduced 60 percent for reproduction.)

Figure 6. 2014 aluminum alloy sample. Representative of results produced in final polishing with "C-RO" and "Finish-Pol" mixed oxide abrasive preparations. 500X. (Reduced 60 percent for reproduction.)

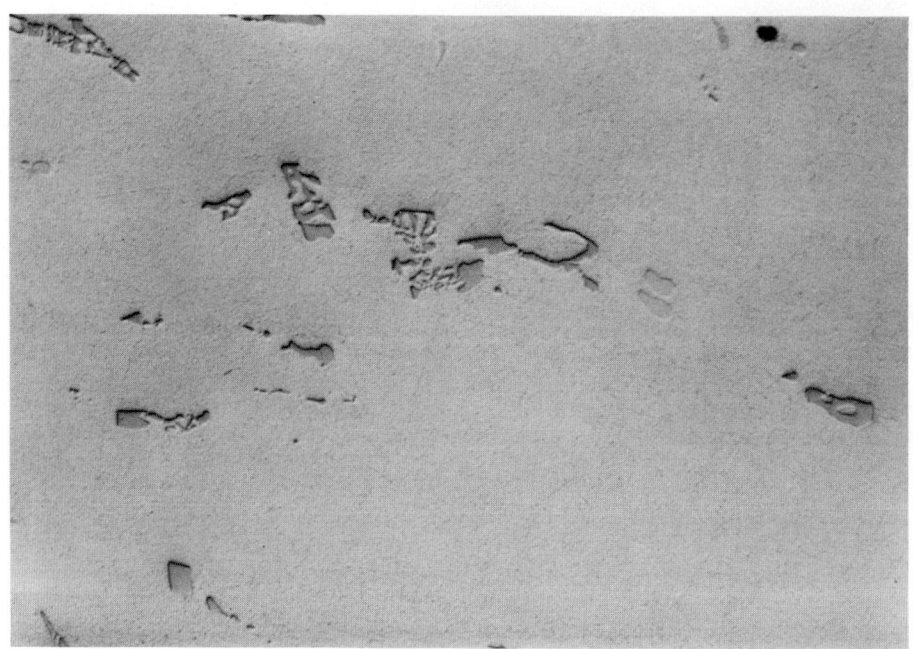

Figure 7. 2021 aluminum sample polished on "Microcloth" using "Finish-Pol" abrasive. Bright field, as polished. 500X. (Reduced 25 percent for reproduction.)

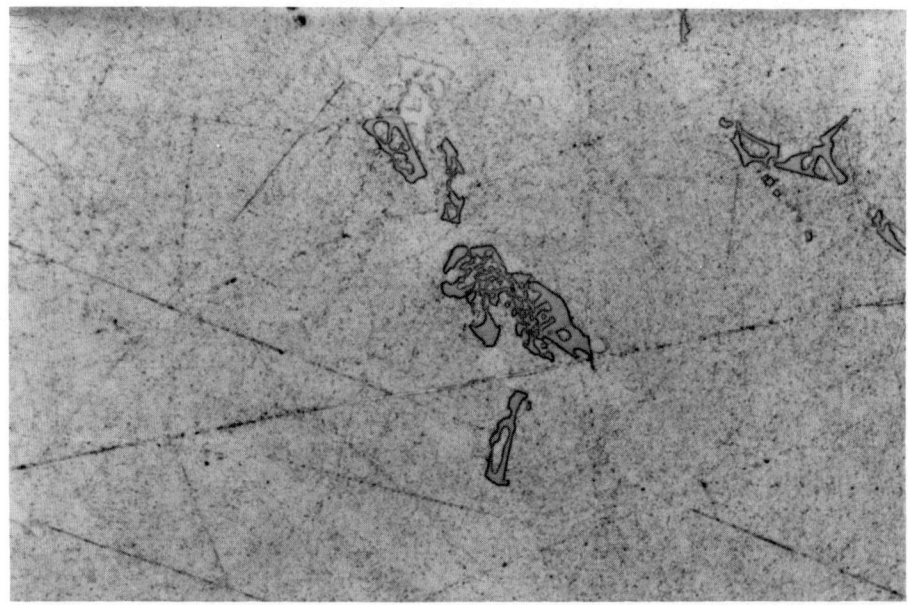

Figure 8. 2021 aluminum sample polished on "Miramet" cloth using "Finish-Pol" abrasive. Bright field, as polished. 500X. (Reduced 25 percent for reproduction.)

THE EVALUATION OF FINAL POLISHING ABRASIVES

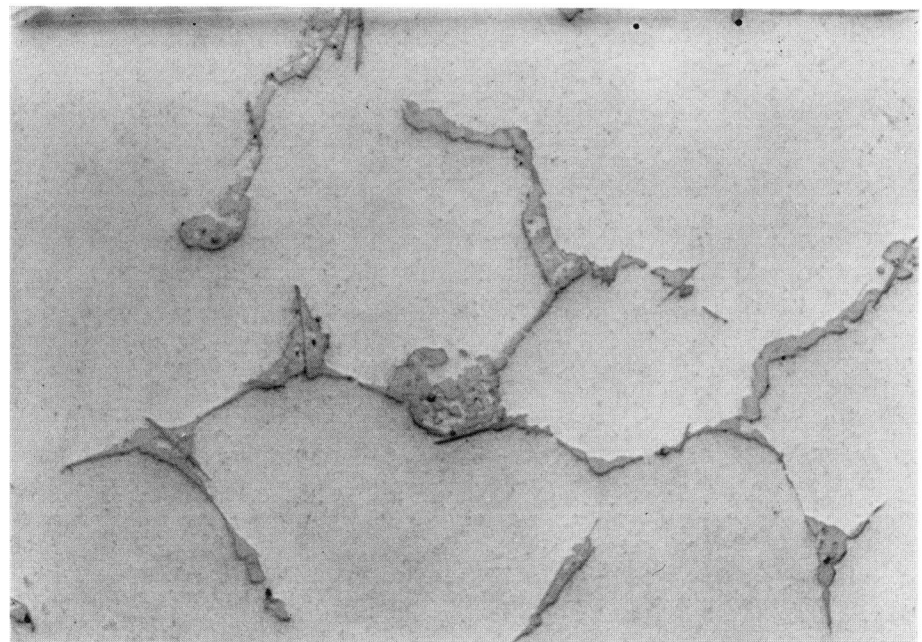

Figure 9. 2014 aluminum sample polished in 40g weight on "Microcloth" using "Finish-Pol" abrasive. Bright field, as polished. 500X. (Reduced 25 percent for reproduction.)

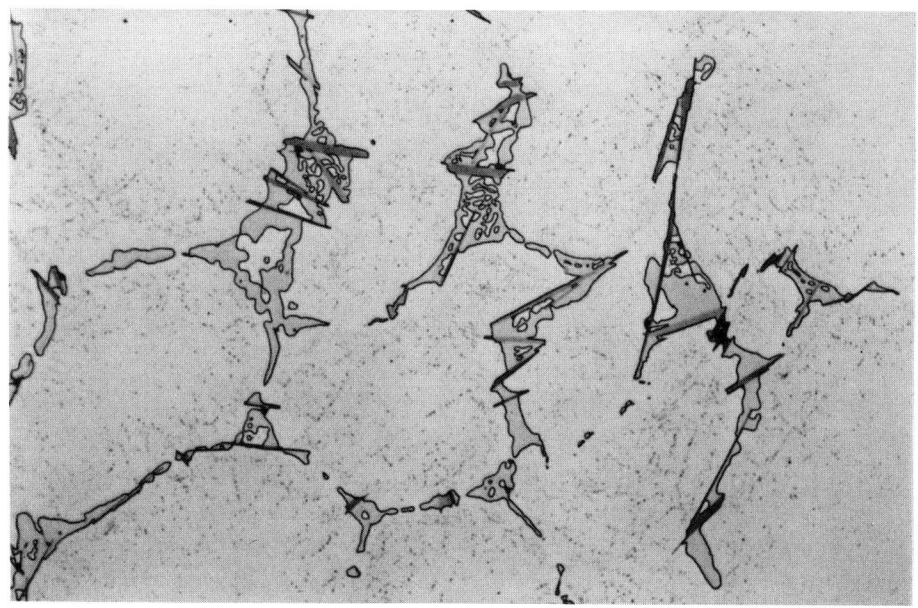

Figure 10. Same as Figure 9 after etching. Note fine details of microstructure. Bright field, etched. 500X. (Reduced 25 percent for reproduction.)

The proposition is put forth that the best abrasive for polishing of all material samples should have uniform particles with a maximum size of between 0.5 and 1.0 microns.

REFERENCES

1. T. G. Gregory and D. R. Schuyler, "SEM Study of Metallographic Polishing Aluminas". Metallography, 5, No. 2, April 1972, pp. 195-199.

MOUNTING, LAPPING AND POLISHING LARGE SIZES AND QUANTITIES OF
METALLOGRAPHIC SPECIMENS

N. J. Gendron

General Electric Company

Schenectady, New York

Since the year 1900, metallography has been practiced at the Materials and Processes Laboratory of the General Electric Company in Schenectady, New York. In the beginning, most samples were small, were cut to size with a hacksaw, filed, sanded on abrasive paper and then polished using various materials on a variety of cloths. Although, through the years, methods, materials and equipment gradually improved, there were limitations, particularly in the mounting of large size specimens, the heat and pressure to be used, and the ability to impregnate porous samples.

In 1950 experiments were begun in this group to minimize these limitations in order that any solid material could be mounted and polished. Prime consideration was given to increasing the size of the mounts that easily could be made. Transparency also was a most desirable objective since it would make identification easier and allow viewing of the whole sample in relation to the polished face. In this regard, in 1946 Kehl and Church (1) published a report on "Room Temperature Casting Resin for Metallographic Mounts". This was a transparent lucite-type mount which was soft and lacked impregnation ability.

To effect an improvement, experiments were conducted for about three years. Concentration was on epoxy resins (2). After many trials, the best of several epoxy systems was put into operation. The formulation selected was composed of 100 grams of Bakelite (Union Carbide Corporation), ERL-2795 epoxy resin, and 10 grams of Hysol (Hysol Corporation) H2-3416 clear hardener. As mixed, this system accelerates in 20 minutes time and becomes a mass of bubbles and shrinks. However, if it is poured into molds with the samples and placed before a room air conditioner (Figure 1) for approximately

Figure 1. Varying sizes of mounts (largest in photo - 3" X 5") being cooled before a room air conditioner. Mounts are set in pans with about 1/4" of water to prevent spilled drops of epoxy from cementing cans to trays.

10 hours, mounts up to 6" in diameter can be prepared which are transparent, shrink resistant and bubble free. During hardening, a mount 6" in diameter will reach no more than approximately 90°F, a temperature which will produce no deterioration in most materials. Since this epoxy system, when mixed, has the viscosity of about 10W oil, it is a simple matter to vacuum impregnate any void which is exposed in the sample.

Using this method, many of the limitations inherent in past metallographic procedures were reduced. Specimens up to 6" in diameter could be readily mounted. Only ambient temperatures were required. Pressure was not applied. Vacuum impregnation was readily accomplished. The mounts produced were transparent and hard. While the hardening time before a room air conditioner might sound quite long, if mounts are poured late in the afternoon and set before a room air conditioner with a timer to shut it off at 3:00 A.M., only about three working hours are required for mounting. On mounts where ambient temperatures must not be exceeded, polishing can proceed after the mold has been removed. For most specimens, particularly where edge preservation is desired, a 150°F two-hour bake will increase the mount hardness. The advantages of transparent, non-

MOUNTING, LAPPING, AND POLISHING LARGE SPECIMENS

shrinking, vacuum-impregnated, large-size samples far outweigh other limiting factors.

With this capability, an unlimited variety of solid materials could be mounted. These include metal, wood, plastics, rubber, teflon, coal, coke, rocks (3), fossils, drilled earth core bars (4), compounds, paper, film, paint chips, teeth, enameled wires, carbon brushes (5) and powder (6), oxide scale, fine metallic powders of all descriptions, cloth, all kinds of very thin foils, and many more materials too numerous to mention.

Since the capability to mount very large quantities of samples in many different size molds was now attainable, a method had to be found to prepare these large quantities and sizes by machine. From 1953 to 1960, all mounts were completed by first mounting in epoxy, surface grinding parallel faces (Figure 2) and finishing on 280-, 400- and 600-grit aluminum oxide paper. Silicon carbide paper was used for titanium, hafnium, zircaloy, high-temperature alloys and hard materials. Sanding was followed by 9-micron diamond and 0.3-micron aluminum oxide.

In 1960 some experiments were conducted to first replace the 280-grit paper sanding operation with a lapping operation. This was done utilizing a large diameter (24") Crane Packing Company

Figure 2. Large hydraulic grinder preparing to grind metal faces of three mounts ground to same height with machine in Figure 5. With proper fixturing, this grinder can do up to 15 - 3" mounts at one time. Total grinding time approximately 15 minutes.

"Lapmaster" (Figure 3). With this operation, all surface grinding effects (Figure 4) were removed, the 280-grit operation was no longer needed, and large mounts could be produced to flatness requirements not attainable with hand sanding. Due to the flatness, the 400- and 600-grit operation was now reduced to approximately five seconds on each paper. For the lapping operation, all mounts are brought to the same height by a small vertical spindle grinder called a "Swisher" (Figure 5).

As the average size of the mounts increased, a machine method following sanding was needed. Since either a meehanite wheel for lapping or a brass wheel for polishing can be used on a Lapmaster, experiments were conducted with the brass-wheel-equipped machine in 1962. To this wheel, a 2.17 drill cloth with medium sizing was affixed. The abrasive used was 30-micron diamond (General Electric "Man-Made"). After a cycle of three to eight minutes (depending on the size of the mount and the amount of cloth used), it was established that all evidence of 600-grit scratches were removed on the wide variety of materials tried. From this condition only about one minute was required to finish-polish the mounts on 9-micron diamond (General Electric "Man-Made") and 0.3 micron aluminum oxide by hand on conventional metallographic polishing wheels.

Figure 3. Three 24" Lapmasters in operation. Furthest machine laps ground surfaces with 14.5 micron Al_2O_3. Center machine removes lap surface with 60 micron diamond on cloth. Forward machine removes 60 micron diamond surface with 30 micron diamond on cloth.

MOUNTING, LAPPING, AND POLISHING LARGE SPECIMENS 125

Figure 4. Various size mounts in lapped condition.

Figure 5. Small Blanchard grinder removing top of large mount to bring to same height as group of others. Up to 200 mils can be removed at one pass on 4" diameter mount in approximately 10-15 seconds.

By 1972 the average mount size had increased to 3-1/2" diameter. Sanding of these large sizes on a 12" metallographic wheel became more and more difficult. Early in 1972 it was decided to eliminate the 400- and 600-grit hand sanding operations. This was accomplished by removing the lapped surface with another lapping machine utilizing a cloth wheel with coarse diamond. Preliminary results using 60-micron diamond (General Electric "Man-Made") on a 24" Lapmaster with static, light loading proved the method to be most feasible. A Lapmaster with pneumatic lifts which could apply a controlled heavier load proved even more efficient. Now, all hand sanding had been eliminated, and all mounts could be machine processed up to the last two operations, which, due to the very flat and uniform surface produced, required about one minute of hand labor to finish. Unit time for complete polishing preparation was now reduced to only 25 percent of the total effort spent for each sample. The remaining time was consumed in etching, consultation with engineers, photomicrography and record keeping. With this method, a total of 25,500 samples were processed in 1972, and the average mount size was 3-1/2" diameter.

The advantages of lapping and lapping machines in a metallographic operation are numerous:

1. Extremely flat, smooth surfaces are produced (7).
2. All previous working due to grinding (tensile stresses) (Table 1) is removed and a shallow worked surface (compressive stresses) is produced. This can be removed easily by diamond polishing. Residual stresses were determined by X-ray diffraction on steel to be about 100,000 psi tensile strength.
3. Sharp edges of samples can be retained.
4. Processing of large quantities of samples can be accomplished with minimal effort. Small batches can also be processed.
5. Preparation of high quality large samples can be made.
6. Outstanding photomacrographs can be made. These are not possible by other methods of sample preparation. (Less than five minutes of hand work is required for a large piece 6" x 6" in size).
7. Specimens which would be transformed or cracked by grinding can be lapped with no subsequent damage.
8. Work not connected with metallography can be generated, such as samples for stress corrosion studies. (The surface produced is the same at all times, eliminating the possibility of different tensile stresses being produced by different operators of surface grinding machines).
9. Semi-automation techniques of metallographic preparation can be used on mounts 1" to 6" in diameter because no time-consuming fixturing is necessary.
10. Metallographic polishing wheels can be refurbished from time to time by lapping, thereby helping to make finish quality easier to attain.

TABLE I
STEEL; 3.93 Ni; 0.33 Cr; 0.33 Mn; 0.22 C. RESIDUAL STRESS MEASUREMENTS ON SURFACES OF FORGED MATERIAL HARDNESS 60 RA (228 BRINELL).

Specimen	Residual Stress, KSI
Surface Ground	+12
Lapped	-50
280 grit Al_2O_3	-18
400 grit Al_2O_3	-18
600 grit Al_2O_3	-18
60 μ Diamond	-36
30 μ Diamond	-25
9 μ Diamond	-16
.3 μ Al_2O_3	-17
Etched w/2% Nital	-16

The principle of electric-powered machine lapping has been in existence since about 1890. As a metallographic tool, it is invaluable. Since the use of this type machine both for lapping and polishing has virtually done away with the "art" of sample preparation, the time saved can now be utilized more effectively in the "art" that is left, that is, etching. Due to the many complex alloys and heat treatments in use today, this has become the new challenge of the future.

Since Sorby (8) in 1863 first viewed a metallic structure under a microscope, metallographers have constantly tried to perfect new and better methods of preparation. Application of the method described here is just another step in the continuing improvement.

REFERENCES

1. Kehl, G. L. and Church, J.S., "Room Temperature Casting Resin for Metallographic Mounts", Metals Progress, 51, 1946.
2. Gendron, N.J., "Cold Setting Plastic Used for Metallographic Mountings", Industrial Laboratories, June 1958 pp. 26-27.
3. Gendron, N.J., "Mounting of Geological Specimens in Clear, Cold Setting Plastic", Economic Geology, 54, 1959.
4. Buoma, Arnold H., Methods for the Study of Sedimentary Structures, John Wiley and Sons, New York, p. 267.
5. Pincus, I. and Gendron, N.J., "Microscopy and Structures of Brush Carbons", Proceedings of Fourth International Conference on Carbon, Buffalo, New York, 1959 (Pergamon Press, 1960).
6. Titus, A.C., "Boron Carbide in the Graphitization of the Lampblack Carbon Brush Constituent", Proceedings of Fourth International Conference on Carbon, Buffalo, New York 1959 (Pergamon Press, 1960).

7. Gendron, N.J., "Lapping Produces Smooth Surfaces on Metallographic Specimens", Metals Engineering Quarterly, February 1973.
8. The Sorby Centennial Symposium on the History of Metallurgy, Metallurgical Society Conferences, October 1963 (Gordon and Breach Science Publishers).

TECHNIQUES OF SPECIMEN PREPARATION FOR METALLOGRAPHY USING A NEW AND UNIQUE AUTOMATED POLISHING PROCESS

K. H. Roth, KENINCO, Toronto, Ontario, Canada and

D. M. R. Taplin, University of Waterloo, Waterloo, Ontario, Canada

INTRODUCTION

This paper describes a new automatic polishing device which has proven itself capable of producing high quality surfaces for microstructural analysis of a wide range of materials. These included soft materials (lead alloys), brittle materials (coal, oxide coatings), sensitive materials (austenitic stainless steel, zinc alloys) and hard materials (carbides, iron ore pellets). It was also found that the polishing characteristics of the system were relatively insensitive to the material polished so that no special polishing "recipes" were required. Results are consistent, predictable and largely independent of operator factors, such as level of skill, experience or motivation. Polishing by this method is very fast; the polishing time is generally under 10 minutes total.

EQUIPMENT AND PRINCIPLES OF OPERATION

Figure 1 shows the basic machine and the polishing accessories, namely, polishing cloths, a "lapping cup" and a storage container. The polishing cloth is 2¼" diameter and is adhered to the bottom of the lapping cup. When not in use, the lapping cup is stored in the container shown at right. The machine shows a lapping cup in the operating position on the top surface. The lapping cup is stationary and is held in place by means of a locating pin. The arm above the cup enters a hole drilled into the specimen and is the means of applying the abrasion motion and load. The machine is set in operation by means of the switch at the right while the load is applied using the central knob. The switch at the left operates a special load-reducing mechanism discussed later.

Figure 1. Basic polishing machine.

The heart of the system is the abrasion motion. This motion, traced in Figure 2, produces multi-directional, yet systematic, abrasion with the following characteristics:
(1) The abrasive slurry is drawn towards the center of the motion, assuring that fresh abrasive arrives continuously at the abrasion interface.
(2) The abrasion vector changes direction rapidly, subjecting each and every point on the specimen surface to essentially identical, multi-directional abrasion. No specific abrasion direction is ever favored at any given time throughout the entire polishing cycle. While the direction of abrasion changes at a rate of 360° in 1/10 sec., the speed of abrasion also changes cyclically between 46 and 80 surface feet per minute.

Figure 2. Trace of abrasion motion.

A most important component of the machine is the lapping cup. From a practical point of view, this cup allows easy cleaning, storing and use. Since cups are small and inexpensive, it is possible to reserve a cup for each specific polishing requirement (i.e., nap surface, 0.3 micron alumina, steel). A cup, once set up, is simply taken from its storage container, fresh abrasive is added, and the cup is used immediately with a new specimen. On completion of that polishing step, the cup is rinsed and returned to storage. While charges can be used several times, it generally proved to be less troublesome to rinse the cups routinely and start fresh. Owing to the small polishing area ($2\frac{1}{4}$" diameter) and because the specimen travels fully within the cup recess, intense mixing action between the specimen and cup sidewalls occurs. This prevents slurry settling and, at the same time, disperses the polishing debris uniformly into the total abrasive mass. Abrasion characteristics consequently remain quite uniform throughout a polishing cycle, primarily because slurries are fully contained and no loss occurs.

The specimen load can be set at any value between 0 and 8 lbs. Furthermore, it can be made to reduce continuously at a predetermined rate from the set value towards 0 by means of an automatic mechanism. This enables the following two capabilities - automatic "skid" polishing and polishing under a continuously decreasing load.

The technique of skid polishing as developed by Samuels (1) is capable of producing ultrafine finishes, considerably better than those achievable using conventional contact polishing. Using a conventional lap, the technique requires skidding the specimen on a layer of abrasive paste sufficiently thick to avoid contact between the specimen surface and the underlying cloth. By using a viscous paste in the lapping cup and applying a low load, the hydrodynamic lift induced by the continuous polishing motion is capable of prolonged automatic skid polishing by floating the specimen off the polishing cloth. The second capability, namely polishing under a slowly reducing load, offers certain special advantages. A polishing cycle generally requires high initial abrasion rates to remove strained material remaining from a previous coarser step. However, once this is removed, importance shifts towards minimizing the introduction of new polishing damage. If the rate of material removal is sufficiently fast to remove strained layers while new damage occurs at a decreasing rate and to a decreasing depth owing to lower pressures and less severe polishing, the final damaged layer will be minimum for that particular abrasive size and material response to abrasion. Since skid polishing occurs at the end of a reducing pressure cycle if the paste is sufficiently viscous, it is possible to have a fully automatic cycle beginning with high abrasion rates, and ending in skid polishing.

POLISHING PROCEDURE

The specimen is prepared by drilling a hole into the mount as close to the surface to be polished as possible. The surface is then prepared by conventional pre-grinding or by automatic pre-grinding on the machine as discussed below. Subsequent to this, various polishing regimes are possible. All, however, use a setup sequence as shown in Figures 3 and 4. The appropriate lapping cup is taken from storage, abrasive and fluid are added to prepare a suitable slurry or paste, the specimen is placed into the cup and the cup is placed on the machine. The machine is started and the appropriate load is applied (with or without the load reducing action). At the end of the predetermined polishing cycle, the cup is removed, rinsed and returned to its storage container. The specimen is cleaned ultrasonically and the above procedure is repeated if required using a second lapping cup which has been reserved for the next stage of polishing. Throughout the polishing stage, cleanliness must be maintained and no "contaminants" can be allowed to enter the abrasion process (i.e., coarser abrasives from previous steps).

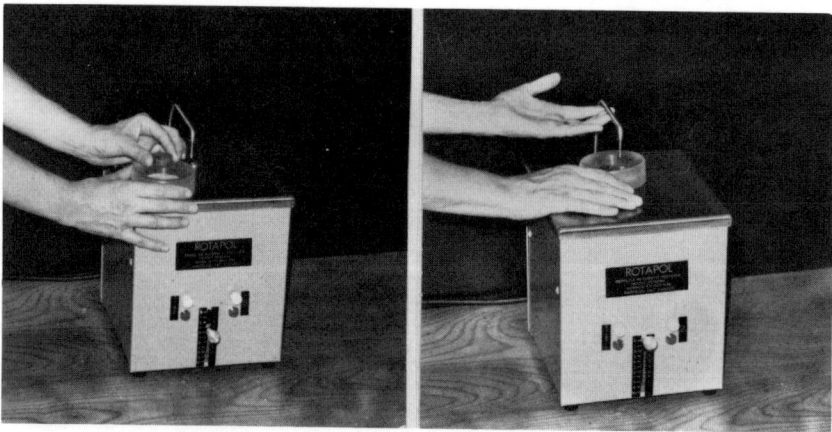

Figure 3. Polishing set up sequence.

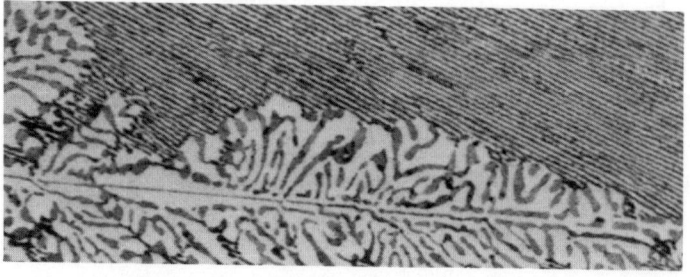

Figure 4. Pb-Sn-Sb eutectic alloy.

RESULTS

It has been found that a wide range of materials can be polished readily by the method using only two polishing stages subsequent to conventional pre-grinding on 600-grit wet emery. The same abrasives and loads are used, while polishing times at each stage are kept to the minimum time required for that stage. The two stages used 0.3-micron alumina in distilled water (1 to 3 minutes) followed by colloidal magnesia in distilled water (2 to 4 minutes). Quite frequently, the first stage alone will produce an adequately polished surface.

In general, the second step is required to remove fine scratches, especially for the softer materials. While this has been found to be the optimum method for all specimens, its use was simply to illustrate the capability of this technique to achieve good results for a variety of dissimilar materials using essentially the same method. Figures 4 through 10 show etched surfaces for a wide variety of materials polished using the above method:

Figure 4 is a Pb-Sn-Sb eutectic alloy. This material exhibits a typical eutectic structure consisting of three distinct soft phases. Conventional polishing requires considerable expertise. Polishing times using the described method were 4 minutes for the first stage and 3 minutes for the second stage using reduced loads.

Figure 5 is a $Fe/FeO \cdot Fe_2O_3 \cdot Fe_3O_4$ alloy. This specimen contains two very different phases - soft ductile iron and brittle, cracked oxide. Polishing was accomplished using 2 min. on 0.3-micron alumina and 5 min. on magnesia. The initial abrasion was 600 grit. Better results are possible using diamond and automatic pre-grinding.

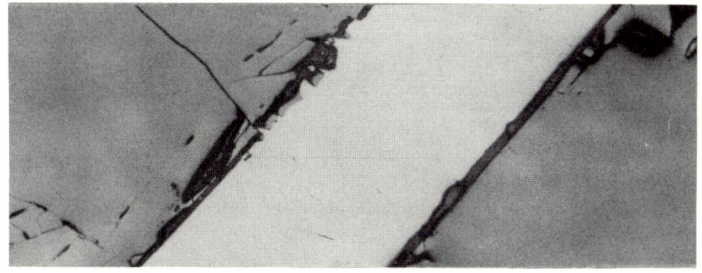

Figure 5. $Fe/FeO \cdot Fe_2O_3 \cdot Fe_3O_4$ on iron.

Figure 6 is of an annealed automotive lead body solder; 92% Pb, 5% Sb, 3% Su, 0.4% As. This is another soft material which is difficult to polish manually. The polishing was done with 2 min. on 0.3-micron alumina and 4 min. on magnesia. The initial abrasion was 600 grit.

Figure 7 is of an Al-Si eutectic alloy. Polishing was done using 2 min. on 0.3 micron alumina and 3 min. on magnesia.

Figure 8 is of a weld puddle in mild steel after introduction of some nickel to change the fluid-flow pattern. This specimen illustrates the results of single stage polishing for macroscopic examination. The polishing was accomplished using only 5 min. on 0.3 micron alumina from a 600-grit abrasion.

Figure 9 is of a zinc (99% pure, as drawn) specimen. This was polished for 4 min. on 0.3 micron alumina followed by 5 min. on magnesia.

Figure 10 is of a pulsed arc weld in mild steel using copper filler metal. This specimen also illustrates single stage macroscopic polishing. The polishing was accomplished using 5 min. on 0.3-micron alumina after a 600-grit abrasion.

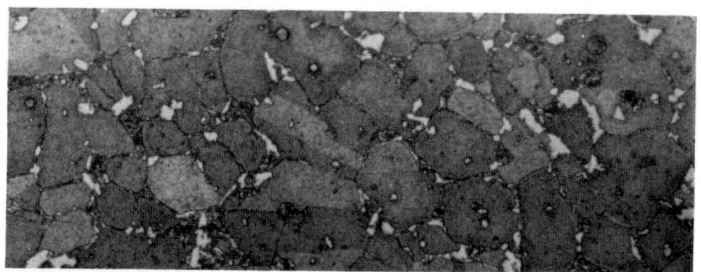

Figure 6. Annealed automotive lead body solder.

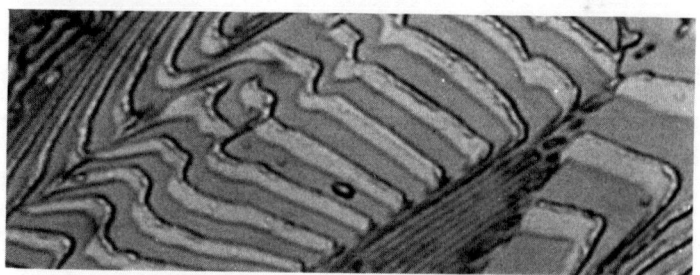

Figure 7. Al-Si eutectic alloy.

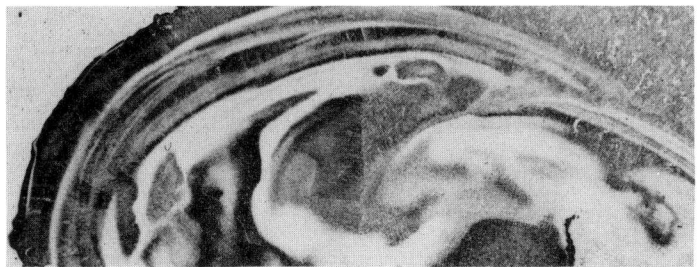

Figure 8. Weld puddle in mild steel.

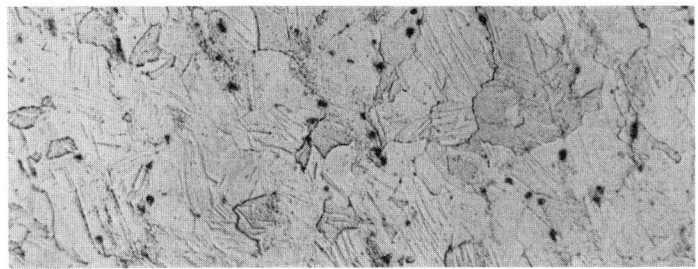

Figure 9. Zinc specimen.

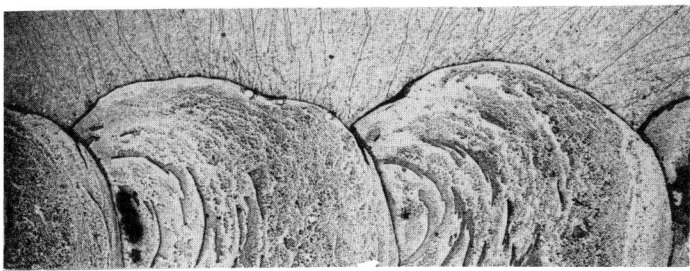

Figure 10. Arc weld in mild steel.

Figures 11 through 14 were polished using diamond and lapping oil in place of 0.3 micron alumina. Figure 11 is of a copper specimen which used 5 min. with magnesia for the second stage. Figure 12 is of a spheroidized steel specimen which used 4 min. on 0.05-micron alumina. Figure 13 is of an aluminum alloy specimen which involved 2 min. on magnesia for the second-stage polishing. Figure 14 is of a Hastelloy alloy specimen which used 5 min. on magnesia for the second-stage polishing. Other materials that have been polished successfully using similar techniques include iron ore pellets, 99.9% Zn, Inconel X-750 and austenitic stainless steel.

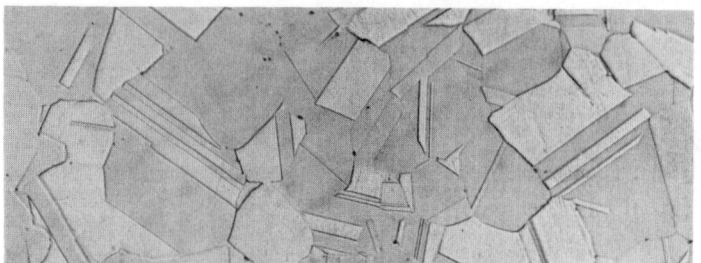

Figure 11. Copper specimen.

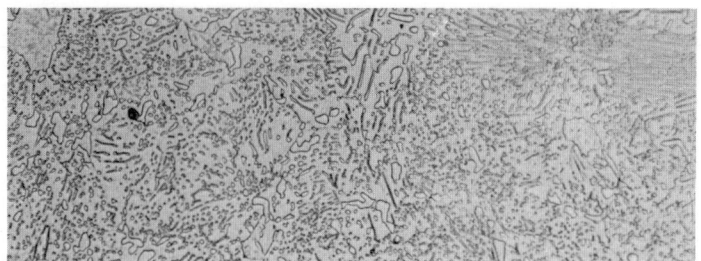

Figure 12. Spheroidized steel specimen.

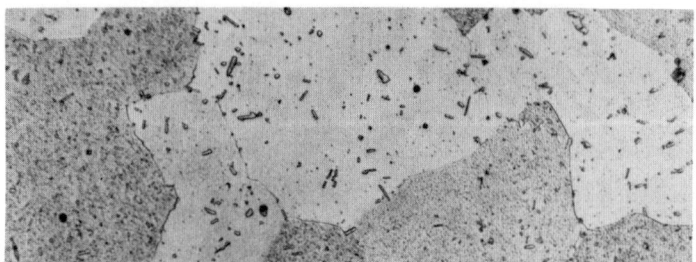

Figure 13. Aluminum specimen.

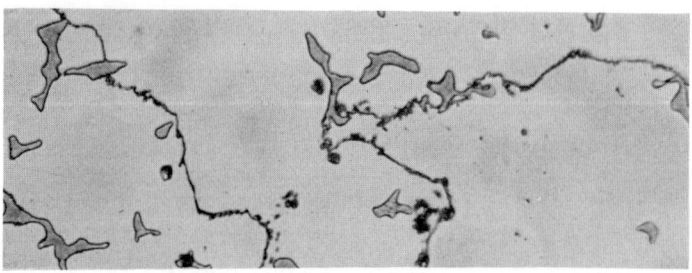

Figure 14. Hastelloy specimen.

Figures 15 through 18 show two special applications and capabilities of this system. Since abrasion is subject to rapid and systematic changes in direction, the tendency to "smear" cavities and cracks is minimized since abrasion into the cavity occurs on the average as much as abrasion into the material. Through the use of hard abrasion surfaces and low pressures, it is possible to reveal minute defects in a specimen and present a clear outline of the defect. The material of Figures 15 and 16 is 99.99% Zn containing high temperatures creep cavities. The cavity morphology is clearly shown in Figure 15, while the higher magnification of Figure 16 (750X) shows the cavity edge. Edge retention is particularly easy to achieve using this system since fine finishes using hard abrasion surfaces can be achieved. Nap cloth is used only for the final stage and for the shortest possible time.

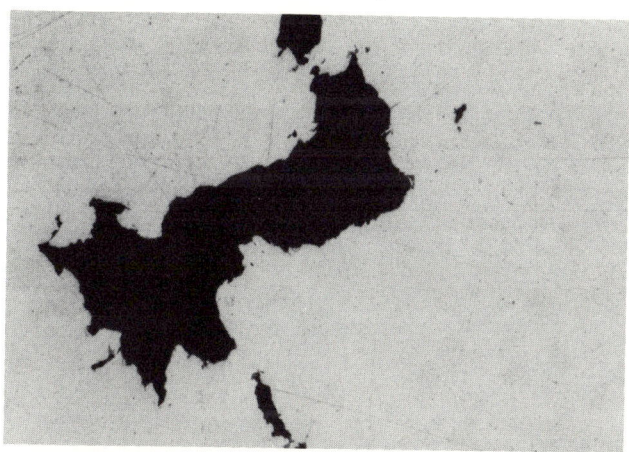

Figure 15. Creep cavities in 99.9% Zn.

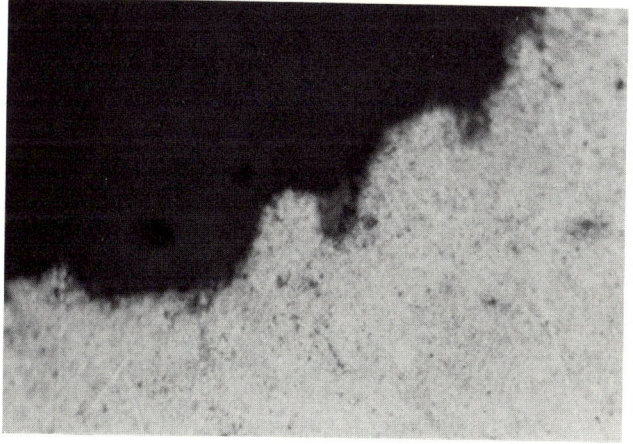

Figure 16. Creep cavity in zinc. 750X (Reduced 10 percent for reproduction.)

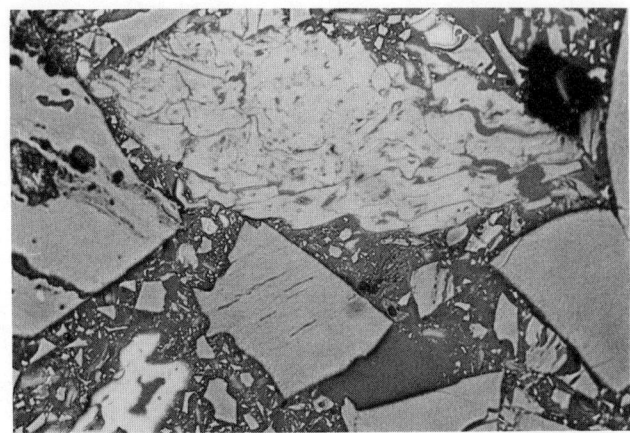

Figure 17. Coal chips impregnated with epoxy.

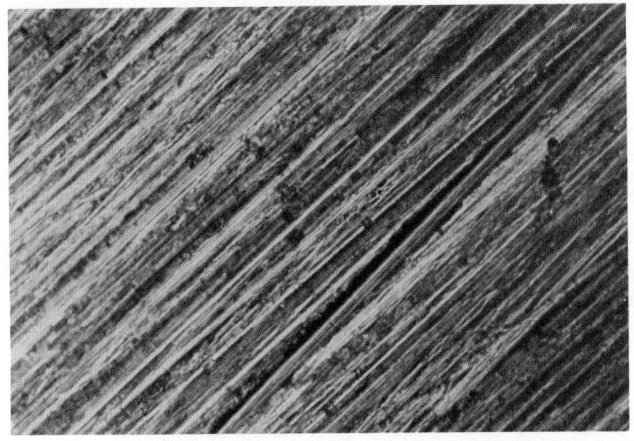

Figure 18. 60/40 brass specimen after 240-grit wet emery grind.

Figure 17 shows the application of this system to a non metallic material - coal chips impregnated with epoxy. The objective here is a scratch free surface with minimum relief for component reflectivity measurement. The sample was polished using the following regime:

(1) Automatic pre-grinding using Texmet (2) and 0.3-micron alumina for one minute.
(2) Preliminary polishing using nap cloth and 0.3-micron alumina for one minute.
(3) Final polishing using colloidal magnesia for one minute.

While these are not optimum conditions (and because of the great variability of coal will not perform equally well on all types of coal) the results obtained, as can be seen, are quite acceptable.

Relief is minor and scratches are not visible at this level of magnification (1000X visual).

AUTOMATIC PRE-GRINDING

It has been found that the machine functions effectively as an automatic pre-grinder. Two different methods may be used:

1. A special cup with 400-grit emery paper adhered to its bottom and submerged in water is used on the machine in place of lapping cups. This single step is capable of replacing 4 stages of conventional pre-grinding. The specimen is placed in the machine, and a minor load is applied. Since the direction of abrasion changes continuously and rapidly as described before, owing to their random orientation, many cutting points are engaged in rapid succession and a high rate of metal removal results. The abrasion debris is flushed from the surface through rapid agitation of the water and abrasion continues in a uniform and regular manner as long as the cutting quality of the abrasive remains adequate. A single stage, using 400-grit emery in this system combines the metal removal rates of 280- and 320-grit emery and the fine grinding property of 600-grit emery. Pre-grinding by this method consistently required less polishing time to remove abrasion damage than 4 stage manual pre-grinding.

2. This method uses no conventional emery paper at all, but automatic pre-grinding is performed in a cup whose bottom has a hard polishing surface such as is used for polishing mineral specimens. The cup is charged with an intermediate grade of alumina, and the specimen is allowed to abrade in the usual fashion. The final surface is considerably better than can be achieved on the finest fixed abrasive surface and for some purposes can be considered to be adequately polished. Due to high metal removal rates, and the uniform and fine nature in which this is done, it is possible to start with a very rough surface (saw cut) and in a single step to prepare a surface equivalent to, or better than, that obtained using conventional pre-grinding followed by rough and intermediate polishing. This method is illustrated in Figures 18 and 19. Figure 18 shows a 60/40 brass specimen ground on 240-grit emery. Figure 19 shows the same surface after polishing for 4 minutes using 0.3-micron alumina on Texmet (2). The single polishing step replaces 3 pre-grinding steps and rough polishings.

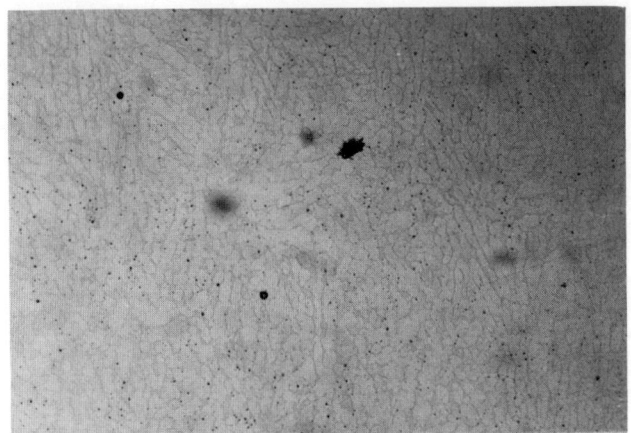

Figure 19. Same as Figure 18 after polishing.

POLISHING IRREGULAR PIECES

Since the polishing motion is relatively slow and abrasion is largely owed to a rapid change in abrasion direction, it has been found possible to polish irregular, unmounted pieces of metal directly without removing sharp edges or mounting in plastic (as illustrated in Figure 20). A hole is drilled in the approximate center of gravity of the metal. Subsequent treatment is identical to that given a conventionally mounted sample. Pieces weighing up to one-half pound have been polished readily and automatically, and without physical danger as is normally encountered by polishing these pieces on a rotating lap. Since the application of pressure is uniform, no opportunity exists for sharp corners to dig into the polishing cloth and to rip it. This technique is particularly suitable in situations where quality checks have to be made and it is not practical to mount samples in a conventional fashion each time a specimen requires analysis.

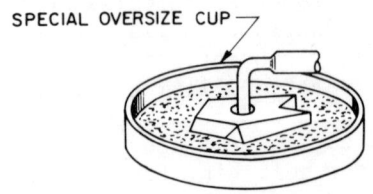

Figure 20. Polishing irregular specimens.

APPLICATION TO RESEARCH

This system has several advantages over conventional techniques, as described above, that make it particularly suitable for use in university research laboratories where graduate students are required to prepare their own specimens. Quite generally, fine effects are under investigation and these require very careful preparation so that they are fully revealed and a minimum of artifacts are introduced during the polishing process. To achieve these aims, however, requires considerable skill and patience on a rotating lap and, not only does polishing itself have to be learned, but frequently a special technique has to be developed which is suitable to the material and the objective in question. This is frequently very time consuming and may reflect on the quantity and quality of metallography performed. (One of the authors, in part, developed this system as a result of 3 months of frustration encountered in attempting to polish 99.99% Zn).

APPLICATION TO MATERIALS EDUCATION

Another potentially valuable use of this system is in education. Since conventional polishing requires considerable skill to achieve a surface revealing fine details of structure, and students generally do not acquire this skill in the short time assigned to lab sessions, first hand analytical experience is difficult to arrange. One key step, that of mechanical preparation, in essence prevents in many cases the major objective, that of material analysis. Since this system is automatic and requires no special experience, each student can readily do his own work without requiring a great deal of individual attention. The actual polishing sequence is relatively simple and foolproof so that the necessary results can be achieved with ease.

SUMMARY

Summarizing, this new machine has the ability to produce a high quality polish on a wide variety of materials, to do so much faster than conventional automatic polishers, and to be almost completely independent of operator shortcomings. It has already proved itself in several laboratories and is expected to become widely accepted in the near future.

ACKNOWLEDGMENTS

Thanks are due to the following for encouragement and assistance throughout the development of this polishing technique: Professor P. Niesen, Professor H. W. Kerr, W. H. S. Lawson and S. Kadela, all of the Department of Mechanical Engineering of the University of Waterloo, Waterloo, Ontario, Canada; Dr. H. J. Westwood, formerly of the University of Waterloo, now with Ontario Hydro, Research Division, Toronto and Dr. J. M. Stuart, formerly of the University of Waterloo, now with Rollmet Inc., Santa Ana, California, U.S.A.

REFERENCES

1. L. E. Samuels, Metallographic Polishing by Mechanical Methods, (Pitman 1967).
2. Proprietary Polishing Surface - Buehler Ltd. (Chicago).

MECHANICAL PREPARATION OF $Gd_3Ga_5O_{12}$ AND VARIOUS ELECTRONIC MATERIALS BY VIBRATORY POLISHING TECHNIQUES

D. Medellin, M. F. Ehman, and G. W. Johnson

Electronics Research Division, Rockwell International

Anaheim, California 92803

INTRODUCTION

The use of gadolinium gallium garnet - $Gd_2Ga_5O_{12}$ "GGG" - single-crystal wafers as substrates for epitaxial thin films requires the preparation of surfaces that are flat, free of scratches and pits, and with minimal residual surface strain. These requirements must be achieved both for chemical vapor deposition (CVD) and liquid phase epitaxial (LPE) deposition of rare-earth magnetic oxides. The procedure described in this paper is basic to the preparation of many types of electronics materials, including ceramic oxide crystals and semiconductors. GGG was selected as an example due to its intermediate hardness and the current interest in the material for magnetic bubble-domain memory devices.

Although several works have been published on the chemical preparation of rare-earth garnet (1-5) surfaces and, more recently, of GGG specifically (6-7), techniques for the mechanical preparation have not been reported.

This paper describes detailed techniques for slicing, lapping and polishing GGG wafers, and the evaluation of the quality of polish by the constancy-of-etch-rate technique.

EXPERIMENTAL PROCEDURES

Crystals

The wafers utilized in this study were sliced from 1"-diameter, core-free, Czochralski-grown, $Gd_3Ga_5O_{12}$ boules with total dislocation densities less than $10/cm^2$

Figure 1. GGG wafers mounted on polishing fixture.

The hardness of the (111) surfaces was 7 on the Mohs scale and 1098 on the Knoop microhardness scale.

Slicing

The boules are first mounted in Do-All No Load Cement* and oriented by x-ray techniques to within ±1 degree of the (111) crystallographic direction. They are then sliced into (111)-oriented wafers with an inside-diameter (I.D.) tensioned-membrane saw using a 4" blade with 340-mesh synthetic diamond abrasive. A standard commercial coolant is used for the slicing. Typical feed speeds are approximately 8mm/min; blade speed is typically 2900 RPM.

Lapping

The thicknesses of the as-sawed GGG wafers are then determined, to facilitate mounting of wafers in groups and reduce lapping times. Any wafers of irregular thickness are individually hand lapped on both sides to provide flat surfaces and uniform thickness. The wafers are mounted on polishing fixtures, as shown in Figure 1, using commercially available low-viscosity waxes. For production runs, six 1/2-in-diameter wafers can be mounted on each fixture, as shown in Figure 1. When several smaller GGG crystals of approxi-

---
*A product of Do-All Company, Des Plaines, Illinois

Figure 2. Mounting pattern for glass spacers and GGG wafers.

mately the same diameter and thickness are to be polished, the slices are mounted equally-spaced around the periphery of the mounting surface of the polishing fixture, as shown in Figure 2. When polishing a small irregular-shape crystal the standard procedure involves the use of three small cylindrical glass spacers, equal in thickness to the GGG wafer. This maintains flat contact between the single GGG wafer, mounted centrally within the triangle formed by the glass spacers, and the polishing surface. It also reduces the amount of edge-breakage of the wafer, avoiding the severe scratching which results from such chipping, and promotes the continuous rotational motion of the sample fixture, necessary for uniform lapping and polishing action. There are two steps in the lapping process, both of which are carried out on Syntron* vibratory polishers. The first step utilizes 9 micron Al_2O_3 in a slurry consisting of 10 ml ethylene glycol, 150 ml H_2O; and 135 gms 9 micron Microgrit** Al_2O_3 on a serrated-grid cast-iron plate, shown in Figure 3. This step is used for approximately 30 minutes, depending upon the severity of saw-marks and initial wedging of the samples. A 5 micron Microgrit Al_2O_3 slurry (using the same proportions described above) is used as the second step, also on the cast-iron plate.

---

* Manufactured by the Syntron Company; Homer City, Pa.
**Al_2O_3 abrasive manufactured by Geoscience, Mount Vernon, N. Y.

Figure 3. Cast iron grid lapping plate.

The loadings on the polishing fixtures are adjusted to 200 g/cm^2 of wafer surface area. The vibrational amplitude of the vibratory motion is between 0.030 in. and 0.035 in. (8) which appears to give the optimum cutting rate with a minimum of surface damage in the wafers. In each set, a separate bowl is used on the polisher for each abrasive size. The mounted samples and fixtures are cleaned ultrasonically between steps to avoid contamination of the subsequent step, since it is essential not to demount the wafers between steps.

Polishing

The polishing process also involves two steps, both of which are carried out in separate Syntron bowls on cloth-covered optically-flat glass plates. The first step (semi-polish) utilizes a slurry of 10 ml ethylene glycol 120 ml H_2O, and 10 g 0.3 micron Linde "A"*** on nylon cloth. The final polish is obtained using a slurry consisting of HT30 Syton+ (0.03 - 0.04 micron SiO_2) in a 1:1 ratio with water, with the pH adjusted to 10.5. The cloth in the latter step is a perforated Pellon cloth.++

***Al_2O_3 polishing abrasive manufactured by Union Carbide Corp., San Diego, CA.
+ A product of Monsanto, St. Louis, MO.
++ A product of Pellon Corp., Lowell, Mass.

The Syton polishing is continued until all scratches from the semi-polish step have disappeared from the surface when it is viewed in the microscope with the Nomarski interference contrast (9) objective at 230X.

A summary of the total lapping and polishing sequence, with times and material removed in each step, is given in Table I.

TABLE I.

MECHANICAL PREPARATION OF GGG SURFACES

Abrasive	Plate/cloth	Time	Material Removed	Slurry Changes
9-micron Al_2O_3	Cast iron plate	Until flat	>0.002"	None
5-micron Al_2O_3	Cast iron plate	30 min	0.002"	None
0.3-micron Al_2O_3	Nylon	16 hr	0.0015"	1
0.03-micron SiO_2	Pellon perforated pad	8 hr	0.0002"	1

The procedures described above deviate from those accepted as the state of the art in several areas, including pressure applied to the samples, amplitude of vibration, polishing cloth selection and slurry concentrations. The pressures utilized in our laboratories are significantly lower than those used for polishing metallurgical specimens. They are useful, however, for several different ceramic oxide crystals ranging in bulk hardness from 3 to 9 (Mohs).

The lower vibrational amplitudes of the vibratory polishers have two distinct advantages: (1) they eliminate bouncing of the polishing fixtures and (2) they provide more intimate contact between the sample surface and the charged polishing pad. The selection of the polishing pads is based upon observations of the motion of the fixtures and samples and abrasive particles on the cloth.

The cloths described hold the abrasives so that the particles are not pushed ahead of the sample. Before a slurry is replenished, a hard-bristle nylon brush is used to raise the nap on the cloth. The slurries are completely changed when depleted instead of just adding more abrasive. This avoids the build-up of a depleted abrasive layer which would reduce the contact between the sample and the charged cloth. The slurry concentrations are chosen to provide the optimum movement of the polishing fixture around the periphery of the bowl counterclockwise while it turns slowly about its own axis with a clockwise motion, while still providing sufficient abrasive for adequate cutting rates.

Etching

The chemical etchant used throughout this study was concentrated H_3PO_4 used at 180°C (3-7). This acid is well known as a dislocation etchant for garnet materials at this temperature, which would normally disqualify it for this type of study because serious deviations in surface areas caused by faceting at dislocation pits would introduce errors in calculating the depth of damage. This was determined not to be a problem in this work, however, since the number of etch pits was very small (less than 10 per wafer) and relatively constant (approximately 5 per wafer). The primary reason for selecting this etchant was the availability of data on the water content, degree of polymerization, and etch rate-temperature relationship (7). A second reason was that in the absence of defects (bulk and surface) this acid acts essentially as a chemical polish, uniformly removing the material in a nonselective manner.

The constancy-of-etch-rate technique, first described by Faust (10), is based upon the observation that a highly strained region etches faster than a region of lower strain. Thus, when a chemical etchant initially attacks a strained surface a high etch rate is measured, but this decreases in proportion to the decreasing amount of strain in successive layers; as the etchant reaches a region in the material which is characterized by the amount of strain present in the bulk material the etch rate becomes constant. Typical curves are shown in Figure 4. For the depth at which the etch rate first becomes constant represents the deepest penetration of the damage produced by the surface processing. From the total weight loss (i.e., removed by etching) per unit area at this point the depth of damage can then be determined.

The acid solutions were contained in Pyrex beakers covered with reflux condensers to avoid excessive water loss. The entire assembly was immersed in a constant-temperature bath controlled to ±0.2C degree. The acid solutions were stirred at 100 RPM by magnetic stirrers to reduce boundary layer interference at the crystal-acid interface. The samples were suspended in the acid by a platinum wire mesh basket.

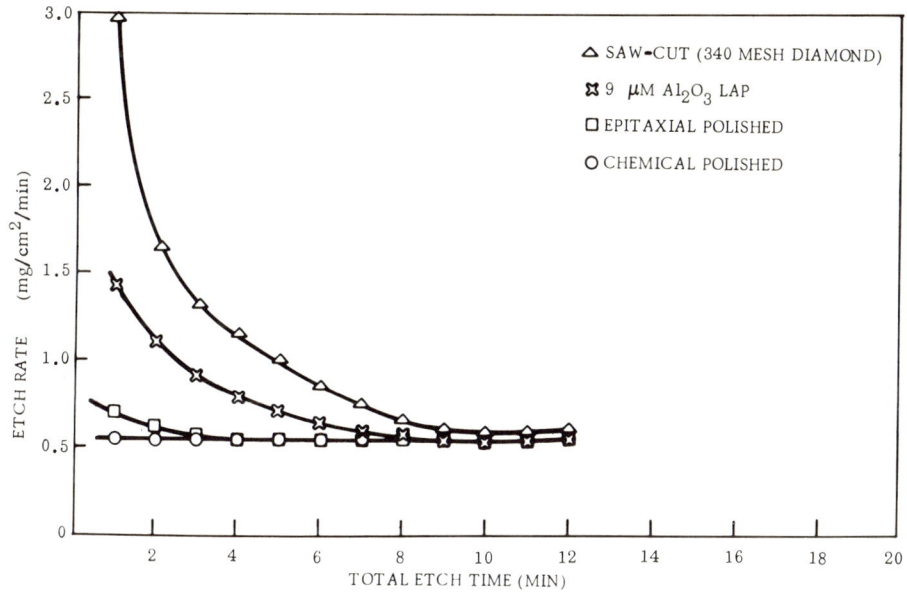

Figure 4. Etch rate vs. etching time plot for GGG depth of damage determination.

SURFACE QUALITY

The polishing process described above produced, on a pilot-line processing basis, scratch- and pit-free surfaces. The technique used to detect polishing pits was to scan the entire wafer surface at 115X using a Nomarski interference contrast objective.

Three different techniques were utilized to detect the presence of a damaged layer. The primary method was that of standard reflection optical microscopy using the Nomarski interference contrast objective. A second technique was that of observing with crossed Nicol prisms the birefringence associated with the strain field around the scratch; this is a standard procedure for mapping the location of bulk defects in the wafers. The third techniquewas x-ray diffraction topography, which was used to detect subsurface damage. The latter technique was not used as a routine quality control step due to the long times required for analysis. We were unable to detect any scratches by any of these techniques.

The surfaces were essentially flat as measured by a Sloan-Dektak Surface Line Profiler*, where the maximum peak-to-valley

---

*Sloan Instruments Corp., Santa Barbara, CA.

height on the lightly undulating surface was 0.1 micron. The average spacing of the undulations was between 100 and 150 micron.

The depth of damage induced by each processing step was evaluated using the constancy-of-etch-rate technique described earlier. Figure 4 shows the etch rate vs total etching time curves for sawcut, 9-micron Al_2O_3 lap, epitaxially polished 0.2 μm Syton, and chemically polished (111) GGG surfaces.

The arrows indicate the points at which it was determined that the etch rates became constant. Since this point cannot be accurately determined, the estimated experimental error or the thickness of the damages layer is given in Table II.

TABLE II.

DEPTH OF DAMAGE (111) $Gd_3Ga_5O_{12}$ SURFACES

Surface Finish	Damaged Layer (microns)
I.D. Saw-cut (180 mesh diamond)	10 ± 1
Lap (9-micron Al_2O_3)	3.5 ± 0.4
Epitaxial Polish (0.03-micron SiO_2)	1.0 ± 0.1
Chemical Polish	0

The etch rates after all damage had been removed reached constant values with the exception of the saw-cut surfaces. In this case, due to the severe damage incurred during slicing, a myriad of faceted, trigonal pits developed.

The real surface area under these conditions was significantly larger than the measured surface area which provided a much larger area for attack of the GGG by the etchant. Consequently, the mass losses per unit area, per unit time as plotted were higher than the actual values.

The chemically polished - at 300°C for one minute - sample was included for two reasons: first, to detect any changes in the reactivity of the acid due to dehydration during an etching run; and

TABLE III (11)

DEPTH OF DAMAGE (microns)

Crystal	Mohs	Orient.	Knoop	I.D.Saw 340 Mesh	SiC Lap 12 micron	Epitaxial Polish
$Gd_3Ga_5O_{12}$ (GGG)	7-3/4	(111)	1098	11	3.1	1.0
BeO	8	$(11\bar{2}2)$	900	15	1.2	0.16
		$(10\bar{1}0)$	1000*	13	1.0	0.15
		$(10\bar{1}1)$	1075	12	0.9	0.14
		(0001)	1250	9	0.8	0.10
$MgAl_2O_4$	8-1/2	(100)	1300	3.8	1.0	0.41
		(111)	1500	3.4	0.82	0.36
		(100)	1700	3.0	0.79	0.22
Al_2O_3	9	(0001)	2050	1.8	0.22	0.09
		$(01\bar{1}2)$	2250	1.6	0.21	0.08
		$(11\bar{2}0)$	2550	1.2	0.20	0.08
		$(11\bar{2}0)$ 6°off	2600	1.2	0.20	0.07
		$(10\bar{1}4)$	2700	1.0	0.18	0.07

*Note that this hardness is dependent on orientation, the average values are reported.

second, to establish a base etching rate should there be any significant variations in the surface areas of the other samples due to nonuniform surface treatment.

Double-crystal x-ray diffractometer analyses were made on the wafers at each step. Depth of damage values determined for the sawcut and lapped surfaces were in agreement within 10% with the etching data. This technique is limited, however, in that only moderate-to-severe damage can be detected. It was not possible to detect the presence of any damage on the epitaxially polished wafers by the x-ray techniques, since variations in lattice parameters (± 0.001Å) masked any distortion of the rocking curve half-widths contributed by the thin damaged layers.

SUMMARY

A vibratory polishing technique has been developed which produces essentially flat, scratch- and pit-free surfaces on (111) gadolinium gallium garnet and other single-crystal wafers. The slightly undulating surfaces have maximum variations in height of 0.1-micron, with an average separation of 100 to 150 micron. The surface damage incurred by GGG wafers during slicing, lapping and polishing is reported. The depth of residual surface damage in GGG wafers after epitaxial polishing was ∼1 micron, as determined by the constancy-of-etch-rate technique. The vibratory technique is essentially the same as that utilized on other microelectronic materials such as silicon, gallium arsenide, beryllium oxide, spinel and sapphire. Table III shows a compilation of the depths of damage incurred by several different materials using vibratory techniques.

REFERENCES

1. J. Hanke, Z-Angew Phys. 14 (1962), pp. 465 and 557.
2. P. V. Klevtsov and V.D. Zamozhskii, Sov. Phys. Solid State 5 (1963), p. 246.
3. R. F. Belt, J. Appl. Phys. 40 (1969), p. 1644.
4. R. Aeschlimann, F. Gassmann, and T. P. Woodman, Mat. Res. Bull. 5 (1970), p. 167.
5. L. K. Schick, J. Electrochem. Soc. 119 (1972), p. 118.
6. A. M. Szaplonczay and H. H. D. Quon, J. Mat. Sci. 7 (1972), p. 1280.
7. D. C. Miller, J. Electrochem. Soc. 120 (1973), p. 678.

8. E. L. Long, J. T. Meador and R. J. Gray, Symposium on Methods of Metallographic Specimen Preparation ASTM Special Publication No. 285 (1960), p. 70.
9. D. Medellin and C.Y. Ang, Advances in Metallography, Twentieth Metallographic Conference, Denver (1966), p. 1.
10. J. W. Faust, Jr., Electrochem. Tech. $\underline{2}$, (1964), p. 339.
11. M. F. Ehman, "Surface Preparation of Ceramic Oxide Single Crystals-Work Damage and Microhardness", Electrochemical Society Meeting, Boston, October 1973.

MAGNETIC ETCHING WITH FERROFLUID*

R. J. Gray

Oak Ridge National Laboratory

Oak Ridge, Tennessee

INTRODUCTION

In the application of magnetic etching as a metallographic tool we should be aware that magnetism is, in its most basic form, an atomic phenomenon. The atoms in most materials have electrons that are paired, with both electrons spinning in opposite directions so the net magnetism is cancelled. Some materials have atoms with one or more unpaired electrons that produce a net magnetic moment. Materials exhibiting strong magnetism are classified as <u>ferromagnetic</u>; other materials displaying significant magnetism are classified as <u>ferrimagnetic</u>. Materials showing weak magnetic response are <u>paramagnetic</u> and materials having no net magnetism are <u>antiferromagnetic</u> (diamagnetic).

The origin of magnetic etching must be attributed to the development of Bitter (1), and independently by von Hamos and Thiessen (2) during the same year - 1931. The original "Bitter patterns" were obtained by sprinkling magnetic powder on the surface of a material in a magnetic field and observing the distribution of the particles. The concept remains today; however, many advancements have been made. Although Bitter (3) improved his technique in 1932, real progress in the early development of the technique for the observation of magnetic patterns was made by McKeehan and Elmore (4). They utilized magnetic particles in colloidal suspension in the successful observation of well-defined and reproducible magnetic domain patterns on the faces of iron crystals. "The magnetization

---

*Research sponsored by the U.S. Atomic Energy Commission under contract with the Union Carbide Corporation.

of a ferromagnetic or a ferrimagnetic material tends to break up into regions called domains, separated by thin transition regions called domain walls" (5). Magnetic etching offers a technique for the microscopic observation of domain patterns.

Further progress has been reported in the interim years (6-10) in the use of the colloid technique for observing magnetic domain patterns and ferromagnetic phases and constituents in the microstructure, although several problems have continued to hamper the progress of magnetic etching. Nearly every report has presented a recipe for making the colloid; however, the anticipated stability of colloids made in the laboratory has generally not been fulfilled. Since the procedure for making the colloid in the laboratory is quite time consuming, a satisfactory colloid may not be readily available, so the technique is not utilized. Some metallographers have reported that specimens have corroded after using some formulas of laboratory-prepared colloid. The magnetic coils reported previously for use in magnetic etching have been a diameter that just fit the right cylindrical specimen mount. Since the observation of a specimen often can extend to 30 minutes or more, the heat generated in the coil can evaporate the aqueous carrier and terminate the examination.

Recent improvements in the magnetic etching have been reported (12-15). These improvements have included the use of a commercial colloid, Ferrofluid (16), and the use of a larger coil magnet to remove the coil heat from the specimen and provide a larger central area where the flux is normal to the specimen surface. This progress report offers some suggestions in simplifying the use of this technique and applying it in the metallographic studies of several materials.

RESULTS

Ferrofluid and Its Application

We continue to find Ferrofluid to be an extremely stable colloid. We are using the first trial bottle that was purchased over three years ago, with no evidence of colloid flocculation. The water-base Ferrofluid is now available for $75 from the supplier in a 3-oz. bottle (catalog No. 2W03) with saturation magnetization of 200 gauss. This measurement is an indication of the colloid concentration which we have found is two to four times the particle density most suitable for magnetic etching. The 50-100 gauss concentration is achieved merely by diluting with distilled water. The lower colloid density usually has been most satisfactory since it reduces "clouding" over the microstructural details by the colloid particles.

Application of the Ferrofluid on the specimen follows the technique shown in Figure 1. We have found the most desirable specimen size to be 1/2-2 cm square. A specimen should be mechanically polished so as to be free of detectable scratches, then lightly electropolished to provide a scratch-free, undisturbed surface. The best mechanically polished specimen retains a sufficiently disturbed surface to preclude sharp magnetic patterns. The mechanically polished surface usually is slightly in relief to the surrounding mount, so a light electropolish must be limited to removing the relief effect and any evidence of the mechanically polished surface. A specimen profile that is lower than the specimen mount will cause the colloid layer to be too thick to relate the colloid pattern to the substrate. The recommended specimen size is a distinct advantage in following these very important preparation steps. The specimen mount should be very dense. An epoxy resin is very suitable; whereas, bakelite is too porous and will absorb the colloid carrier.

The dispensing syringe (see Figure 1) should be small. The plastic tuberculin-type syringe with a 1 cc capacity is easily controlled when dispensing the required small volume, less than one drop -- approximately 5 microliters. We have found the cover glass, or cover slip, should be the No. 0 type (22 mm^2 x 0.085-0.12 mm thick). An objective designed for viewing a specimen through a

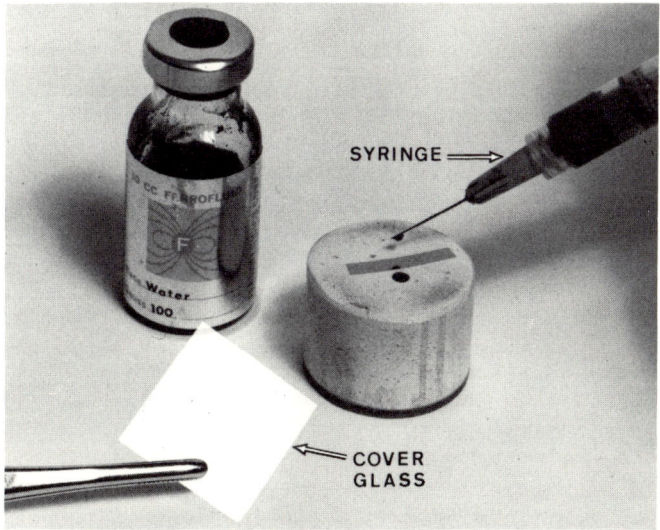

Figure 1. Application of Ferrofluid. Less than 1/4 drop (approximately 5 microliters) of the iron colloid is applied to the specimen. The cover glass is positioned over the colloid to form a thin fluid layer between the glass and specimen surface.

cover glass might be thought desirable, however, metallographic objectives seldom carry this classification, and our work has been conducted on a Bausch and Lomb Research metallograph with very satisfactory results.

After positioning the ultrasonically cleaned cover glass over the colloid and specimen, the colloid should remain under the perimeter of the cover glass. If the colloid fluid boundary extends beyond the perimeter of the cover glass, the glass should be removed, the specimen surface cleaned with detergent and the process repeated. By confining the colloid within the perimeter of the glass, the fluid surface tension secures the cover glass in place on the specimen mount. Generally, microscopic examination of magnetic patterns are made at high magnifications using immersion oil. If the cover glass is not held in place by the surface tension of the colloid, the glass will float between the two fluids and the examination becomes difficult.

Optical Examinations

An iron filings display of the flux pattern of the $11\frac{1}{2}$ cm inside diameter magnet coil and the position of the coil on a metallograph are shown in Figure 2. A schematic display is shown in Figure 3 with the various components in place for the examination. Another schematic demonstration is shown in Figure 4 to display the response of the colloid to ferromagnetic martensite and delta ferrite in an austenitic matrix.

In conventional etching a metallographic microstructure is in a static state. A very noticeable change is observed by the metallographer in the application of magnetic etching. Although the colloid particle size is less than 300 Å and is not resolvable with the optical microscope, motion is faintly observed at 1000X. This motion is due to Brownian movement - the random thermal agitation produced by impact of the colloid particles with molecules of the liquid. Motion is also present if the magnet is energized and de-energized with specimens that are fully or partially ferromagnetic. Masses of colloid particles collect over a ferromagnetic phase or constituent in a magnetic field, then disperse in a variety of patterns when the magnet is de-energized. This particle mobility can offer a most interesting and informative display.

An example of magnetic domain patterns as observed on delta ferrite in an austenitic matrix is shown in Figure 5. Details on the study of delta ferrite in a CF8 casting have been reported (17).

Interpretations of the response of the colloid to a ferromagnetic phase or constituent in a paramagnetic matrix must not be based merely on the respective activity or non-activity of the par-

MAGNETIC ETCHING WITH FERROFLUID

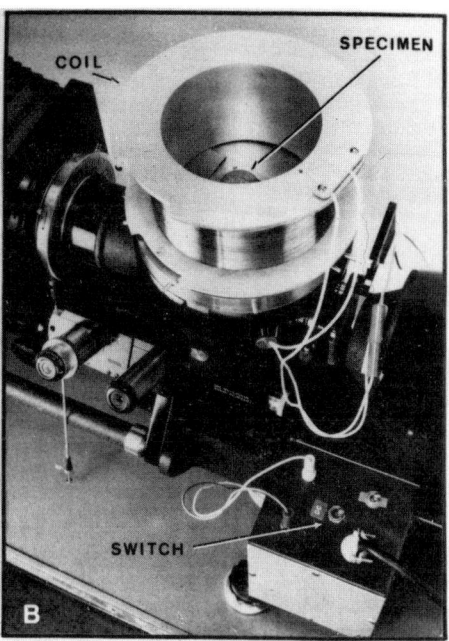

Figure 2. A - Magnetic Etching Coil with Iron Filings Display
          B - Coil and Specimen in Position on Metallograph Stage
The aluminum spool, 1/8" wall, 4½" ID, 2-7/8" high, has 800 turns of #18 AWG, double glass covered copper wire. The coil energized by direct current supplied from a rectifier and is operated at 4.0 amp, 37 volts. Current polarity is controlled by a switch.

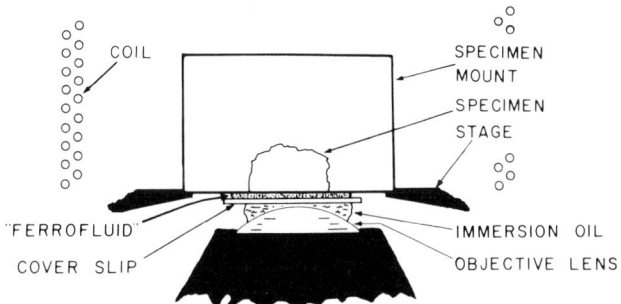

Figure 3. Schematic display of specimen and coil in position on the metallograph stage.

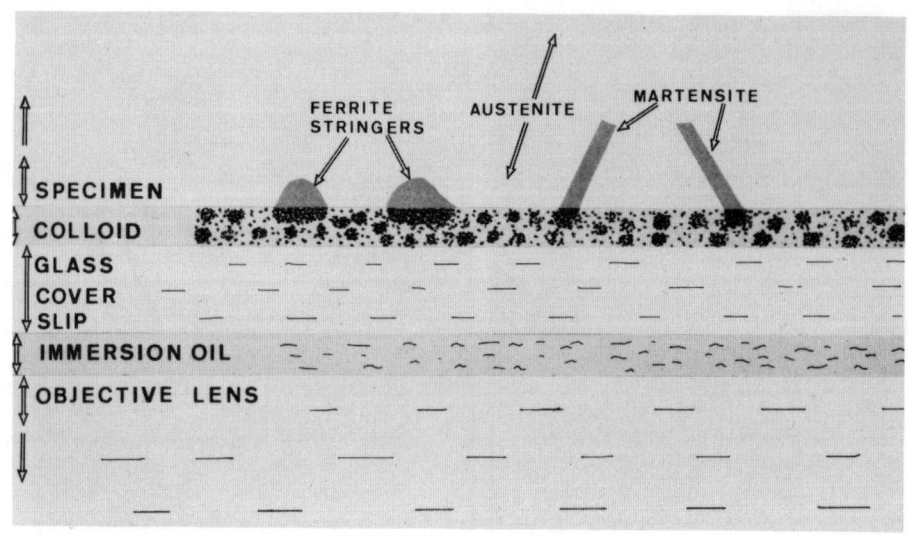

Figure 4. Schematic display of the response of the colloid particles to ferrite, austenite, and martensite.

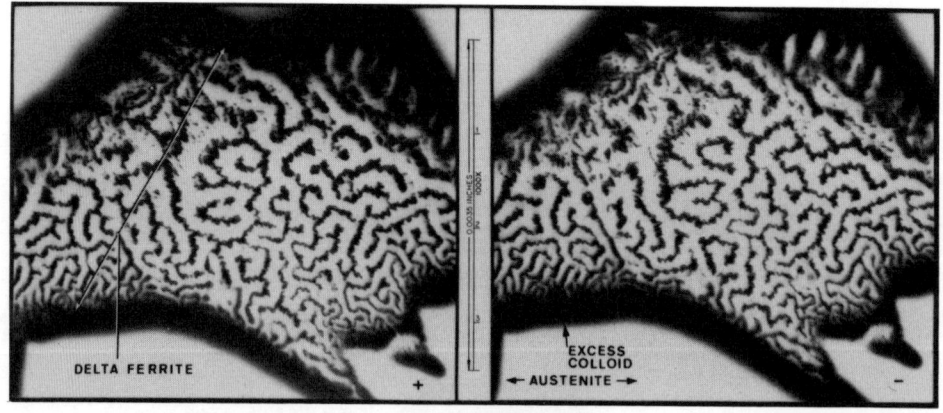

Figure 5. "Coral" type domain patterns on delta ferrite. The magnetic delta ferrite in paramagnetic austenite offers an ideal display of the sensitivity of magnetic etching.

ticles. If the matrix is paramagnetic austenite with some ferromagnetic ferrite or martensite present, the relative amount of these two constituents is quite important. If the principal microstructure is paramagnetic austenite with a minor amount of a ferromagnetic material, the colloid response will follow the pattern seen in Figure 6. Although the austenite is paramagnetic, the colloid particles respond to form "colloid colonies." The inexperienced eye might first misinterpret this activity over the austenite to indicate a ferro- or ferrimagnetic substrate. These four photomicrographs show the colloid attracted to a ferromagnetic ferrite stringer with a majority of the colloid collected in colonies over the austenite. These colloid colonies display restricted Brownian movement in the magnetic field. The size of the colonies and the confinement of the colloid over a paramagnetic area are inversely related to the magnetic flux.

An example of the colloid sensitivity is shown in Figure 7. This microstructure represents a sufficient distribution and size of delta ferrite to cause the colloid particles to line up from one phase to another in ribbon-like patterns in the absence of a magnetic field. By rotating the metallograph stage 90 degrees, the colloid bands would break up and regroup in the north-south direction.

Magnetic etching is a very sensitive technique to detect the formation of strain-induced martensite in some types of austenitic stainless steel. Microstructural comparisons of the undeformed shoulder and test areas of a fatigue test specimen of the type shown in Figure 8 can be very informative. Two examples are shown in Figures 9 and 10. Two type 301 stainless steel fatigue specimens were heat treated prior to their fatigue tests, so one specimen contained a ferrite-austenite microstructure (Figure 9) and the other test specimen was fully austenitic (Figure 10). The magnetic etching displayed the morphological distribution of the ferromagnetic constituents in the microstructure as related to prior heat treatment and the fatigue tests. Apparently the transformation to martensite (during the fatigue test) in the specimen heat treated at a lower temperature occurred primarily at the grain boundaries. The fully austenitized specimen displayed primarily intragranular transformation to martensite.

A type 304 stainless steel fatigue specimen containing some ferrite stringers in an austenitic matrix was post-test examined near the shoulder where deformation would be minimal and in the test region (see Figure 8) where deformation causing strain-induced martensite transformation would be extensive. Photomicrographs of these two areas are shown in Figures 11 and 12. The microstructure showing minor amounts of transformation (Figure 11) also shows a ferrite stringer. The presence of this stringer had no apparent

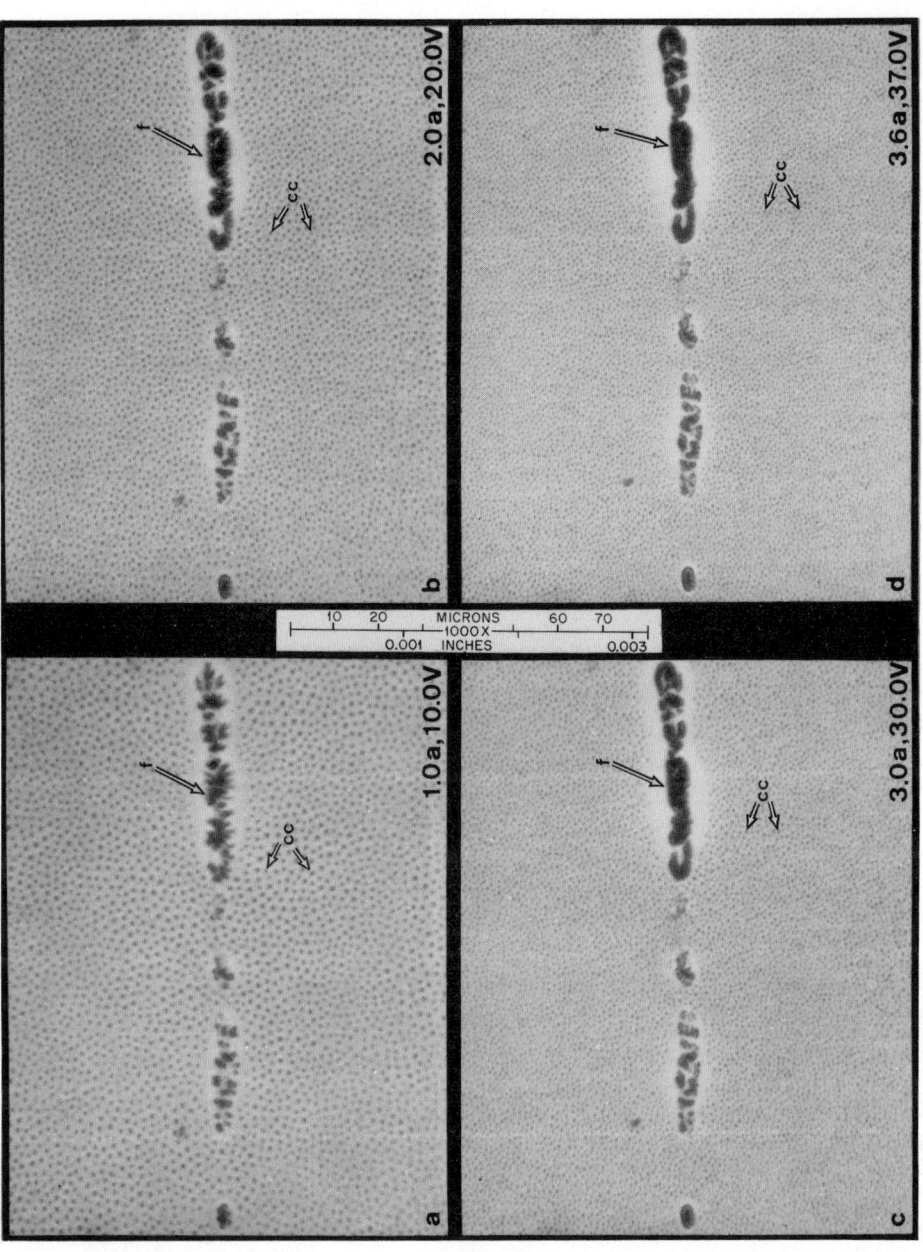

Figure 6. Colloid patterns on paramagnetic austenite and ferromagnetic ferrite with increasing magnetic flux. Note the influence of the magnetic flux on the size and distribution of the colloid colonies (cc) and the confinement of the colloid particles over the ferrite (f).

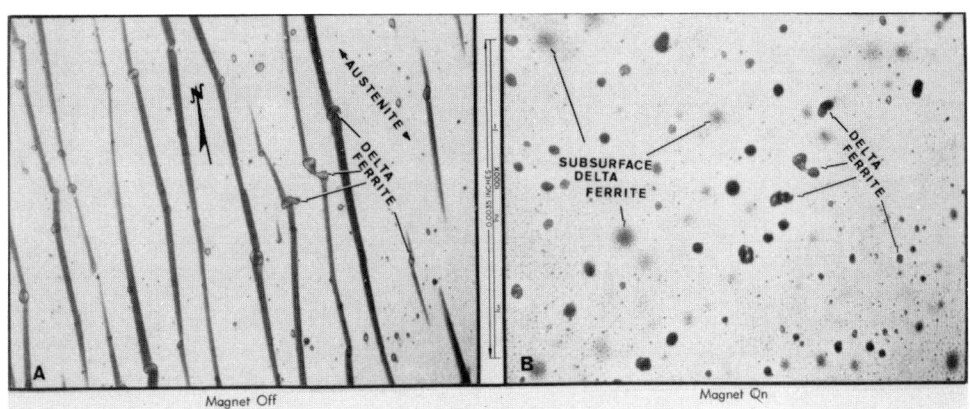

Figure 7. Demonstration of colloid response to weak magnetic field. A type 304 stainless steel specimen was heat treated at 1950°F (1066°C) for one hour and water quenched to convert the ferrite to austenite. Some ferrite remained and served to align the colloid in bands as per the magnetic field of the earth.

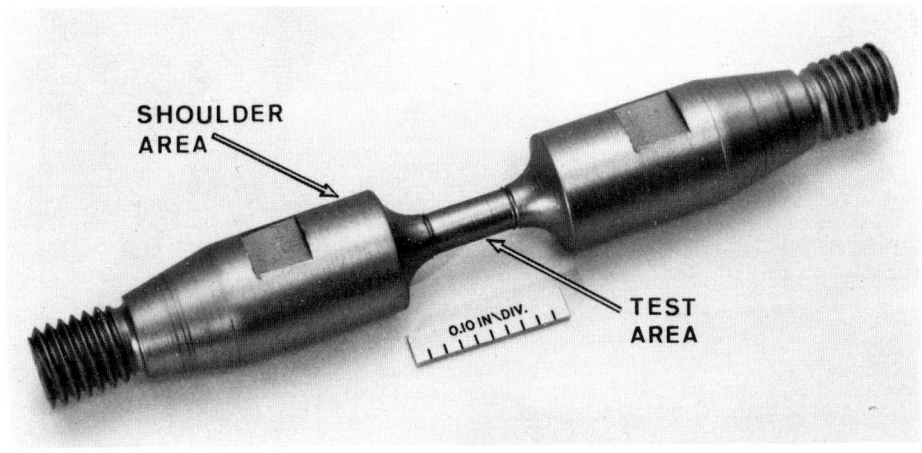

Figure 8. Fatigue test specimen. Microstructural comparisons are made of the shoulder and test areas to evaluate the effects of the fatigue tests.

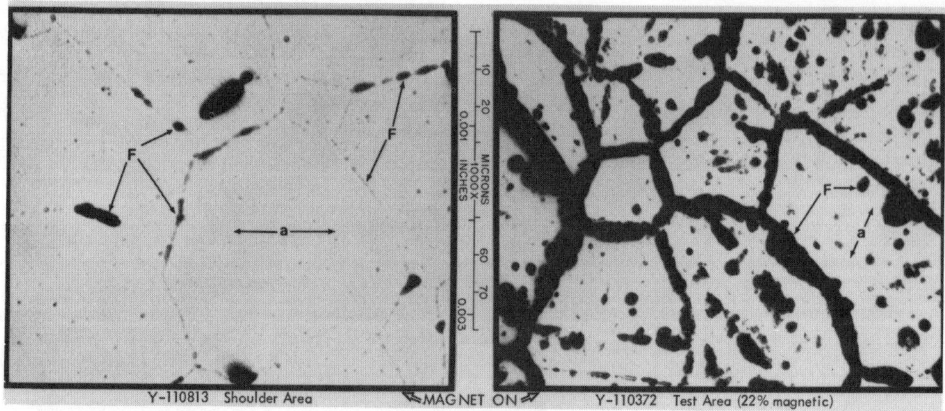

Figure 9. Strain-induced martensite in type 301 stainless steel as revealed by ferromagnetic etching. Specimen was annealed at 1750°F (954°C) for one hour prior to fatigue test. Shoulder area is austenite with grain boundary ferrite; the test area is grain boundary ferrite and strain-induced martensite in austenite. A major part of the transformation to martensite in the test area has occurred at grain boundaries.

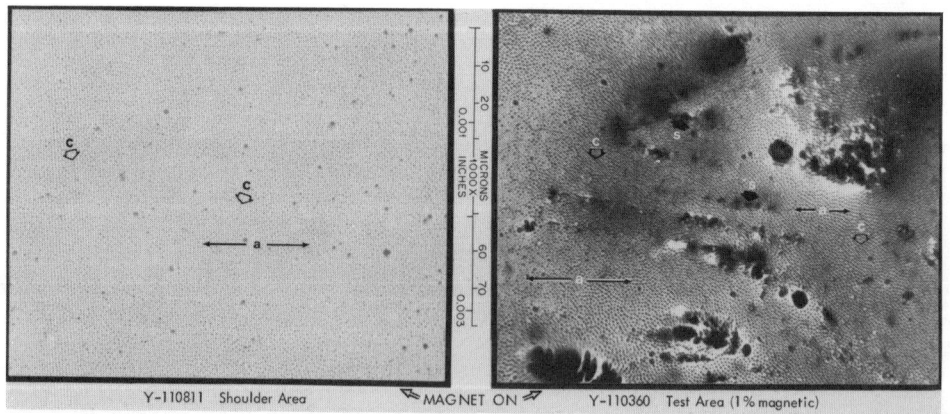

Figure 10. Strain-induced martensite in type 301 stainless steel as revealed by ferromagnetic etching. Specimen was annealed at 2000°F (1093°C) for one hour prior to fatigue test. Shoulder area is austenite; the test area is strain-induced martensite in austenite.

MAGNETIC ETCHING WITH FERROFLUID

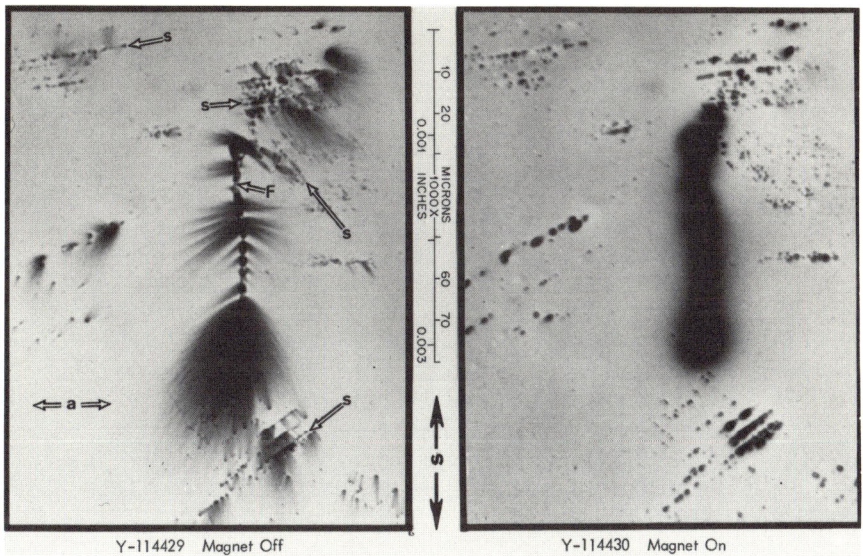

Figure 11. Strain-induced martensite in type 304 stainless steel fatigue specimen as revealed by magnetic etching. This region of low density shear (s) is near the shoulder. The ferrite stringer (F) has no apparent effect on the transformation of the austenite (a) to martensite (s).

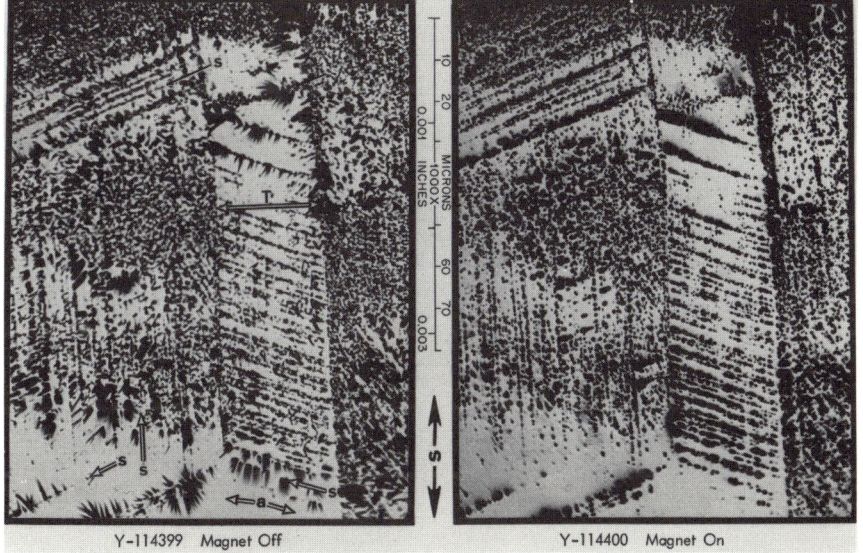

Figure 12. Strain-induced martensite in type 304 stainless steel fatigue specimen as revealed by magnetic etching. Region of high density shear (s) is in the gauge area. Note the twin (T) boundaries. Some retained austenite (a) can be seen.

influence on the nucleation of the strain-induced martensite. The microstructure in the test area shows high density martensite transformation in a region containing annealing twin boundaries that were present prior to the fatigue test.

A type 304 stainless steel tensile specimen tested to failure was magnetically examined from the undeformed shoulder area to the highly deformed area near the fracture, and photomicrographs were made of selected areas, as shown in Figure 13. The heterogeneous transformation of the austenite to martensite is vividly displayed even very near the fracture.

The magnetic microstructures of a type 308 stainless steel weldment are shown in the weld zone in Figure 14. This specimen had been heat treated to transform the ferromagnetic ferrite to a paramagnetic sigma. Minute transformation to sigma can be seen after a 100 hour treatment and is very evident after the 1000 hour treatment.

The detection of a ferromagnetic constituent in a Fe-16 Cr-8 Ni-2 Mo weldment by magnetic etching is shown in Figure 15. This 1.6-cm thick weldment is shown in cross section in Figure 15a. The root and second pass display a cored ferrite structure in an austenitic matrix. The subsequent passes were made by a weave (oscillation) process and contain an acicular ferromagnetic structure in an austenitic matrix. This acicular structure was assumed to be martensite, although x-ray diffraction (18) indicated no tetragonality representative of martensite, so a more positive identification is underway. Considerable tempering of the first three or four weld passes is evident from the presence of domain patterns seen in Figure 15c. Such patterns are not seen in microstructures with a high dislocation density. The delta ferrite and acicular structure can be observed in considerable detail with magnetic etching. Further verification of the observed microstructure by another technique, such as chemical etching, has been only moderately successful, however, phase contrast microscopy has been a complementary application. This microscopy approach is ideal for enhancing the resolution of microtopographical surfaces. The same final preparation--an electropolished surface--on the same specimen was used for phase contrast microscopy, and the results are shown in Figure 16. The cored ferrite, acicular structure, and austenite are well defined. The phase contrast microscopy approach was not intended as an aid in the identification of the acicular structure, but merely to verify the patterns observed by the magnetic etching technique.

The most dramatic application of magnetic etching that we have found has been to show the presence of a ferromagnetic phase in a contrasting paramagnetic matrix. However, interesting results

MAGNETIC ETCHING WITH FERROFLUID

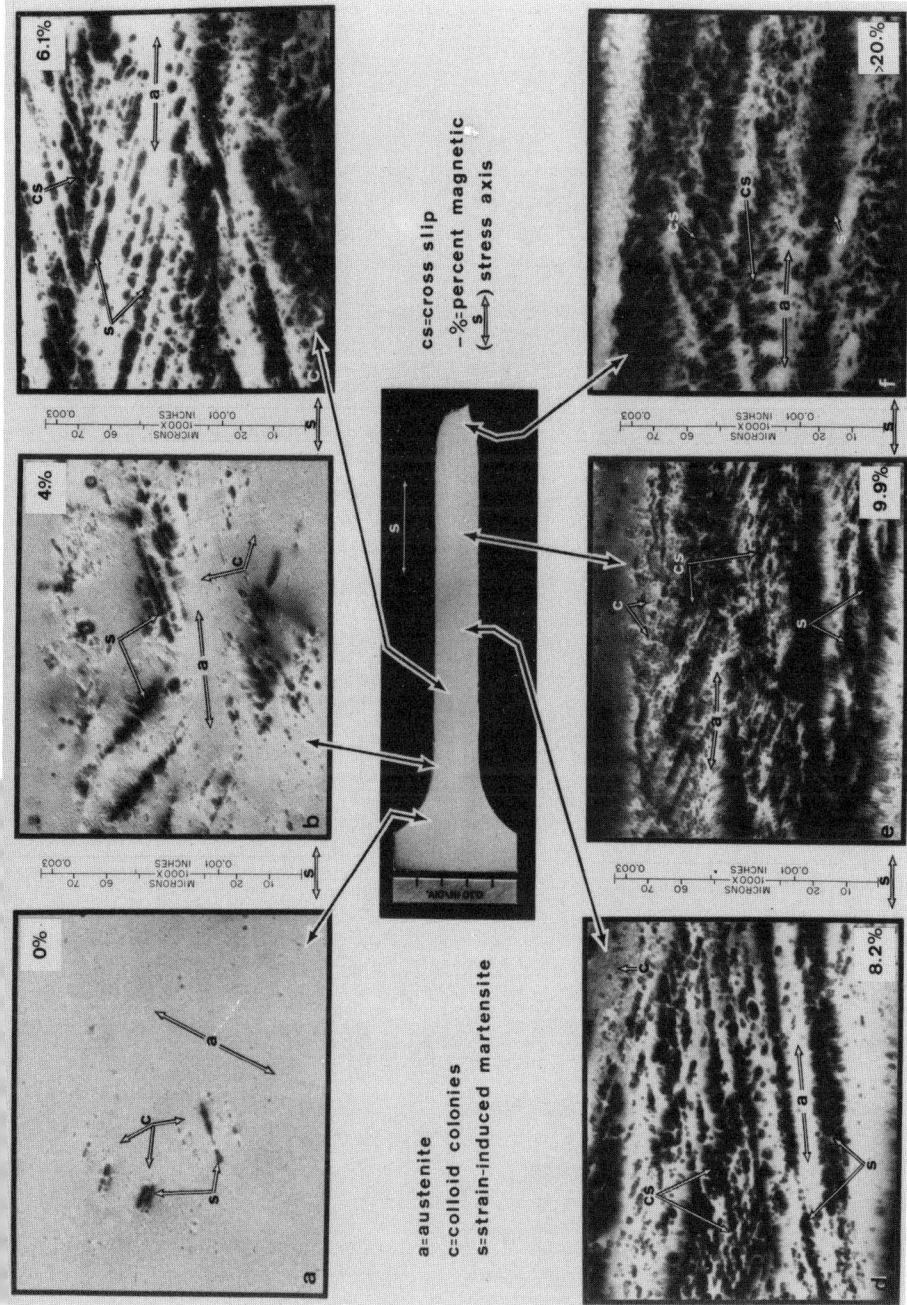

Figure 13. Detection of strain-induced martensite in type 304 stainless steel tensile specimen by magnetic etching.

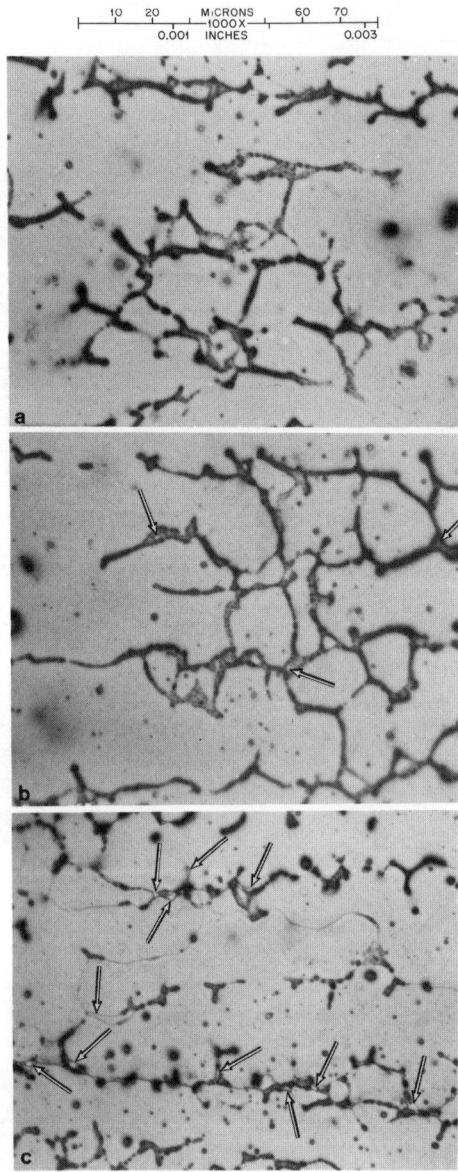

Figure 14. Microstructure of type 308 stainless steel weldment
   a. As received
   b. 100 hours at 1100°F
   c. 1000 hours at 1100°F

Ferromagnetic ferrite attracts colloid. Arrows point out paramagnetic sigma phase thermally transformed from ferrite. Transformation is easily detected after the 1000 hour treatment.

MAGNETIC ETCHING WITH FERROFLUID 169

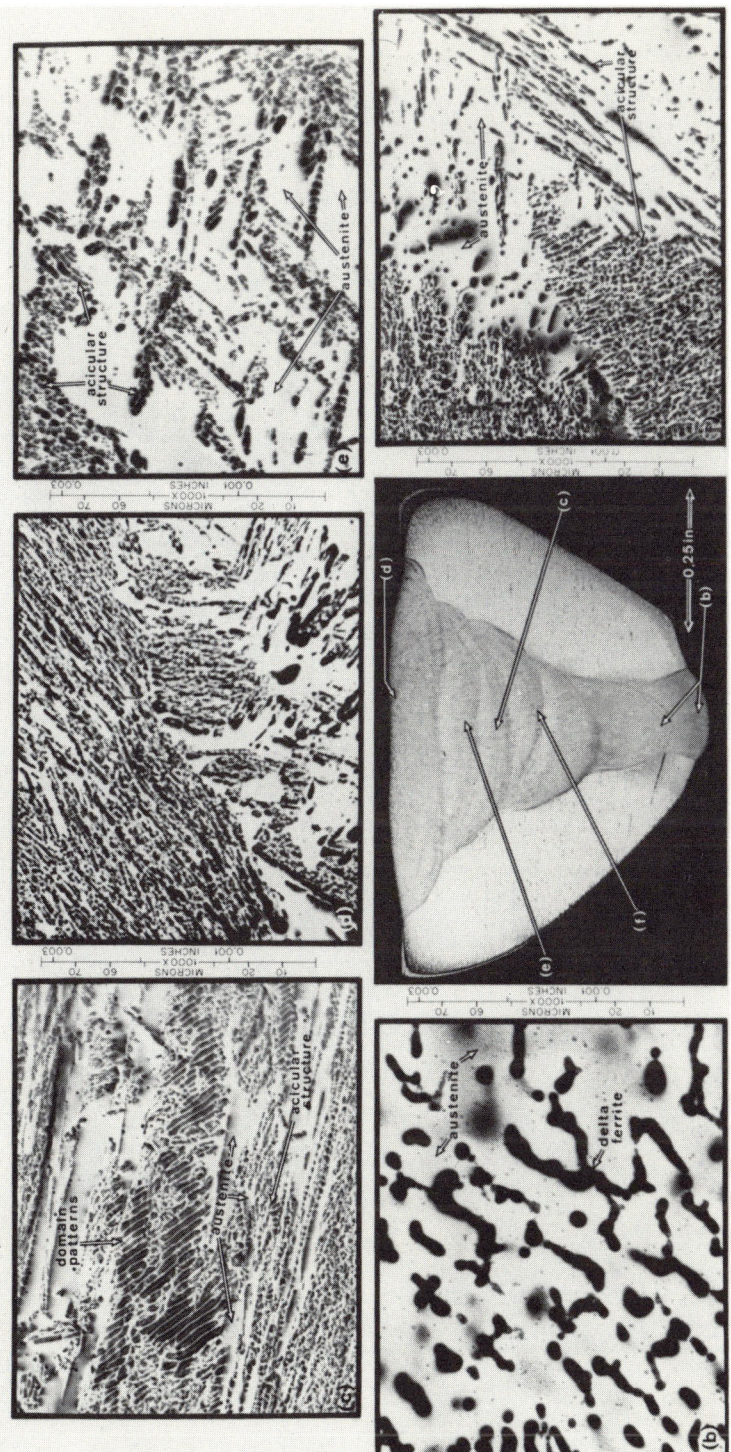

Figure 15. Detection of delta ferrite and an acicular structure in 16 Cr-8 Ni-2 Mo weldment by magnetic etching. A photomacrograph (a) of a cross section of the weldment after chemical etching is shown. Photomicrographs (b-f) show microstructural patterns produced by colloid particles (<300Å) in a magnetic flux. Root pass and second pass (b) were made manually and contain ferromagnetic delta ferrite in a paramagnetic austenitic matrix. The third-seventh passes (c-f) were made by weave (oscillation) process and show a ferromagnetic acicular structure in a paramagnetic austenitic matrix. Tempering effects by the last passes influenced the formation of a more ordered lattice in the acicular structure which caused the colloid to form domain patterns (c).

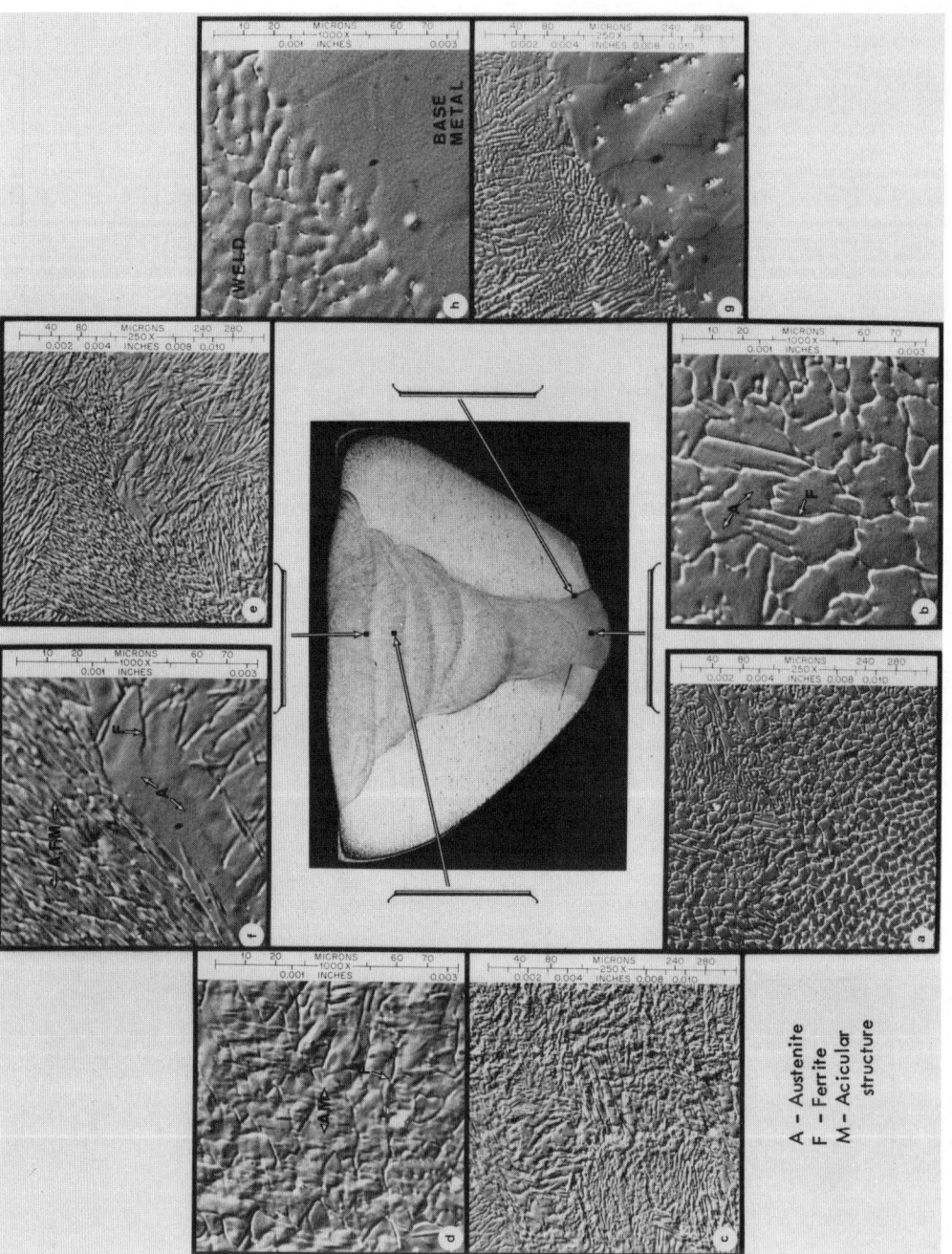

Figure 16. Photomicrographs of 16 Cr-8 Ni-2 Mo weldment with phase contrast microscopy.

A – Austenite
F – Ferrite
M – Acicular structure

were found in the examination of a type 416 stainless steel specimen that contained ferromagnetic stringers (ferrite) in a ferromagnetic matrix (martensite). In this case, there are two different ferromagnetic constituents present. The results are shown in Figure 17. The ferrite stringer displays a more ordered domain pattern than the pattern over the martensite.

Domain patterns on iron oxide specimens are shown in Figures 18 and 19, and patterns on lithium-iron specimens are shown in Figures 20 and 21. Since these materials were ceramic and final electropolishing was not possible, the patterns are shown on a mechanically polished surface and do not display the usual sharp contrast of an electro-polished surface.

Iron-2.75-4.50 weight % silicon sheet steel is of great value as a primary component in electrical transformers and motors. Grain oriented sheet steel is highly desirable to reduce the operating noise level initiated by magnetostriction, which varies with the flux density. Transformer efficiency can be increased up to 30% by the use of grain-oriented sheet steel (19). Of the three principal crystallographic directions - (100), (110), and (111)-magnetization is easiest in iron silicon sheet in the (100) direction. Magnetic domain patterns on this material with a cube-on-face preferred texture can be seen in Figures 22 and 23. The pairs of pho-

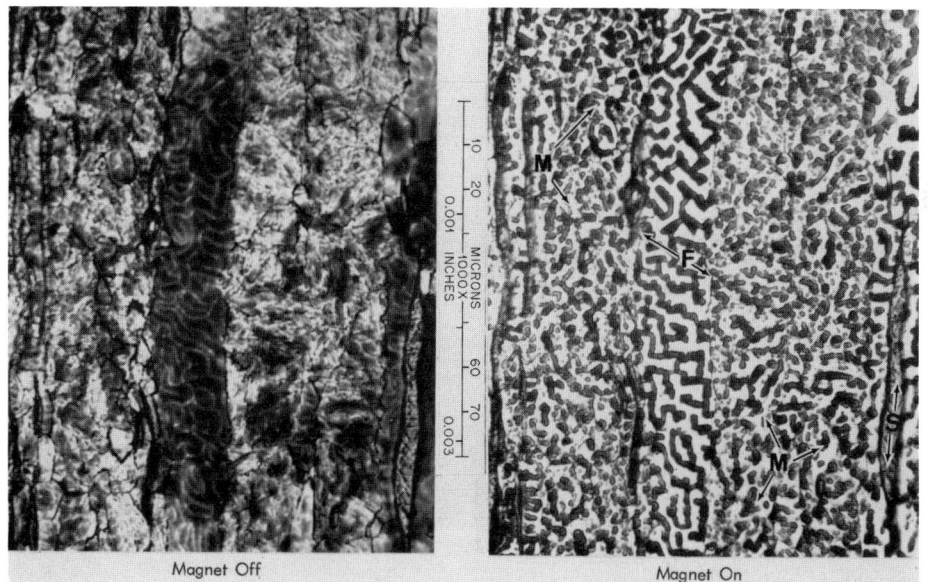

Figure 17. Magnetic etching of type 416 stainless steel. Note the irregular pattern on martensite matrix (M) and sharp patterns on ferrite stringer (F). A sulfide stringer (S) can be seen at lower right.

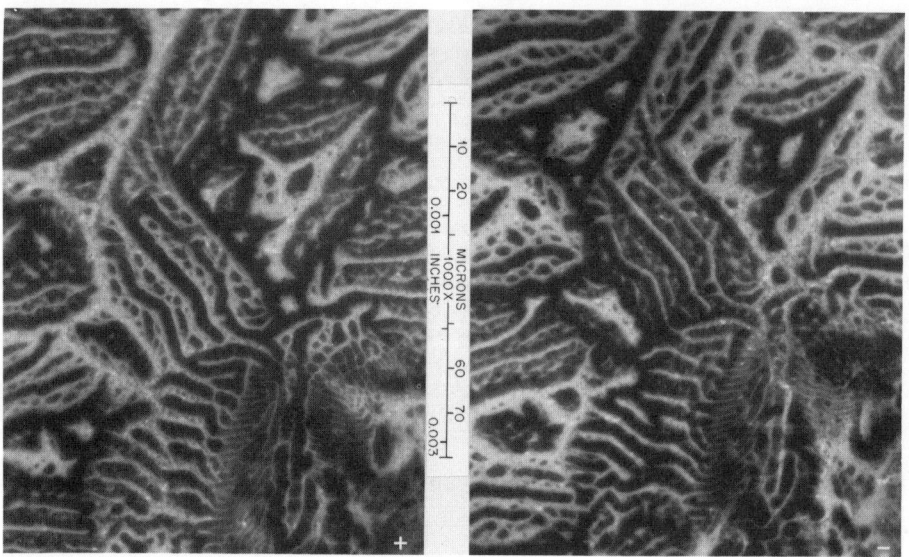

Figure 18. Magnetic domain patterns on Fe_3O_4. Note the nearly reversed patterns with change in polarity of direct current to the coil.

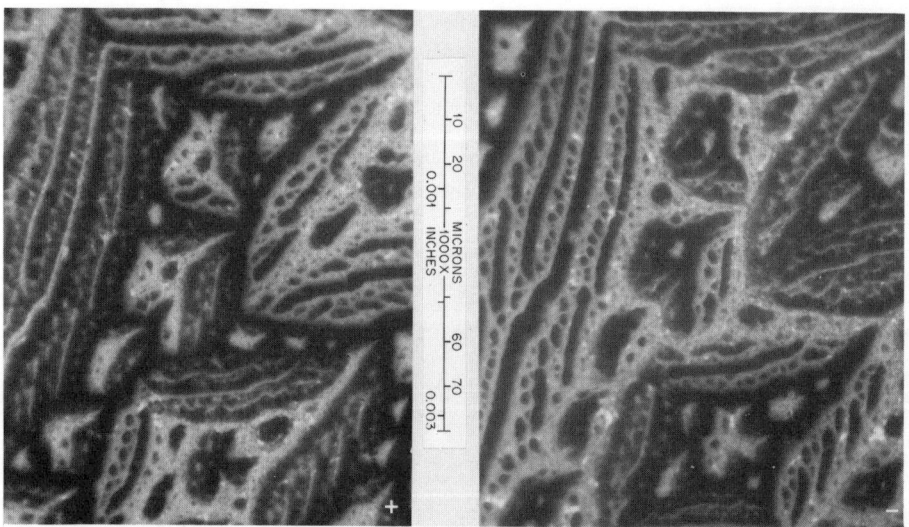

Figure 19. Magnetic domain patterns on Fe_3O_4. Mechanically polished. Note the nearly reversed pattern with change in polarity of direct current to the coil.

MAGNETIC ETCHING WITH FERROFLUID 173

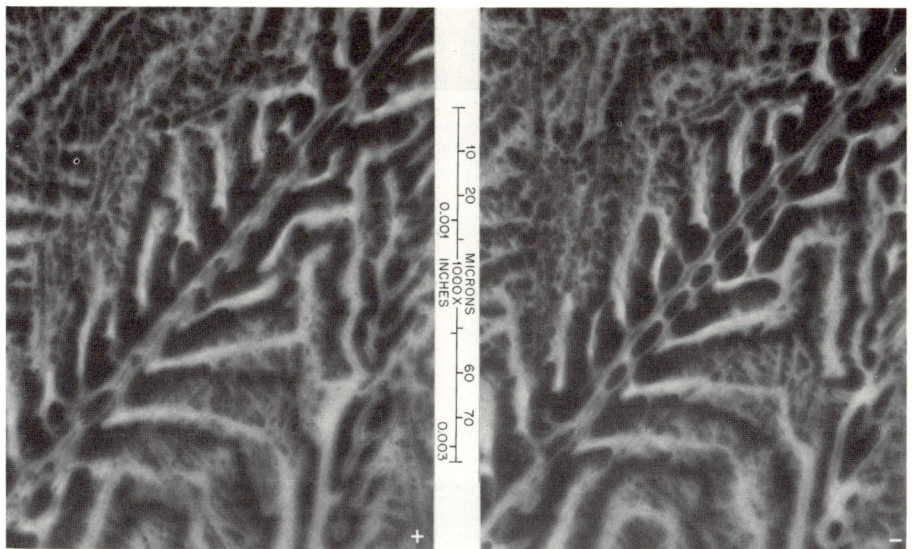

Figure 20. Magnetic domain patterns on $Li_2Fe_{2.8}O_4$. Mechanically polished. Note the nearly reversed pattern with change in polarity of direct current to coil.

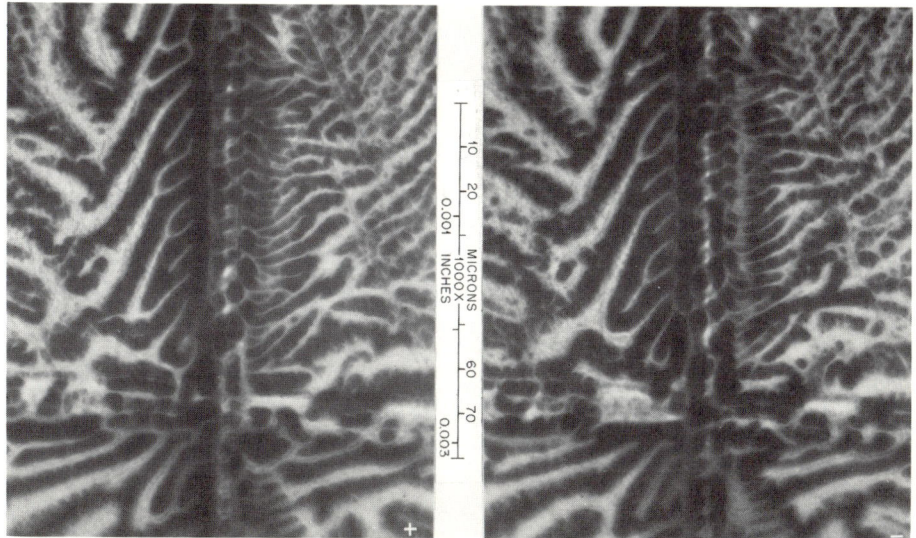

Figure 21. Magnetic domain patterns on $Li_2Fe_{2.8}O_4$. Mechanically polished. Note the nearly reversed pattern with change in polarity of direct current to coil.

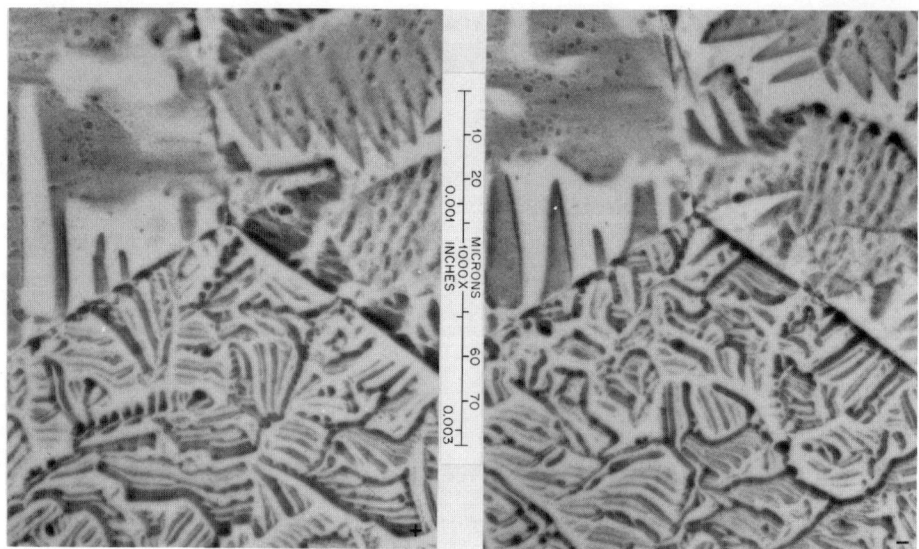

Figure 22. Magnetic domain patterns on iron-silicon sheet-steel. Triple point of grain boundaries. Note these grains do not display reversed patterns usually observed in a majority of specimens with change in direct current polarity. Electrolytically polished.

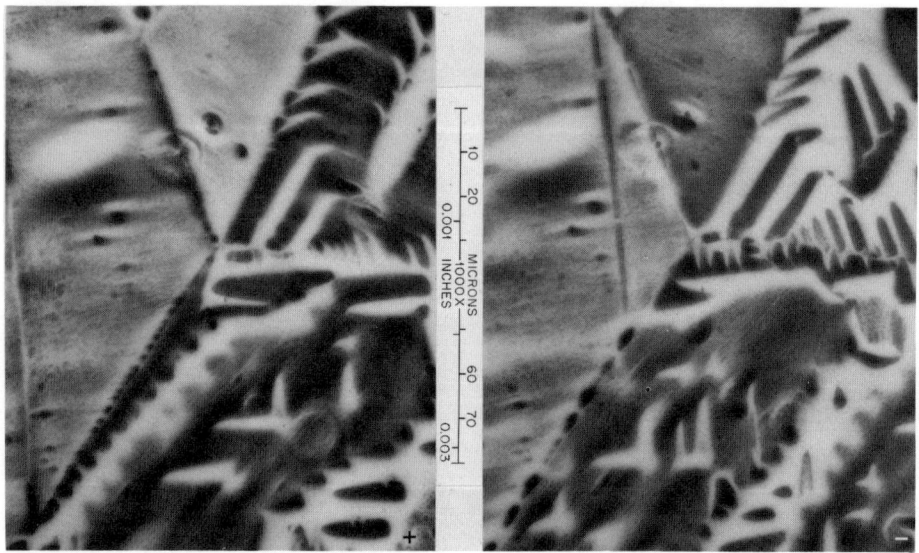

Figure 23. Magnetic domain patterns on iron-silicon sheet-steel. Triple point of grain boundaries. Note some intragranular patterns are reversed with change indirect current polarity to the coil. Electrolytically polished.

tomicrographs display grain triple points of the same microscopic areas with the microstructural variations (formed by the colloid patterns) attributed to changes in crystallographic orientations. Pattern changes are effected on each site if the polarity of the direct current to the coil is reversed. In most ferromagnetic materials a reverse in polarity of the current induces a direct reversal of the pattern, as demonstrated in the delta ferrite shown previously in Figure 5. Close study shows the domain patterns in some of the grains in Figures 22 and 23 do not follow this trend. Photomicrographs of the interior of the grain in the lower right of Figure 23 are shown in Figure 24, and the reversed pattern just described can be seen. The explanation for some grains displaying the reversed pattern and some grains not following this trend is not known.

CONCLUSIONS

Magnetic etching is one of the more potent tools in the metallography laboratory. The availability of a commercial colloid and the design of a larger magnetic coil greatly increase the potential for this application over previous uses. The observations of microscopic ferromagnetic phases and constituents are now limited to the resolution of the optical microscope. Delta ferrite, strain-

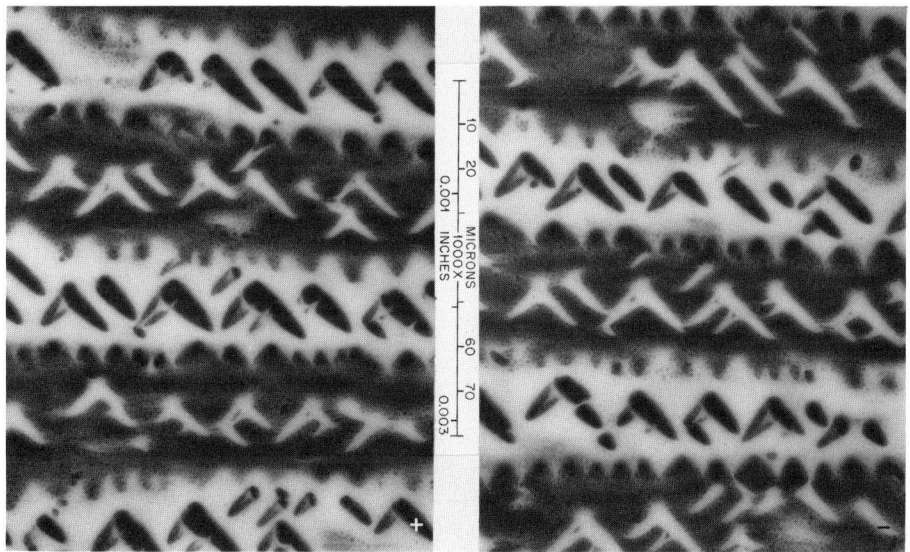

Figure 24. Magnetic domain patterns on a single grain of iron-silicon sheet steel. The reversed pattern with the change in polarity of the direct current to the coil is evident. Electrolytically polished.

induced martensite, and several other ferromagnetic constituents can be seen in a totally different mode that is complementary to conventional chemical and electrolytic etching techniques. Paramagnetic sigma phase isothermally transformed from ferromagnetic ferrite can be detected in a more positive manner than the questionable identification by chemical staining.

The full technological significance of domain patterns for metallographic evaluations remains to be resolved. The morphological relationship of the patterns to the crystallographic orientation of the substrate is significant since definite pattern changes across grain boundaries in iron-silicon sheet steel are shown.

Magnetic etching is a very fascinating tool. Much of this attraction is generated from the dynamic display as the colloid patterns are formed. As metallographers become more aware of this method of studying some ceramics, metals, and alloys, it will assume its rightful place in the laboratory.

ACKNOWLEDGEMENTS

Grateful thanks are extended to E. S. Bomar and R. S. Crouse for their suggestions in reviewing this paper and to Mary Combs for her assistance in preparing the manuscript.

REFERENCES

1. F. Bitter, "On inhomogeneities in the magnetization of ferromagnetic materials," Phys. Rev. 38 (1931) p. 1903.
2. L. von Hámos and P.A. Thiessen, "Über die Sichtbarmachung von Bezirken verschiedemen ferromagnetrsihen Zustandes festen Korpen," Z. Phys. 71 (1931) p. 442.
3. F. Bitter, "Experiments on the nature of ferromagnetism," Phys. Rev. 41 (1932) p. 507.
4. L. W. McKeehan and W.C. Elmore, "Surface magnetization in ferro magnetic crystals," Phys. Rev. 46 (1934) p. 226.
5. Handbook of Chemistry and Physics, 48th Edition, 1967-68, p. F-96, The Chemical Rubber Co., 18901 Cranwood Parkway, Cleveland, Ohio 44128.
6. H. S. Avery, V. O. Homerberg, and E. Cook, "Metallographic Identification of Ferro Magnetic Phases," Metals Alloys 10 (1935) pp. 353-355.
7. W. C. Elmore, "Ferromagnetic Colloid for Studying Magnetic Phases," Phys. Rev. 32 (1938) pp. 309-310.
8. E.A.M. Harvey, "Metallographic Identification of Ferro-Magnetic Phases," Metallurgia 32 (June 1945) pp. 71-72.
9. P.F. Weinrich, "Mciroferrographic Technique," Australasian Engr. (November 1948), pp. 42-44.

10. George F. Fisinai, "Magnetic Oxide 'Etchant'," Metal Progr. 17 (October 1956) pp. 120-122.
11. R. Carey and E.D. Isaac, eds., Magnetic Domains and Techniques for Their Observation, Academic Press, New York, 1966.
12. R.J. Gray, "Revealing Ferromagnetic Microstructures with Ferrofluids," Proceedings of the International Microstructural Analysis Society, Denver, Colo. Sept. 21-23, 1971, International Microstructural Analysis Society, Northglen, Colo. 80221.
13. R. J. Gray, Revealing Ferromagnetic Microstructures with Ferrofluid, ORNL-TM-368, Oak Ridge National Laboratory, Oak Ridge, Tenn. (March 1972).
14. R.J. Gray, "The Detection of Ferromagnetic Phases in Types 304 and 301 Stainless Steels by Epitaxial Ferromagnetic Etching", Microstructural Science, edited by R.J. Gray and J. L. McCall, American Elsevier Publishing Co., Inc. New York, 1973.
15. R.J. Gray, "The Detection of Ferromagnetic Phases in Types 304 and 301 Stainless Steels by Epitaxial Ferromagnetic Etching," Metallographic Review 2 (2), 1973.
16. Ferrofluid, Ferrofluidics Corp., 144 Middlesex Turnpike, Burlington, Mass. 01803.
17. Francis F. Lucas Award 1971, Metal Progress, 100 (5), 1971.
18. H. L. Yakel, private communication.
19. "Magnetical Soft Materials," ASM Handbook, Volume 1, 8th edition, (1961) pp. 795-797.

DECORATIVE ETCHING*

R. J. Gray, R. S. Crouse, and B. C. Leslie

Oak Ridge National Laboratory

Oak Ridge, Tennessee 37830

INTRODUCTION

Metallography has often been described as both a science and an art. The scientific approach requires an understanding of the chemical reaction of the specimen with an environment (liquid, gas, etc.) that reveals the structure for observation with the microscope. Additional scientific knowledge is needed to relate the observed structure to the composition, history of the material, and phase diagrams. The artistic role in metallography is human contrivance, ingenuity, knack, and sometimes even a "magical" skill for producing results that are aesthetic and informative. Some metallographers are scientifically oriented and others display a pronounced artistic talent; but undoubtedly the dedicated and accomplished--the good--metallographer must display nearly equal measures of both attributes. The opening thrill of seeing the structure come into focus after following the careful steps of specimen preparation approaches or equals the personal enjoyment of an artist who has completed his assignment. This introductory enjoyment must be followed by the equal challenge of ferreting out the minute details in the microstructure that disclose the information and meet the obligations of metallography.

We usually consider that metallography had its beginning with Sorby in 1863, but Cyril Stanley Smith (1) describes the complicated etching of swords, armor, and gun barrels that took place over 500 years ago. Metallography in its very earliest sense could have

---

*Research sponsored by the U.S. Atomic Energy Commission under contract with the Union Carbide Corporation.

been the observation of the characteristics of a fractured surface to control the composition of bronze for a primitive axe.

Unless the term "etching" is further clarified, it is usually assumed to be a chemical attack into the surface of the material. Structural details are revealed in part by a process of selective removal of the structure from the surface downward (2). The resultant microtopographical surface is viewed with a microscope and the observed microstructure is the result of the reflection characteristic of this surface. Undoubtedly, this technique for revealing the microstructure will continue in popularity. Some specimens lend themselves, however, to a nearly opposite approach for microstructural study, which we have designated as "decorative etching". Numerous techniques for this approach for revealing the microstructure have been developed and are described.

The reader is definitely penalized in observing these patterns and several other colorful microstructures in this report because, for economic reasons, the reproductions are in black and white. A report (3) displaying some colorful oxidation patterns and several other aesthetic patterns is available.

TYPES OF DECORATIVE ETCHING

The standard chemical attack method of revealing the microstructure can produce a surface that can be misleading in interpretation, or the chemically attacked topographical surface may need to be re-emphasized by selective coloration to enhance the constituents in the microstructure. Although the term decorative etching may be new, some of the techniques have been used for many years and are merely re-categorized. Decorative etching can be defined as "techniques for revealing the microstructure by adding something to the specimen surface that is controlled by or related to the substrate. This control or regulation by the substrate is attributed to crystallographic orientation or chemical composition (or both) and helps to disclose the microstructure with informative and often aesthetic results." We have expanded decorative etching to include the examination of an as-polished surface with sensitive tint or plane polarized light to reveal the microstructures in spectacular colors.

Alkali Chloride Crystallization

Numerous examples of oriented crystal growth have been reported. The reported (4) nucleation of snow in supercooled clouds is accomplished with remarkable efficiency with silver iodide due to the near fit of the ice and iodide lattices. Johnson (5,6) has demonstrated oriented crystal growth by separate additions of chlorides of sodium, potassium, rubidium, and calcium on metal single crystals of known orientation. Laboratory procedures also described epitaxial growth of chloride on polycrystalline metallic substrates. The significant benefit of this technique is to observe substrate crystallographic orientation by the orientation of the epitaxial crystals.

Selective Oxidation (Thermal)

The oxidation of a metallographic specimen can be an approach for revealing the microstructure. Sometimes the results are informative and colorful. To develop a better understanding of this method of revealing the microstructure, we refer to the reported (7-9) investigations on oxidized single crystal copper spheres. A single crystal sphere exhibits many crystallographic planes on the specimen surface. Thermal oxidation rates were determined on the copper spheres by measuring the thickness of the oxide film as a function of time using a polarizing spectrometer. The relative rates of oxidation of the principal faces were found to be (100), (210), (110), (311) in decreasing order. Oxidation was done at

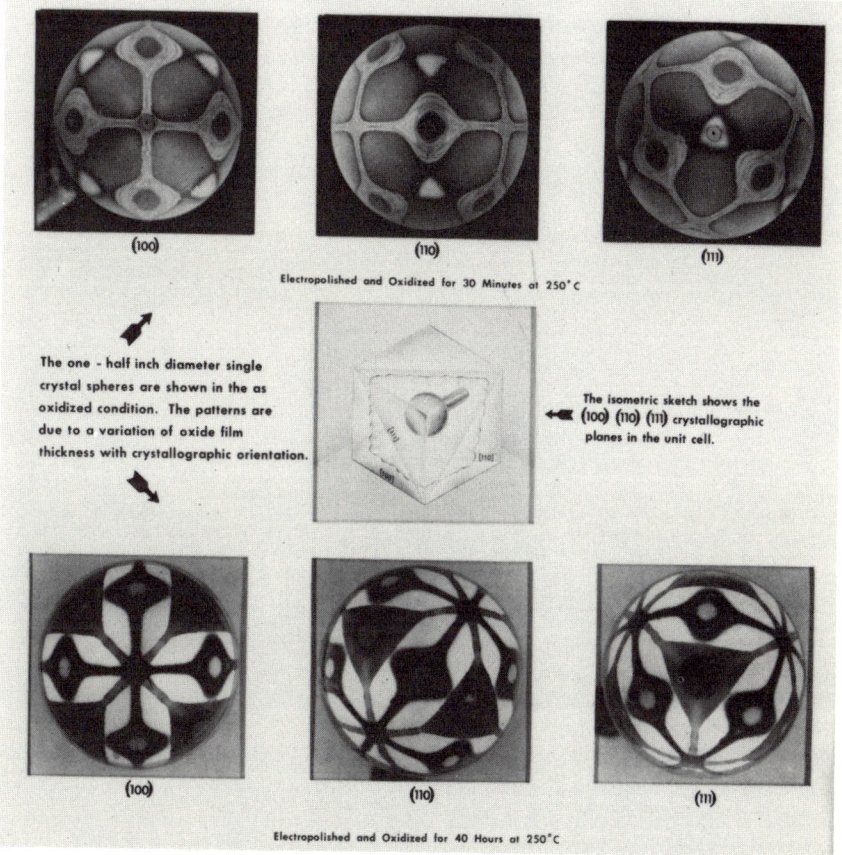

Figure 1. Oxidized single crystal copper spheres.

various temperatures in the range of 70-178°C. Photomicrographs of oxidized single crystal copper spheres are shown in Figure 1.

By observing the surface of the single crystal copper sphere it is possible to better understand the subsequent patterns on the oxidized surface of a polycrystalline copper specimen. The comparison of these two specimens is shown in Figure 2. The polycrystalline copper was electropolished, (10) then heated on a hot plate until the desired coloration was observed. A single crystal copper sphere was oxidized in a carefully controlled closed system so contamination of the oxide layer was minimized or, in some instances, carefully controlled so the environmental conditions were well known. The patterns on the sphere are very sharp and the interference colors are well defined.

The oxidation of single crystals of materials other than copper has been applied in research and development studies. An example of the selective oxidation on a single crystal niobium sphere has been reported (11). A photograph of a 1.25 cm sphere oxidized 24 hr at 400°C is shown in Figure 3. The interference oxide film is tightly adherent in the (100) region; whereas, the (110) and (111) regions exhibit a breakaway oxide formation. Pawel, et al (12) have demonstrated selective platelet oxidation parallel to a particular set of (320) planes of a tantalum single crystal as shown in Figure 4. A comparison of the oxidation patterns on cubic single crystals of copper and copper with 0.1 wt% aluminum, and hexagonal

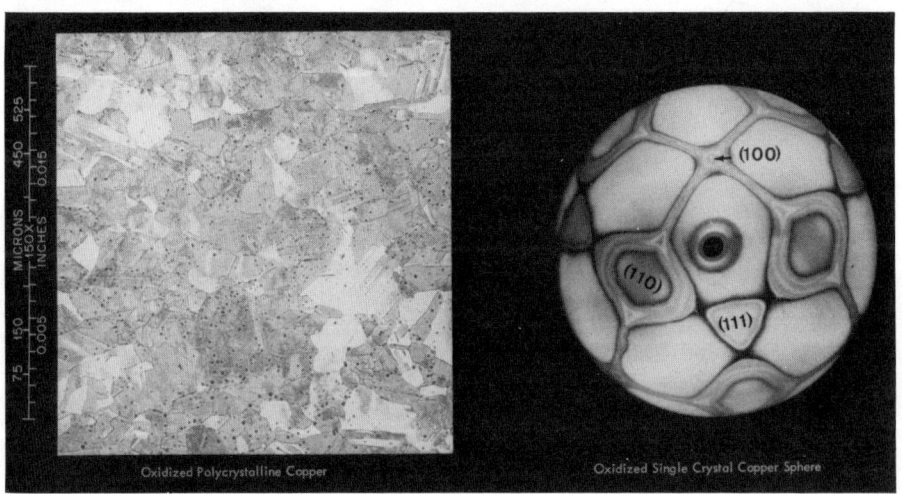

Figure 2. Oxidized polycrystalline copper and oxidized single crystal copper sphere. The polycrystalline copper specimen was electropolished and heat tinted on a hot plate. The single crystal copper sphere was oxidized at 250°C for 30 min at 1 atm O_2.

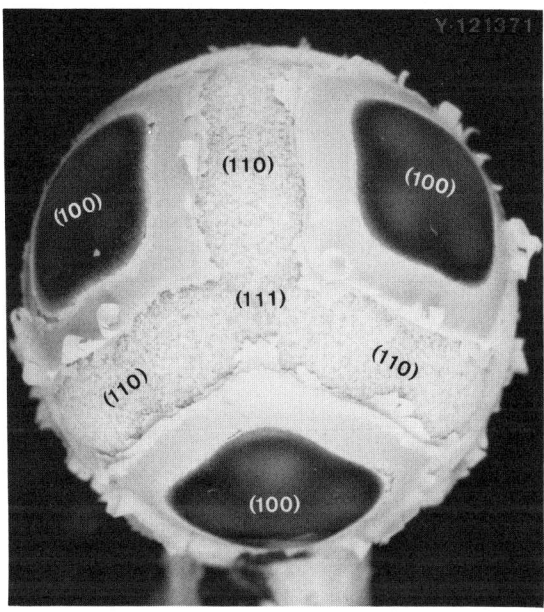

Figure 3. Oxidized surface of single crystal niobium sphere. Crystallographic regions are identified. Note the characteristics of the oxide as related to the substrate orientation.

Figure 4. Tantalum single crystal specimen oxidized at 500°C. Selective platelet oxidation of (320) planes. One set of planes is nearly parallel to specimen surface. Other sets form at angles to the specimen surface.

close-packed crystals of zirconium and hafnium oxidized in a controlled system is shown in Figure 5. With this background, it is possible to develop a basic understanding and appreciation for the patterns obtained in heat tinting a metallographic specimen, as shown in Figure 6. The microstructures displayed in this category have been generated by heating the specimen to well above ambient temperatures.

We cannot dismiss the oft used etching technique in metallography of allowing a specimen to stand at room temperature for selective oxidation. Such an example is shown in Figure 7. A uranium-zirconium alloy (U-14 at.%Zr) was oxidized for 40 min at 900°C. Several conventional etching techniques were applied to reveal the characteristics of the oxide-metallic interface with no success. After exposing the specimen to ambient atmosphere for 48 hr, however, a thin zirconium-rich layer with slender finger-like penetrations into the bulk oxide could be seen at 2000X.

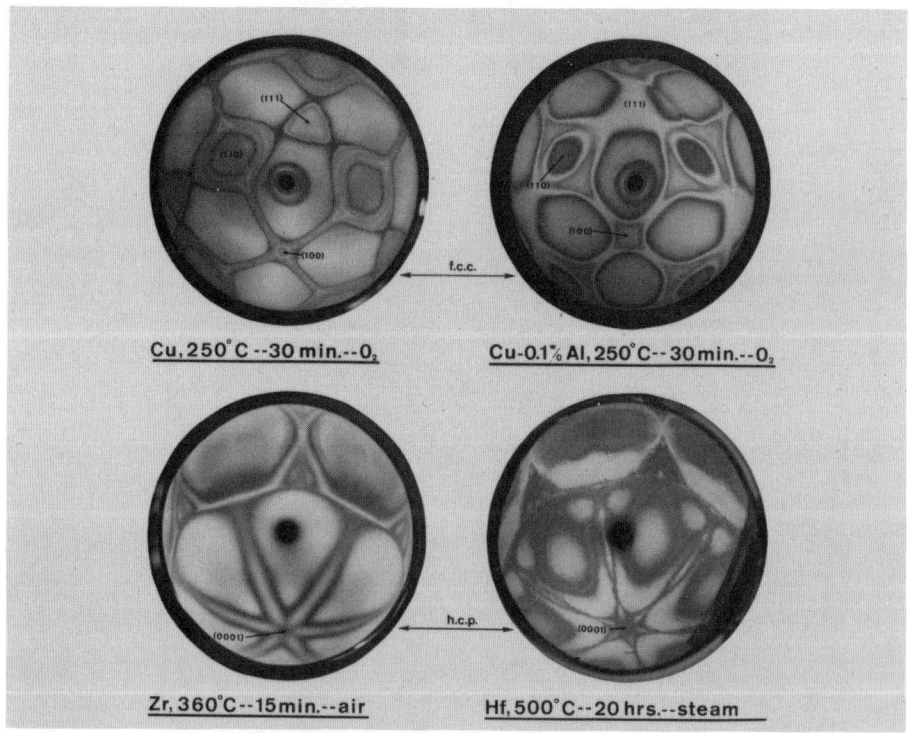

Figure 5. Oxidized single crystal spheres. Note the change in the oxidation patterns of high-purity copper and copper with 0.1 wt% aluminum. Hexagonal close-packed crystals of zirconium and hafnium exhibit very different patterns than the cubic copper and copper-aluminum crystals.

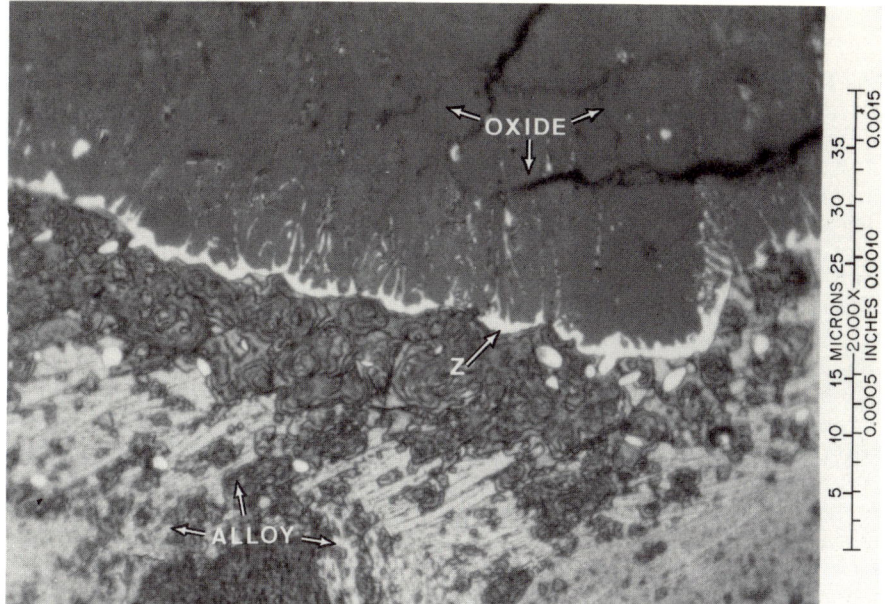

Figure 6. Heat tinted braze joint. Mallory 1000 (bottom) bonded to a cermet (top) with a nickel-palladium braze.

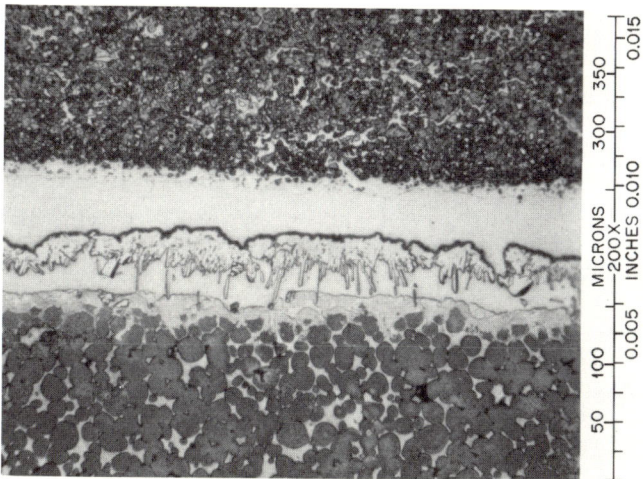

Figure 7. Cross section of oxide-uranium-zirconium alloy (U-14 at. %Zr) after 40 min at 900°C. Polished specimen surface was exposed to ambient temperatures. The alloy (bottom) was non-uniformly oxidized and a thin zirconium-rich layer (arrow Z) was resolved.

Selective Staining-Oxidation (Chemical)

The previous classification is also a chemical reaction, but it identifies etching techniques that are thermally initiated. The etching techniques described in this category are initiated in liquid solutions.

Some etching by chemical staining or oxidation also is performed by heating the etching solution, however, the etchant usually is at 100°C or less. Several materials can be selectively chemically stained or oxidized by immersing the specimen in an oxidizing etchant. Uranium carbides can be treated in this manner. The techniques for preparing uranium carbide specimens and the colorful results obtained have been reported (3,13,14). Examples of these microstructures are shown in Figure 8. We have always considered the colorful stain that develops on uranium carbide to be an oxide film of varying thickness displaying interference colors, the film thickness being regulated by the crystallographic orientation of the substrate. To investigate this hypothesis of the oxide film, a uranium carbide (UC_2) stain-etched specimen was analyzed by Auger electron spectroscopy (15,16). This analyzing technique is specifically designed for thin film analyses. A specimen surface is bombarded with a primary beam of high energy electrons in an ultrahigh vacuum (10^{-10}torr). Identification of the components of the film is related to the spectral energies of the excited electrons. Surface depths of about the first 10 Å can be analyzed. By employing ion sputtering to remove surface layers, analyses of greater depths are accomplished. The results (17) of this investigation revealed the film has much oxygen but also carbon and apparently a relatively small amount of uranium. The film contains considerable amounts of water (hundreds of monolayers). Continuation of the analysis showed a subsequent decrease in total carbon and an increase in oxygen. The composition of the surface layer may be a complex combination of carbon, oxygen, and hydrogen, perhaps even an organic layer. Auger spectroscopy with the application to analyze surface depths of a partial monotonic layer is a very powerful tool in the efforts to understand the chemical and physical phenomena associated with some aspects of decorative etching.

If selective staining is used to reveal the microstructure of a specimen, the metallographer must always be aware of the advantages gained from plane polarized light examination as well as conventional bright field. The addition of the decorative layer may introduce an optically active surface that is additionally informative with polarized light. An example of this approach is demonstrated with a uranium carbide specimen that had been heat treated to produce partial transformation to uranium sesquicarbide (U_2C_3), see Figure 9.

DECORATIVE ETCHING

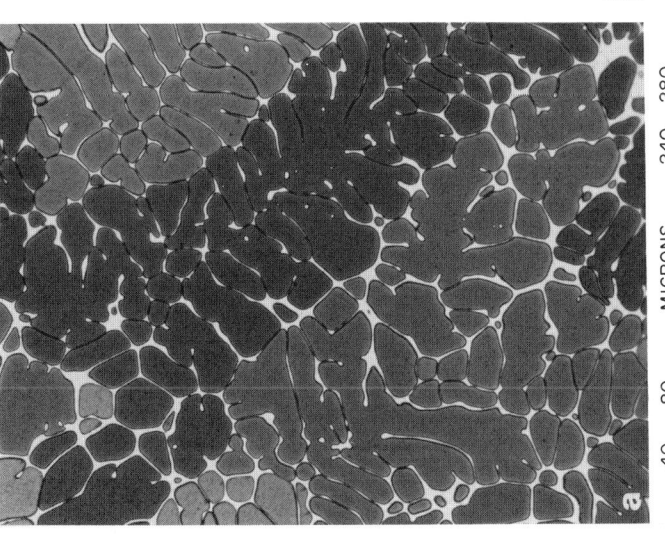

(a) Uranium–4.01 wt % carbon uranium monocarbide with alpha (light) uranium.

(b) Uranium–9.16 wt % carbon uranium dicarbide.

(c) Pyrolytic-carbon coated dicarbide microspheres. Polarized light.

Figure 8. Microstructures of uranium carbide. Etchant: Equal parts of water, nitric acid, and glacial acetic acid.

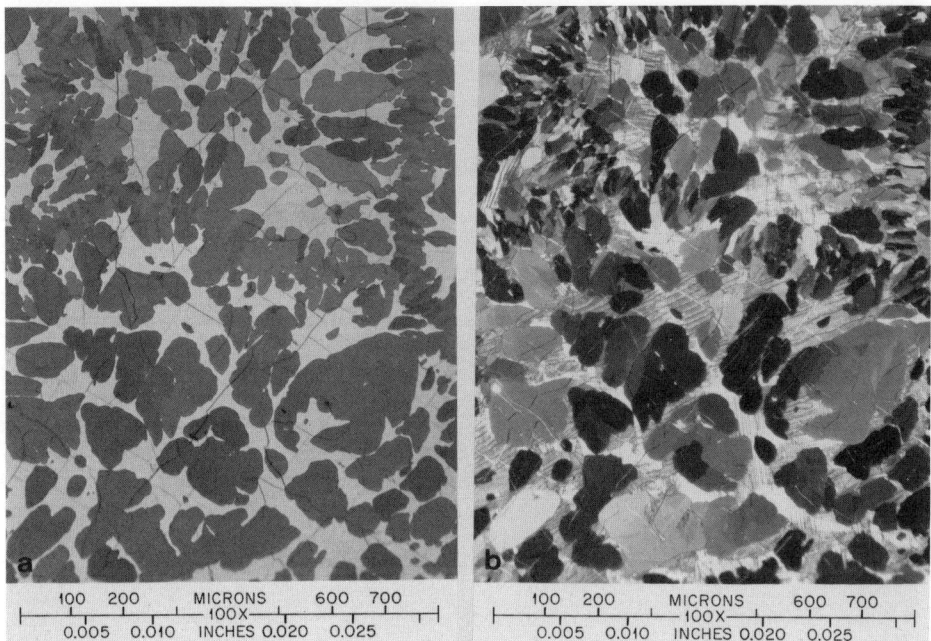

Figure 9. Uranium carbide heat treated for partial transformation to uranium sesquicarbide (U_2C_3). Uranium-7.44 wt% carbon, as cast, then heat treated 60 hr at 1600°C. The uranium sesquicarbide is selectively stained (dark gray) in bright field (a). Polarized light (b) greatly enhances the microstructure. Etchant: Equal parts of water, nitric acid, and glacial acetic acid.

For some materials the decorative etching approach can provide an identifying color that is related to chemical composition. Microstructures of copper phosphorus alloys are shown in Figure 10. The alpha (α) copper phase is clearly defined as a copper-red color in the primary phase and in the eutectic; whereas, the Cu_3P phase, containing 14.24-14.96 wt% copper, is white.

Another example of chemical staining to reveal compositional changes is shown in Figure 11. Crouse has reported (18) this colorful technique in the identification of oxides, nitrides, and carbides of niobium.

Some materials can be decoratively etched with spectacular, colorful results. A TZM alloy (Mo-0.5 Ti-0.08 Zr-0.015 C) weld is shown in Figure 12.

A single crystal zirconium specimen was subjected to a partial shear test. The subsequent deformation produced many matching planes that carried distinguishing colors. The color staining can be re-

DECORATIVE ETCHING

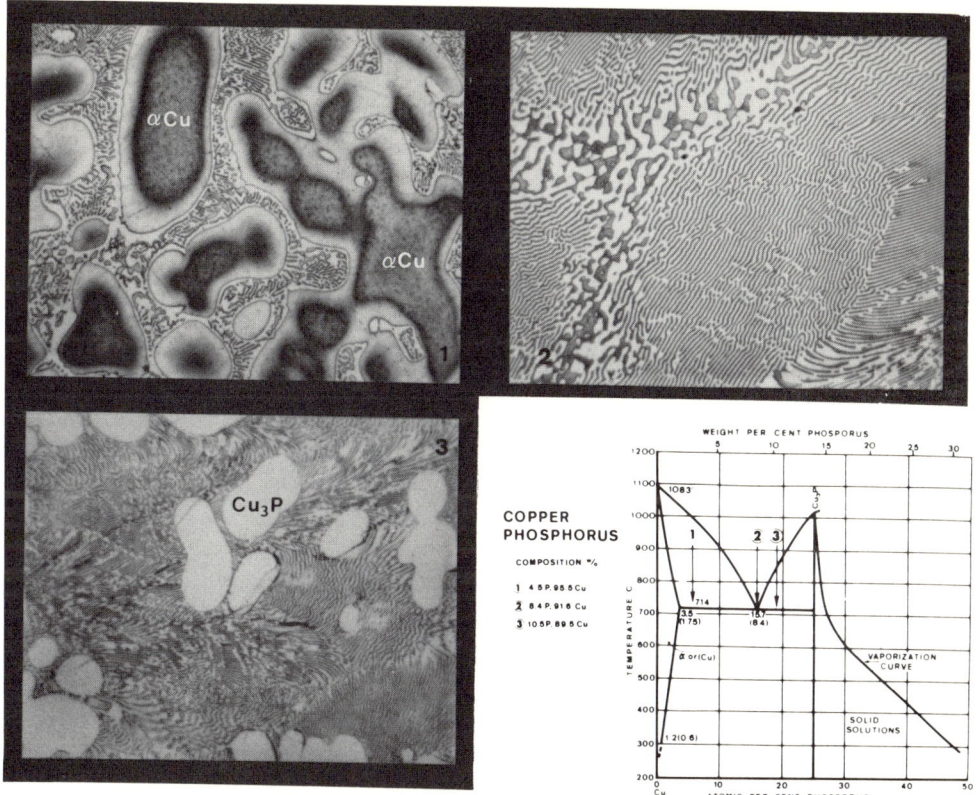

Figure 10. Microstructures of three copper-phosphorus alloys. The definitive color etching shows the α copper and the Cu_3P phases in both the primary and eutectic forms. Color etchant: Waterbury etch: 40 g chromium trioxide, 9 g ammonium chloride, 36 ml nitric acid, 30 ml sulfuric acid; dilute to 1 liter with water.

lated to crystallographic orientation, see Figure 13. Beryllium diboride, shown in Figure 14, displays a spectrum of color. The identification of zirconium hydride in zirconium by chemical staining has been reported (19). This light yellow phase is shown in Figure 15, and offers another example of the detection of a specific phase by color.

Some decorative etching techniques may serve only to protect a polished specimen surface from reacting to moisture in the air or in optical immersion oil. This hygroscopic reaction must be retarded during specimen preparation and the examination. The tech-

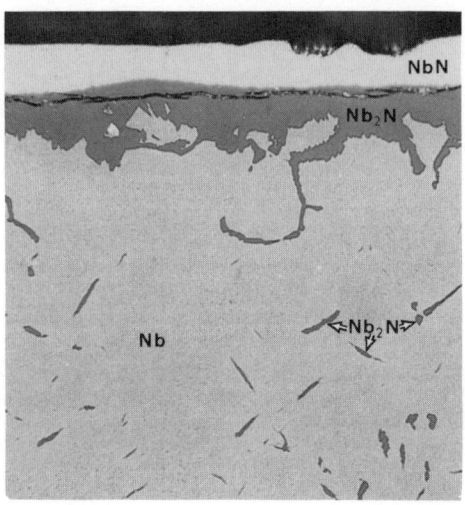

Figure 11. Color stained nitride layers on niobium. The nitrogen-rich NbN is stained yellow and the Nb_2N is reddish brown. The base metal, niobium, is light blue. Heat treated 1365 hr at 1200°C in nitrogen atmosphere. Electrostaining solution: 40 g oxalic acid, 40 g citric acid, 40 ml H_3PO_4, 80 ml lactic acid, 280 ml H_2O, and 480 ml C_2H_5OH.

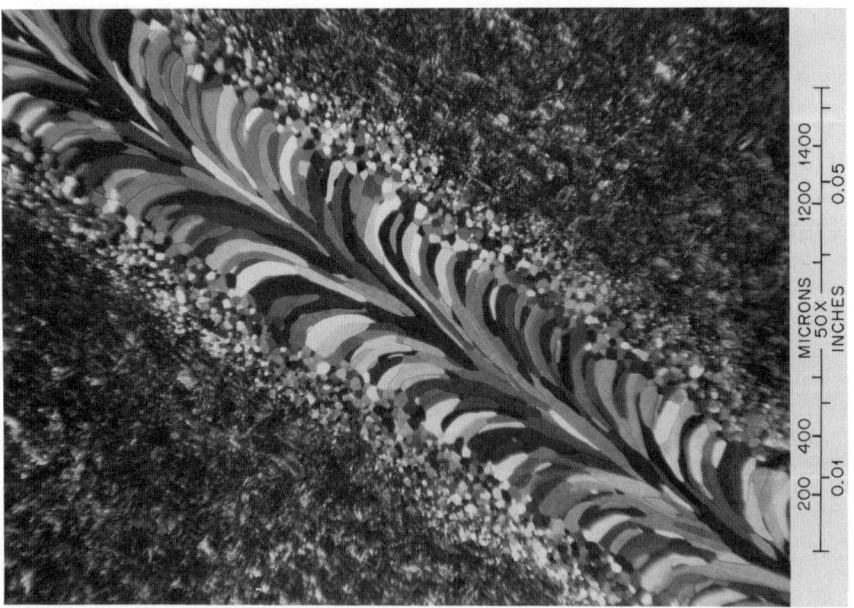

Figure 12. A TZM alloy weld in cold worked matrix. Chemical staining solution: 70 ml H_2O, 20 ml (30%) H_2O_2, and 10 ml H_2SO_4.

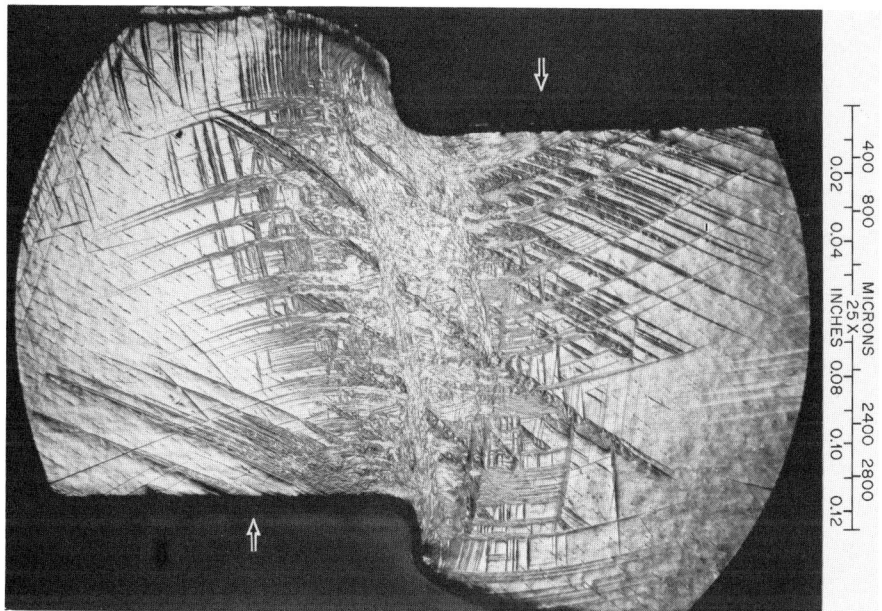

Figure 13. Deformed zirconium single crystal. Stresses were applied to the single crystal as indicated by arrows. Although the color definition is not detectable in a black and white reproduction, the color staining technique provides a more meaningful display of deformation. Electrostaining solution: 20 ml glycerine, 60 ml ethyl alcohol, 35 ml water, 10 ml lactic acid, 5 ml phosphoric acid, and 2 g citric acid.

Figure 14. As-polished (as-cast) beryllium diboride. Many orientations of this cast structure are shown with the application of sensitive tint.

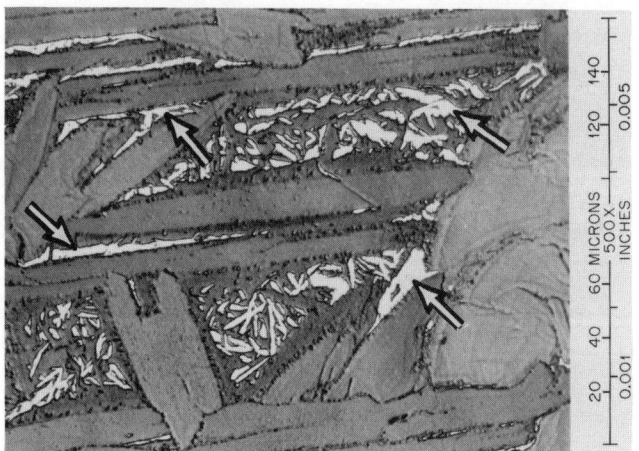

Figure 15. Zirconium hydride precipitate in zircaloy II alloy. The zirconium hydride (arrows) is stained light yellow. This zirconium specimen contains 870 ppm H_2. Electrostaining solution: 20 ml glycerine, 60 ml ethyl alcohol, 35 ml water, 10 ml lactic acid, 5 ml phosphoric acid, 2 g citric acid.

nique for the preparation and etching of thorium dicarbide and thorium-uranium dicarbides of greater than 30 weight % ThC_2 has been reported (20). Specimen preparation in a glovebox containing sufficient desiccant to maintain a low humidity, followed by passivation etch of the surface in a 1:1 HNO_3-H_2O solution allows examination in the open air. The results are shown in Figure 16. Some surface attack takes place in this technique, but the distinct advantage in this technique is the passivation of a normally moisture-reactive surface to allow microscopic examination to proceed for a sufficient time in the open to complete the examination and photomicrography.

Anodizing

A most impressive color display is experienced in the metallographic examination of anodized aluminum. A. Hone and E. C. Pearson (21) and Perryman (22) introduced the use of electrolytic anodization to cause normally isotropic aluminum to exhibit an optically anisotropic surface with plane polarized light. They report that the optical activity is due to furrow-like surfaces of the anodized film on the aluminum. The film develops many characteristic patterns that are related to the crystallographic substrate. We have found this technique to be extremely helpful in the examination of high-

DECORATIVE ETCHING

purity aluminum and aluminum alloys with limited amounts of a second phase in the microstructure. To better understand the characteristics of this anodized surface, Long and Bridges (23) replicated the anodized surface of 99.9% aluminum and observed the replica with the electron microscope. The results are shown in Figures 17 and 18. The variations in the furrow-like pattern are well defined. No attempt was made to relate furrow patterns with known crystallographic orientations. Auger spectroscopic analysis of the anodized surface was made and reported (24) to be an aluminum fluoro-oxide layer with three times as much fluorine as oxygen, the fluorine being from the hydrofluoric acid in the anodizing solution.

Optical examinations of anodized aluminum and aluminum alloys are made with polarized light using a sensitive tint retardation plate (25), which is so named because it provides a very sensitive means of detecting double refraction or birefringence. Optical anisotropy response is seen by color changes rather than the gray shades from white to black customarily seen in viewing an anisotropic surface with conventional (crossed nicols) polarized light technique.

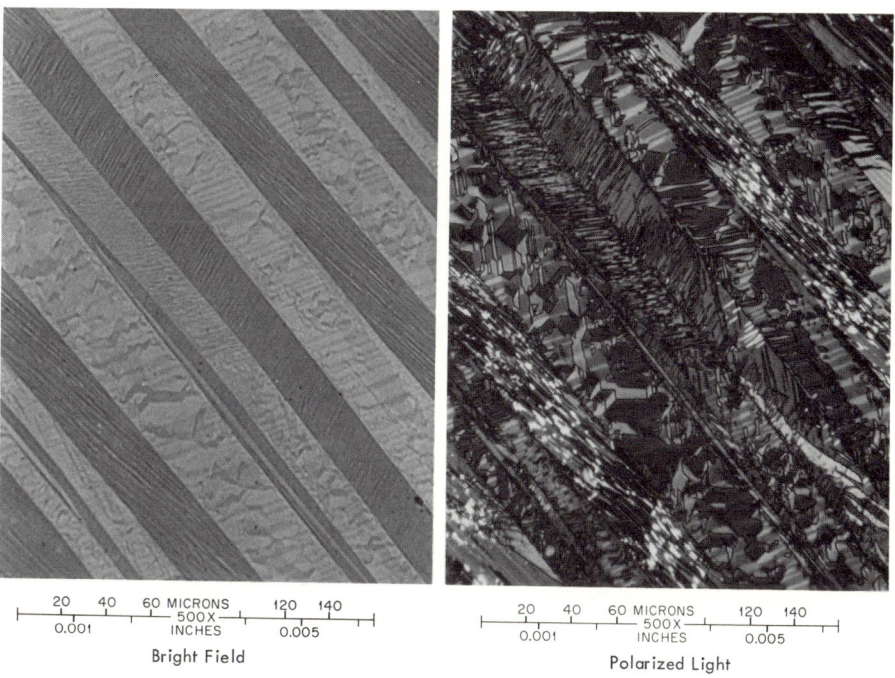

Figure 16. Application of passive decorative etching technique on thorium dicarbide. The polished surface of this material is etched by moisture in the air. A passive film (oxide?) is formed by immersing specimen in a 1:1 HNO_3-H_2O solution.

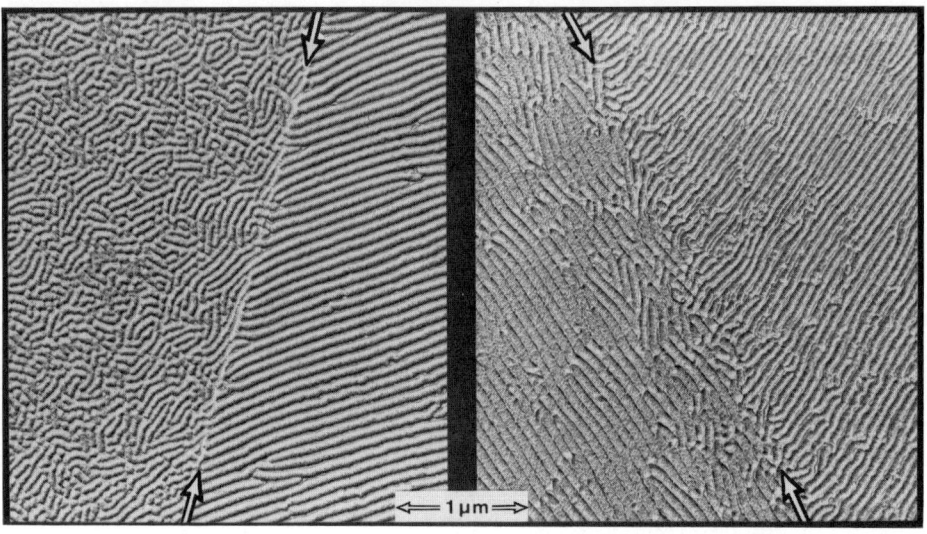

Figure 17. Replicated surface of anodic layer on aluminum. Arrows indicate grain boundaries. Note pattern variations influenced by crystallographic orientation of substrate. Electron micrographs, carbon replica, palladium shadow (30,000X).

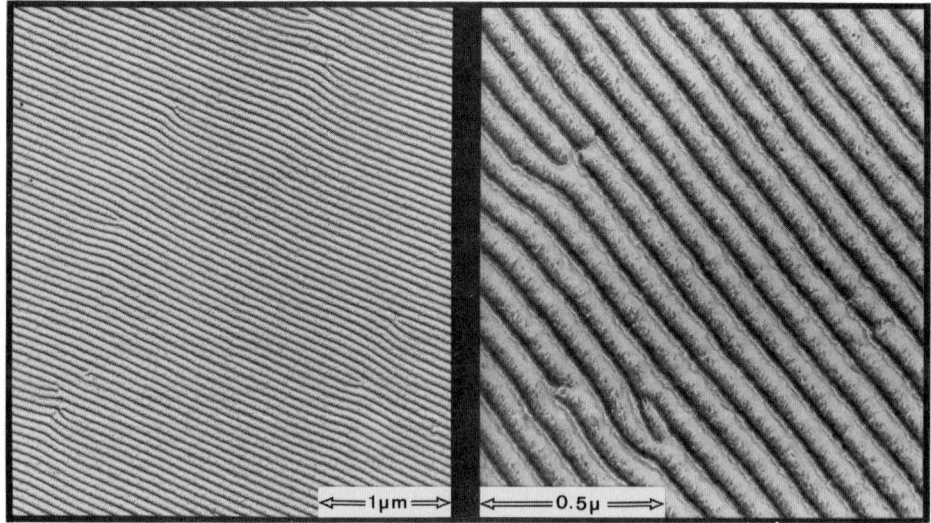

Figure 18. Replicated surface of anodic layer on aluminum. Electron micrograph, carbon replica, palladium shadow (30,000X and 84,000X).

The Bausch and Lomb metallograph is equipped with a "fixed" sensitive tint plate which we have found to handicap the examination of weakly birefringent surfaces. If the sensitive tint is rotatable about its optical axis, the sensitivity is greatly improved. Modifications to the rotatable turret of a Bausch and Lomb Research I are shown in Figure 19. The color-contrast response to this sensitive tint rotation is very delicate, so a control for very minute movement is highly desirable.

The sensitive tint plate is not so accessible in the rotatable turret of the vertical illuminated plate for the Bausch and Lomb Research II, so a rotating device is practically impossible to install. However, if the cement securing the sensitive tint plate is removed (we recommend this modification be done by a Bausch and Lomb service engineer), the brass collar holding the sensitive tint plate can be rotated with a wire by removing an objective from the objective turret, as shown in Figure 20. A performance evaluation of the rotatable sensitive tint is demonstrated in Figure 21. Although the spectral definition is lost in a black and white reproduction, some improved grain contrast displayed in the use of the stationary and the rotating sensitive tint plates can be seen.

Figure 19. Vertical illuminator turret for Bausch and Lomb Research I metallograph. Four positions of turret plate are as follows:
  a. rotatable sensitive tint $\frac{\lambda}{2}$ plate. Note the finely threaded screw to control rotation,
  b. bright field control $\frac{\lambda}{4}$ plate to convert incoming plane polarized light to circularly polarized light (bright field),
  c. original opening for fixed sensitive tint $\frac{\lambda}{2}$ plate,
  d. opening for plane polarized light.

Figure 20. Rotation of sensitive tint plate in Bausch and Lomb Research II Metallograph. The end of a wire (w) is inserted through an objective opening (o) and the sensitive tint plate (st) is rotated. The sensitive tint plate is cemented in place in a factory issued instrument.

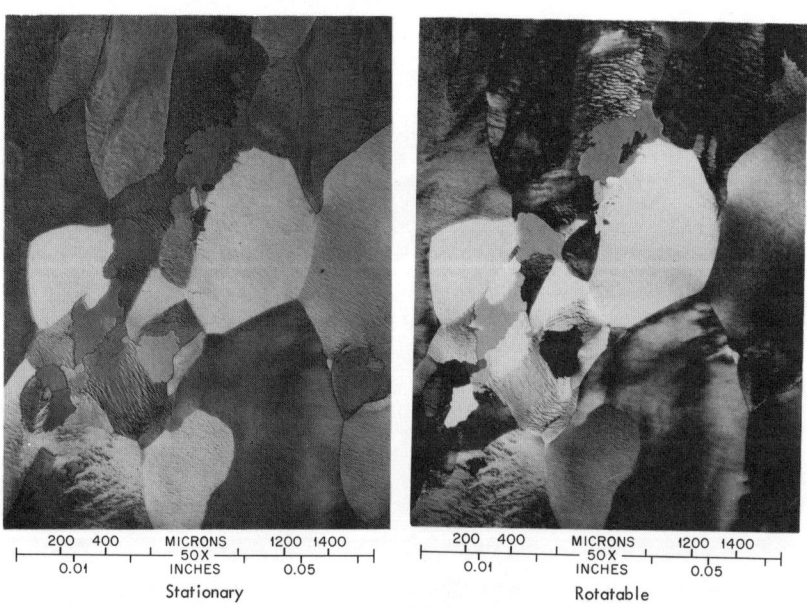

Figure 21. Demonstration of the microstructures of identical fields as seen with stationary and rotatable sensitive tint plates. Specimen: high-purity aluminum with 5 ppm gold. Electropolished and electroanodized.

Electropolishing solution:
   15 parts ethanol
   1 part perchloric acid
   (caution: explosive mixture)

Electroanodizing solution:
   70 parts orthophosphoric acid (85
   2.5 parts distilled water
   26.5 parts ethylene glycol
   1 part hydrofluoric acid (48%)

DECORATIVE ETCHING

Examples of the microstructures of anodized aluminum, with different histories, are shown in Figure 22; and an aluminum-2% uranium alloy, after various treatments, is shown in Figure 23. These photomicrographs (Figures 22 and 23) have been reported in color (3).

Vacuum Deposition

Enhancement of the microstructure through the use of an interference film was introduced by W. Pepperhoff (26). Stiegler and Gray (27) have reported the use of this technique. Titanium dioxide was vacuum deposited on an as-polished specimen. The presence of the film produced constructive or destructive interference as the microscope illumination reflected from the air-film interface and the film-specimen interface. An example of the results is shown in Figure 24. A significant bonus in this approach for etching a specimen is the detection of surface connected voids. The interference color of the epoxy resin mount can serve as a reference to identify the plastic entrapped in the subsurface voids and during the specimen mounting operation.

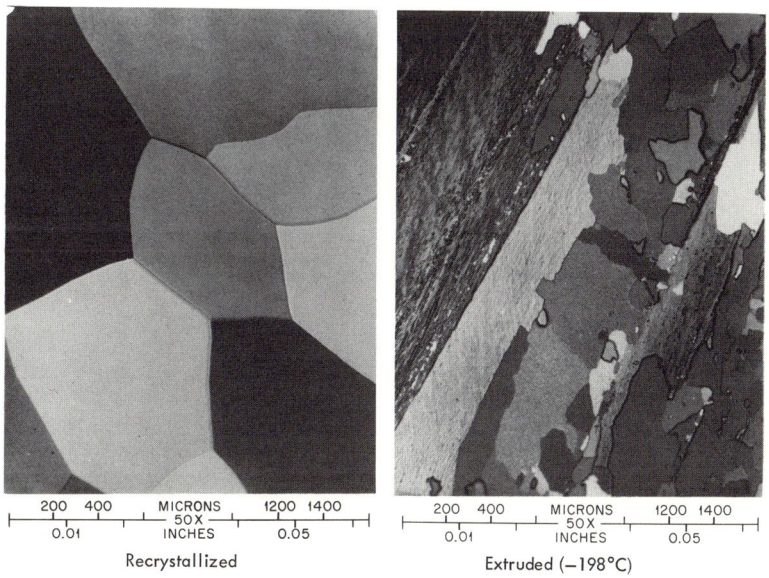

Recrystallized   Extruded (−198°C)

Figure 22. Anodized microstructure of high-purity aluminum. Grain orientations were defined in color by the application of rotatable sensitive tint plate. The extruded microstructure shows severely worked grains contiguous to recrystallized grains. This microstructural phenomenon is attributed to the extrusion temperature.
Electropolishing solution: Electroanodizing solution:
  15 parts ethanol   70 parts orthophosphoric acid (85%)
   1 part perchloric acid   2.5 parts distilled water
(caution: explosive mixture)   26.5 parts ethylene glycol
                                     1 part hydrofluoric acid (48%)

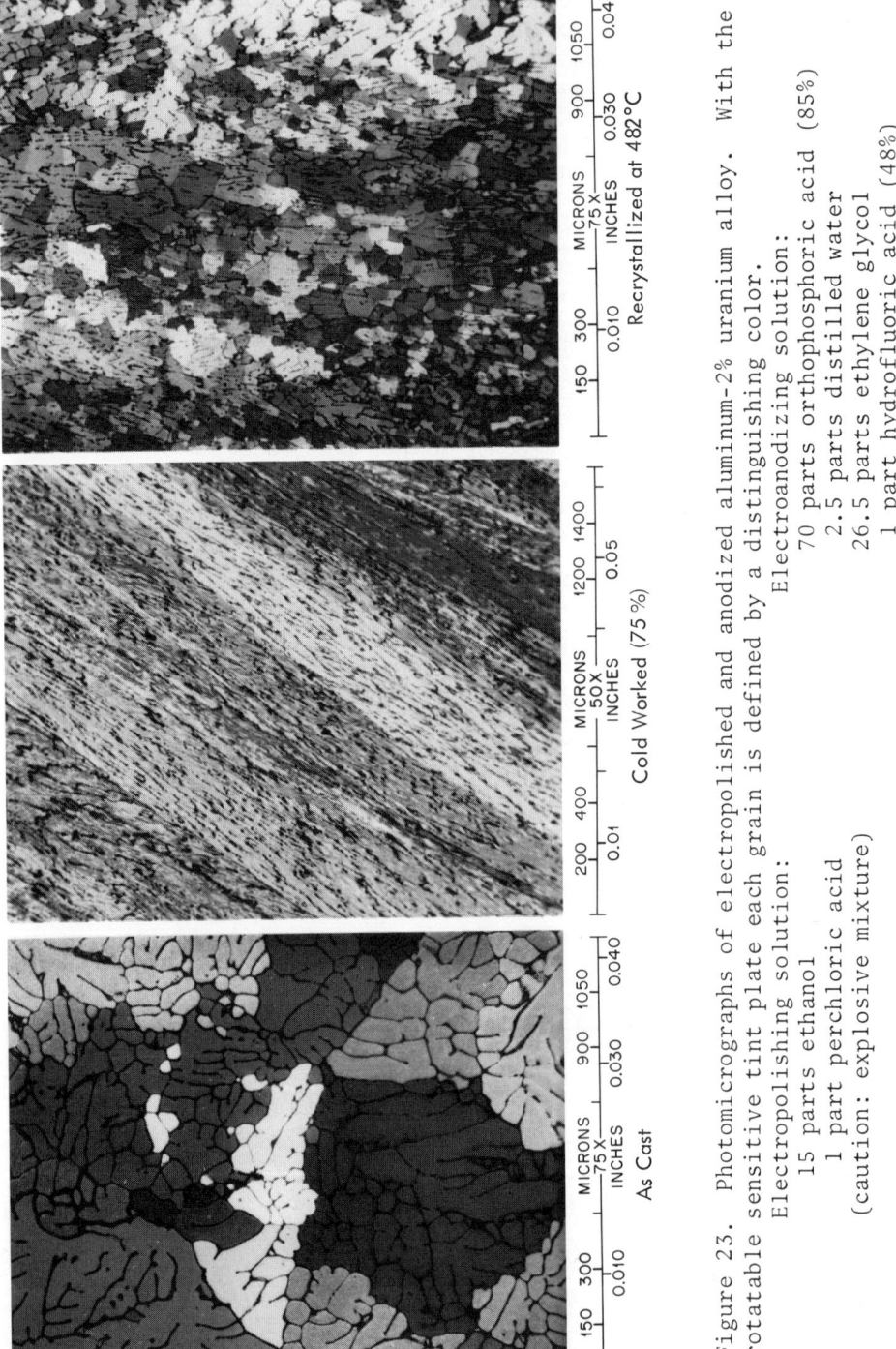

Figure 23. Photomicrographs of electropolished and anodized aluminum-2% uranium alloy. With the rotatable sensitive tint plate each grain is defined by a distinguishing color.
Electropolishing solution:
  15 parts ethanol
   1 part perchloric acid
(caution: explosive mixture)
Electroanodizing solution:
  70 parts orthophosphoric acid (85%)
   2.5 parts distilled water
  26.5 parts ethylene glycol
   1 part hydrofluoric acid (48%)

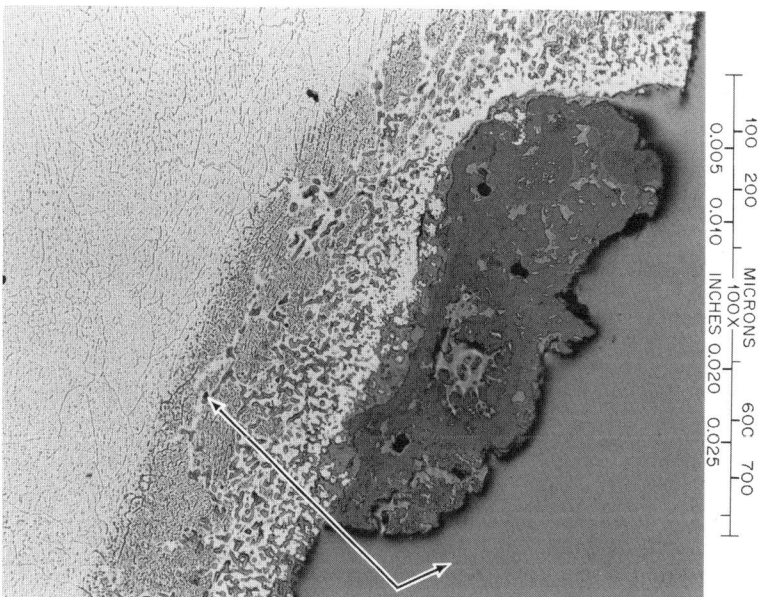

Figure 24. Microstructural discrimination by film deposition. Type 316 stainless steel exposed to air for 3000 hr at 1600°F (871°C). TiO$_2$ was vacuum deposited on the surface of the metallographic specimen. Two oxide films are noticeable and the penetration of the epoxy resin can be seen by comparing colors of areas shown by arrows.

Plastic Overlay

Improvement of the optical contrast of a metallographically etched surface by an approach less complicated than the vacuum technique reported above has been reported by Staub and McCall (28). This technique involves only the application of a thin layer of Parlodian (29). Before and after comparisons of the addition of this interference film are shown in Figure 25.

Elemental Printing

Sulfur, phosphorus and oxide printing to reveal the distribution of inclusions containing these elements in steel have been known to metallographers for many years. Details of these techniques have been reported by Kehl (2). An example of sulfur printing is shown in Figure 26. The 6" diameter area of a 6" x 18" tensile specimen was surveyed for sulfur inclusions. The inset sketch shows the location of the sulfur print relative to the entire tensile specimen.

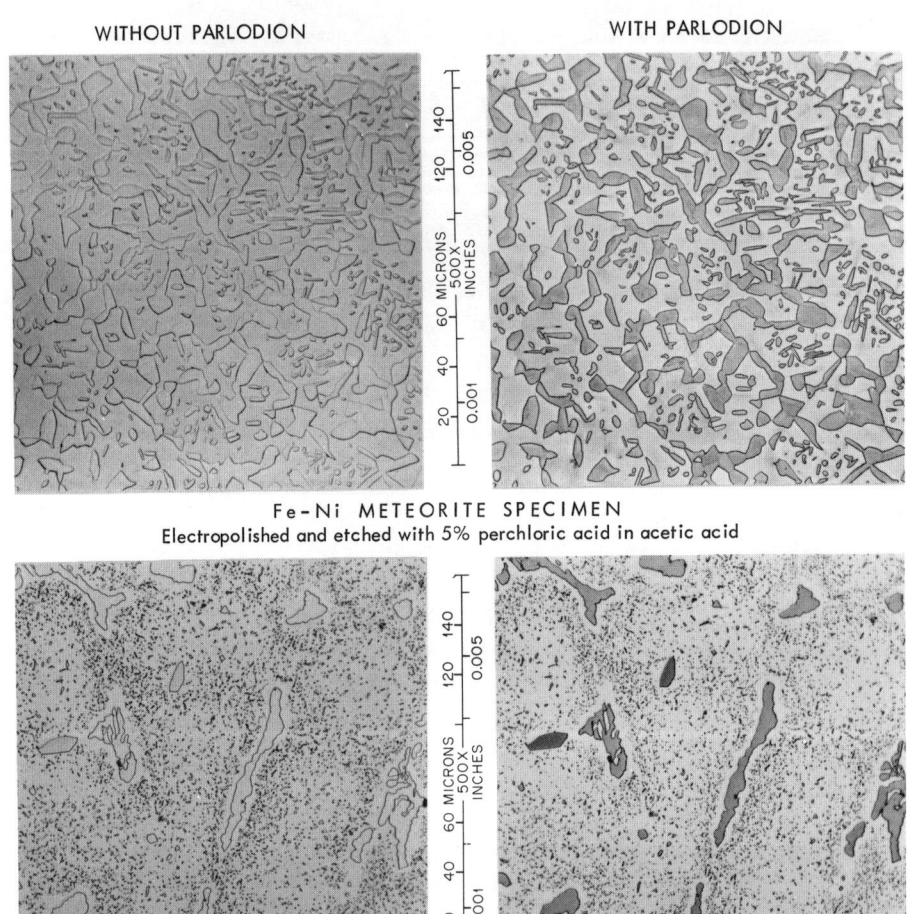

Figure 25. Fe-Cr-Ni alloy specimen. Mechanically polished and etched with glyceregia. Demonstrations of influence of amorphous Parlodian coating on specimen to increase microstructural contrast. (Courtesy of J.L. McCall).

DECORATIVE ETCHING

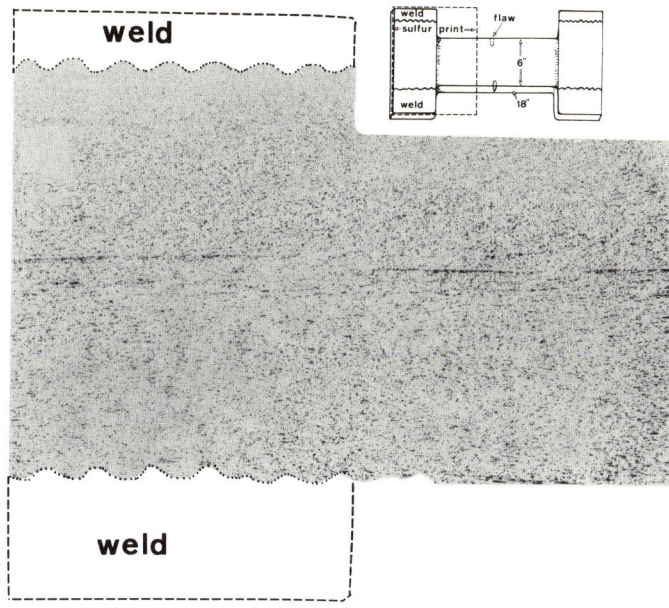

Figure 26. Sulfur print of a section through HSST flawed tensile specimen 6" x 18".

Autoradiography

This technique of registering radioactivity emitted from natural or induced radioactive isotopes in a surface film has carried a prominent role in metallography, particularly since the beginning of the nuclear age about 30 years ago. Only a small attempt will be made to report on the many reported accomplishments on autoradiography.

Considerable care and safety precautions should be exercised in any metallography laboratory involved in the preparation and examination of materials containing radioactive isotopes. One reported (30) technique involves the use of a radiosensitive liquid emulsion film that can be applied directly to the specimen surface, as shown in Figure 27. The application of the film with a glass rod is demonstrated in the photo inset. The film must be thin to be able to relate simultaneously the microstructural areas bearing variable radioactivity as captured by the emulsion and the true microstructure of the substrate.

A metallographic study of some experimental mixtures of plutonium and uranium in the form of microspheres demonstrates another autoradiographic technique and is shown in Figure 28. Varying ratios of plutonium and uranium are present in each sphere. Since

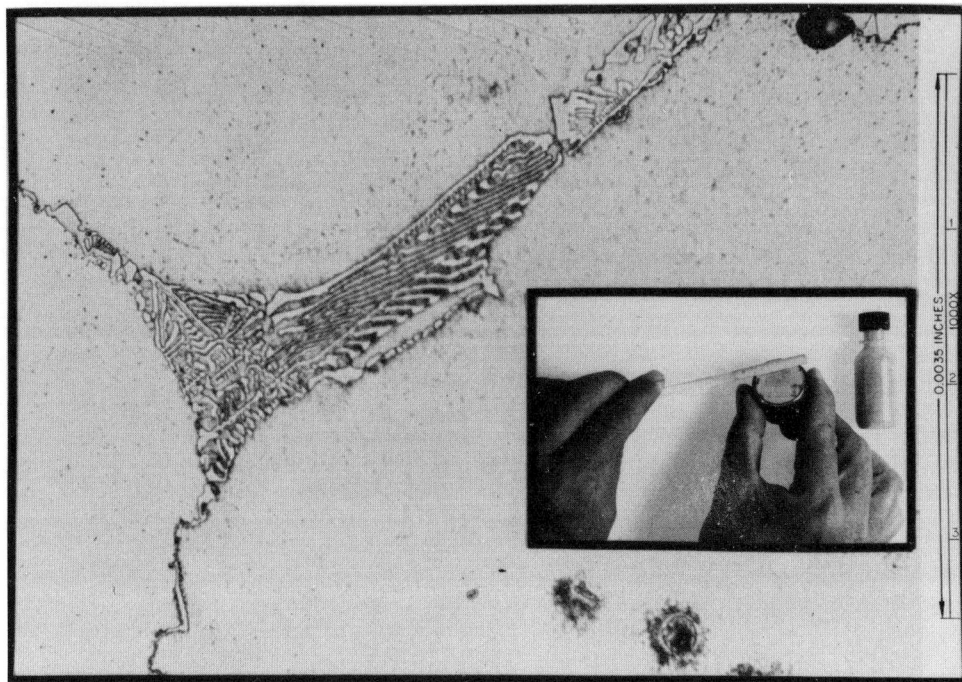

Figure 27. Autoradiograph showing a "script-type eutectic" in Hastelloy N. The liquid emulsion is spread thinly over the specimen in a dark room. After exposure and development, the areas exposed to a radioactive source are dark. In this autoradiograph the ^{14}C radioactive source has been rejected in the formation of the eutectic, hence the clear component is not a carbide. Emulsion: NTE, exposure: 312 hr, developer: D-19 (1:1) 5 min.

plutonium is a high alpha emitter, a technique, reported by Gruber (31), for registering alpha emission density is used. A film of cellulose nitrate (32) is placed in intimate contact with the polished surface. During the contact period, alpha particles bombard the plastic film. After a sufficient exposure, the film is etched with NaOH and examined by transmission microscopy. The high specific alpha activity of plutonium offers easy differentiation between plutonium-rich areas and those with less plutonium. A photomacrograph and an autoradiograph of the same microspheres are shown in Figure 28.

Magnetic Etching

Colloidal magnetic etching has been improved greatly in recent years. Because of the many applications of this technique, a more thorough presentation has been reported (33).

DECORATIVE ETCHING 203

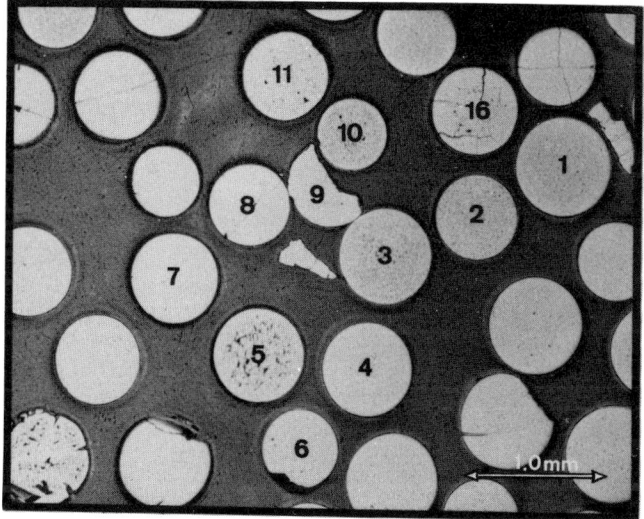

PHOTOMACROGRAPH

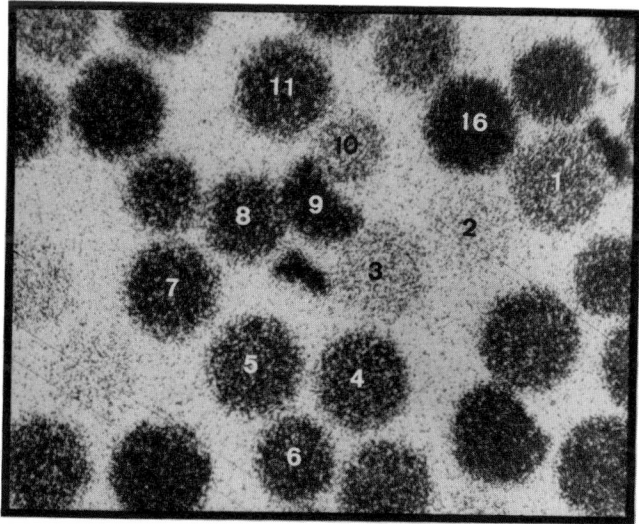

ALPHA AUTORADIOGRAPH

Sphere No.	1	2	3	4	5	8	9	10	11	16
Plutonium, wt %	18	4	9	19	20	23	30	12	22	38
Uranium, wt %	82	96	91	81	80	77	70	88	78	62

Weight percent heavy metal as determined by electron microprobe analyses.

Figure 28. Distribution of plutonium and uranium in mixed oxide microspheres.

CONCLUSIONS

Application of a thermal or chemically formed film, a selective stain, an electrolytic anodic film, a vacuum deposited or an amorphous overlay, relatively simple elemental printing, the complicated procedure of autoradiography, or the examination of an optically active specimen with sensitive tint or plane polarized light are a host of metallographic techniques that can be classified as "decorative etching". Usually in applying this technique the metallographically prepared surface receives a film or overlay that is related to the substrate in some manner so that it becomes the principal or a supplementary means of distinguishing the microstructure. The conventional approach for etching a specimen to produce a microtopographical surface will remain the primary basic etching technique; however, the metallographer should keep in mind the many decorative methods that often reveal the microstructure in an aesthetic way. Scientific and meaningful microstructures in color can be abstract art in its finest sense to non-metallographers. In contrast to the often intangible meaning of abstract art, the photomicrograph with color definitions carries an asset in communication to a layman or scientifically trained personnel in other disciplines.

ACKNOWLEDGMENTS

We are indebted directly to M. D. Allen, H. R. Gaddis, H. R. Tinch, E. H. Lee and to all the other metallographic personnel for their assistance in technique developments. Our photographers were invaluable in solving the many photographic problems. We specifically thank J. V. Cathcart, R. E. Pawel, G. M. Slaughter, D. A. Canonico and G. M. Goodwin who provided material related to their research projects. Helpful recommendations were offered by J. M. Leitnaker and H. E. McCoy in reviewing the paper. Special thanks go to Mary Combs for her preparation of the manuscript.

REFERENCES

1. Cyril Stanley Smith, A History of Metallography, University of Chicago Press (1960).
2. George L. Kehl, The Principles of Metallographic Laboratory Practice, McGraw-Hill Book Co., inc., New York City (1949).
3. R. J. Gray, "Present Status of Metallography", Fifty Years of Progress in Metallographic Techniques, STP-430; American Society for Testing and Materials, Philadelphia, Pa. (April 1968) pp. 17-62.
4. B. Vonnegat, Chemical Review 44, (1949) p. 277.

5. G. W. Johnson, "Some Observations on the Epitaxy of Sodium Chloride on Silver", J. Appl. Phys. 21 (10), (Oct. 1957) pp. 1057-1062.
6. G. W. Johnson, "The Epitaxy of Alkali Chlorides on Metals", J. Appl. Phys. 22 (6), (June 1951) pp. 797-805.
7. F. W. Young, J. V. Cathcart, and A. T. Gwathmey, "The Rates of Oxidation of Several Faces of a Single Crystal of Copper as Determined with Elliptically Polarized Light", Acta Met. 4 (2), (1956) pp. 145-163.
8. J. V. Cathcart, G. F. Petersen, and C. J. Sparks, Jr., "Lattice disregistry in very thin oxide films on copper", Mem. Sci. Rev. Met. 62, (May 1965) pp. 11-16.
9. J. V. Cathcart, G. F. Petersen, and C. J. Sparks, "Oxidation Rate and Oxide Structural Defects", Surfaces and Interfaces Chemical and Physical Characteristics, Burke, Reed, Weiss Ed., Syracuse University Press, Syracuse, N. Y. (1967).
10. W. J. McG. Tegart, The Electrolytic and Chemical Polishing of Metals, Pergamon Press, New York City (1959) p. 61.
11. R. E. Pawel, J. V. Cathcart, and J. J. Campbell, "Break-Away Phenomena in the Oxidation of Columbian Single Crystals", in Columbium Metallurgy (eds. D. L. Douglass and F. W. Kunz), Interscience Publishers, New York (1961) pp. 667-684.
12. R. E. Pawel, J. V. Cathcart, and J. J. Campbell, "Oxide Platelet Formation in Tantalum Single Crystals", Acta Met. 10 (2) (1962) pp. 149-160.
13. R. J. Gray, W. C. Thurber, C. K. H. DuBose, "Preparation of Arc-Melted Uranium Carbides", Metal Progress, 74 (1) (July 1958) pp. 65-70.
14. R. W. McClung, E. S. Bomar, and R. J. Gray, "Evaluating Coated Particles of Nuclear Fuel", Metal Progress, 86 (1) (July 1964) pp. 90-93.
15. "Applications of Auger Electron Spectroscopy", Staff Report in Research and Development (July 1973).
16. D. F. Stein, R. E. Weber, and P. W. Palmberg, "Auger Electron Spectroscopy of Metal Surfaces", J. Metals (Feb. 1971).
17. R. E. Clausing, Oak Ridge National Laboratory, Metals and Ceramics Division, P. O. Box X, Oak Ridge, Tenn., private communication.
18. R. S. Crouse, Identification of Carbides, Nitrides, and Oxides of Niobium and Niobium Alloys by Anodic Staining, ORNL-3821 (July 1965).
19. M. L. Picklesimer, Anodizing as a Metallographic Technique for Zirconium Base Alloys, ORNL-2296 (April 1957).
20. T. M. Kegley, Jr., and B. C. Leslie, "Metallographic Preparation of Dicarbides of Thorium and Thorium-Uranium", J. Nucl. Mater. 13 (2) (1964) pp. 283-287.
21. A. Hone and E. C. Pearson, Metal Progress 53 (1948) p. 363.

22. E. C. W. Perryman, "The Examination of Metal Surfaces", Polarized Light in Metallography, ed. by G. K. T. Conn and F. J. Bradshaw, Butterworths Scientific Publication, (1952) pp. 70-89.
23. E. L. Long, Jr. and W. H. Bridges, Metals and Ceramics Division, Oak Ridge National Laboratory, private communication.
24. R. E. Clausing, Oak Ridge National Laboratory, private communication.
25. G. K. T. Conn and J. N. Naish, "Polarized Light and Anisotropic Materials", Polarized Light in Metallography, op. cit., pp. 14-16.
26. W. Pepperhoff, Sickbarmachung on Gefugestrukturen durch Interferentz-Aufdampfschichten", Des Naturwissenschaften 16 (1960) p. 375.
27. J. O. Stiegler and R. J. Gray, "Microstructural Discrimination by Deposition of Surface Films", pp. 11-17 in Advances in Metallography (ed. by R. J. Jackson and A. E. Calabra), Technical Papers of the 20th Metallographic Conference, The Dow Chemical Company, RFP-658 (Oct. 1966).
28. R. E. Staub, Jr. and J. L. McCall, "Increasing the Microscopic Contrast of Phases with Similar Reflectivities", Metallography 1 (1),(Sept. 1968) pp. 153-155.
29. Trade name for a highly purified form of nonexplosive cellulose nitrate obtained from Mallinchrodt Chemical Works, St. Louis, Mo.
30. R. E. Gehlbach, M. D. Allen, and J. D. Braun, "Autoradiographic Study of Carbon-14 in Hastelloy N", in Proceedings, First Annual Technical Meeting, International Metallographic Society, Inc., Nov. 11-13, 1968, Denver, Colorado (1969) pp. 113-118.
31. W. J. Gruber, "Autoradiography of Irradiated Nuclear Ceramic Fuels", Proceedings of Fall Meeting of the American Ceramic Society, Pittsburgh, Pa., Oct. 6-9, 1968.
32. Kodal special film, Type 106-01-A, nonsilver radiation recording film, Eastman Kodak Co., Rochester, N. Y.
33. R. J. Gray, "Magnetic Etching with Ferrofluid", Symposium, Specimen Preparation Techniques for Optical and Electron Metallography, International Metallographic Society-American Society for Metals, Beverly Hills, Calif., Sept. 23-24, 1973, ed. J. L. McCall and W. M. Mueller.

PRACTICAL APPLICATIONS OF CATHODIC VACUUM ETCHING

F. M. Cain, Jr.

Burrell Corporation

Pittsburgh, Pennsylvania

INTRODUCTION

Cathodic vacuum etching, referred to as CVE throughout this paper, has been defined as a physical process that reveals the microscopic structure of polished samples due to the selective removal of atoms by positive ion bombardment.

There are several publications that deal with the theory and some that deal with the interpretation of microscopic structures that have been prepared by the technique. A number of these are included in the list of references (1-37). Consequently this paper will limit attention to the practical applications of CVE.

The appearance of a number of materials when prepared by nonstaining chemical or electrochemical etching is very similar to that when prepared by CVE. The samples prepared by CVE will show differences however, which are of an advantage to both the electron and the light metallographer. Chemical stains and etching acid residues are sometimes troublesome to the electron microscopist. They are difficult to remove and therefore may introduce artifacts in replicas of microstructures viewed at high magnifications. Such problems are avoided by CVE.

Deep etching for light or electron microscopy by chemical or electrochemical methods usually widens and deepens grain boundaries. On the other hand, deep etching by CVE selectively attacks entire exposed grain surfaces and thus reveals planar grain boundaries and other structural detail not previously available to the light metallographer.

EQUIPMENT

A review of available literature reveals a number of papers on CVE by investigators using laboratory made equipment and only two publications using commercially manufactured equipment (1,2). Both commercial units were manufactured by NUMEC of Apollo, Pennsylvania. NUMEC manufactured a standard laboratory model, a glove box model, a combination etcher-evaporator and a remote model. The manufacture and sale of the complete line of NUMEC Cathodic Vacuum Etchers has been taken over by Burrell Corporation, Pittsburgh, Pennsylvania.

Most of the laboratory-made units described in the literature were limited to etching unmounted samples. After a sample had been mounted and polished it was broken out of the mount before it could be etched. An additional requirement was the need for the polished face of the sample to be reasonably parallel to the face to be affixed to the cathode. A couple of the laboratory-made units could handle mounted samples as could the NUMEC units.

The metallographic samples prepared specifically for this paper were etched using a Burrell Corporation Model 5CVE, Cathodic Vacuum Etcher, Figure 1. A closeup of the base plate, bell jar, cathode, plastic hold down cap, teflon aperture and a mounted sample are shown in Figure 2. The photomicrographs of the structures etched on the Burrell unit were made on Polaroid PN55 film using a new B & L Research II Metallograph, Figure 3. The combination of zoom magnification change, xenon illumination and the built in exposure meter greatly simplified the photography.

SAMPLE PREPARATION

Mounting

Samples to be etched on the Burrell unit were mounted such that the mount served as both an electrical and thermal conductor. In addition the face of the mount served as an electrical insulator. Such a mount could be prepared using P-Mold sections shown in Figure 4. The outer shell and flat washer-shaped sections of the P-Mold were warm pressed using red bakelite. The conducting core sections containing a blend of ball-milled red bakelite powder and copper powder also were warm pressed.

In using a P-Mold the outer bakelite shell is positioned on the base of the mounting press mold assembly. The sample is inserted inside the shell and also placed on the base. The bakelite washer is then inserted inside the shell and brought to rest on the sample. The copper-bakelite core is then inserted in the shell

PRACTICAL APPLICATIONS OF CATHODIC VACUUM ETCHING 209

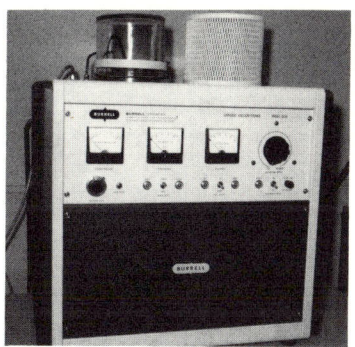

Figure 1. Burrell Corporation Standard Model 5 CVE, Cathodic Vacuum Etcher. This model is adaptable for remote use.

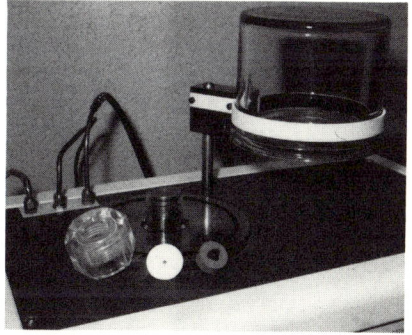

Figure 2. View of base plate and bell jar showing cathode, hold down cap and teflon aperture.

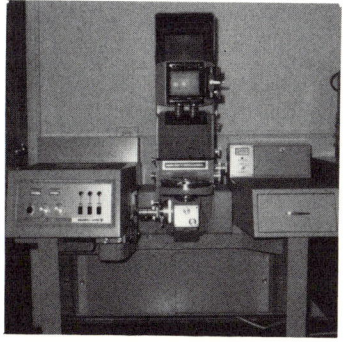

Figure 3. Bausch & Lomb Research II Metallograph.

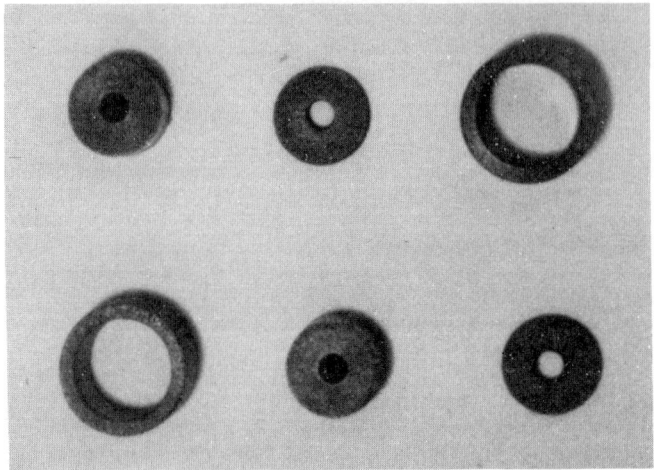

Figure 4. P-Mold sections used to mount samples for the Burrell Corporation CVE. These warm pressed, green dense sections work only in 1¼" molds.

and temperature (270-300°F) and pressure (4500 psi) are applied the same as in mounting any sample in bakelite. A longitudinal cross section of the mount would appear similar to the sketch, Figure 5.

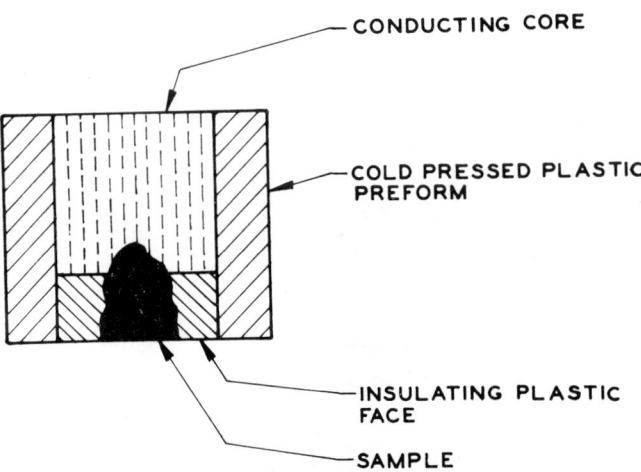

Figure 5. Sketch showing cross section of sample mounted for CVE using bakelite P-Mold or diallyl-phthalate preform.

Another very suitable approach to mounting samples to be cathodic vacuum etched is the use of a preform of the type shown in Figure 6. This type of preform can be made on most any mounting press. A steel washer 1¼" O.D. x 3/4" I.D. x 1/16" thick and a steel insert slightly under 3/4" O.D. x 1¼" long with a rounded end can be machined. The steel washer is positioned on the base of the mounting press mold assembly and the flat end of the insert is placed inside the hole in the washer. This assembly is lowered into the mold to a depth so that the top of the insert is approximately ¼" below the top of the mold. Diallyl-phthalate powder is then poured into the mold around the insert level with the top of the mold. A pressure of 4500 psi is applied and the mold assembly is heated to 150-170°F and held for two minutes at temperature. When the pressure is released and the assembly is ejected from the mold, the washer can be removed from the insert and the insert can be removed from the preform. The thin layer of plastic that forms about the round nose of the insert can be broken away and the preform can then be reinserted into the mold of the press for mounting a sample. With the preform resting on the base of the mold assembly the sample also can be placed on the base and diallyl-phthalate powder poured into the mold to a level that leaves a little of the top of the sample showing. A diallyl-phthalate-copper powder mixture then is added to fill the remaining space in the preform. A molding temperature of 280-290°F and a pressure of 4500 psi should

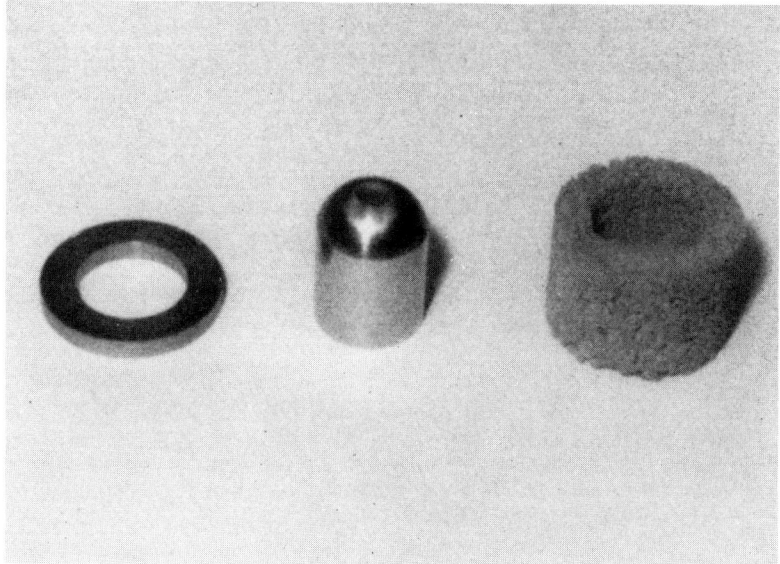

Figure 6. Molding accessories and 1¼" O.D. preform made of diallyl-phthalate powder.

then be applied as for normal mounting with this type of powder. A longitudinal cross section of this mount would again appear similar to the sketch in Figure 5.

Duran and Fisher (1) mounted samples for the NUMEC etcher in Shell Epon 820 resin mixed with a catalyst, diethanolamine. Conductivity through the mount was made possible by cementing a ½" diameter x 3/4" long aluminum rod to the sample with silver conducting epoxy adhesive.

Polishing

The quality of the polish required for CVE is generally similar to that required for chemical or electrochemical etching.

Chemical polishing (3), a technique that precludes the need of mechanical or electrolytic polishing for some materials, gives sample surfaces suitable for CVE. The technique produces a specular, scratch-free surface by the swab application of the proper chemical bath to abraded or machined surfaces. In certain instances the sample can be immersed in the bath without swabbing.

Attack polishing (4-6), a method that utilizes the advantage of simultaneous acid dissolution and abrasive action, also gives sample surfaces suitable for CVE for certain materials. The prescribed chemical solution and abrasive are applied to the final polishing lap.

As in the handling of strong chemical solutions for any purpose, care should be exercised to avoid skin contact by the solution when performing either chemical or attack polishing.

Sample preparation by rotating laps, vibro-met, auto-met and electrolytic polishing also produce surfaces suitable for CVE. A number of materials smear and flow during mechanical polishing and thus require repolishing and reetching to develop the quality of microstructure desired when etching with chemicals or the electrochemical technique. When preparing these materials for CVE the same care must be exercised.

A vibratory polisher at low amplitude for approximately six hours was used by Duran and Fisher (1) to prepare uranium and uranium alloys for CVE.

The sputtering or cathodic vacuum etching characteristics of some metals result in a combination polished-etched surface. This is evident when etching a sample surface that has been abraded only to a 600-grit SiC soft finish. Examples of this observation are illustrated in Figures 7, 8 and 9. Strong (7) ranks a number of

TABLE I. ETCHING CONDITIONS

Material	Pressure μ	Voltage KV	Time Minutes	Figure
This Paper				
Tungsten	150	2	1½	7
Molybdenum	150	3	3	8
Tantalum	150	4	1	11
Stainless Steel	75	3.5	1½	13 & 14
Phosphor-Bronze	40 (air)	3.5	3/4	15 & 16
Carbon Steel	100	3.5	1	17
Silver-Cadmium	100 (air)	3.5	1	19
Aluminum	100	3	1	20
Tungsten Carbide	150	3.5	1½	28
Uranium	300	3.5	½	30
Uranium dioxide	100	3	6	33B
Plutonium dioxide	100	3	1½-2	34 & 35
Aluminum dioxide	120	4	60	36
Zirconium	150	4	1½	42
Referenced Investigators				
Ward (13)				
TZM(Mo-Ti-Zr)	low	6	30	.55 ma
Mo-Ni Clad	low	5.8	30	.95 ma
ThO$_2$ Particles in Mo	low	6.8	45	.80 ma
Duran and Fisher (1)				
U-Cb-Zr alloy	200 millitor	1	20	1 ma
U-10 Mo alloy	200 millitor	1	20	1 ma
Rexer and Vogel (2)				
Zr-Ta-Cr	100	2.5	4	3.5 ma/cm^2
Ti-Va	100	3	4	3.1 "
Ta-Cr	100	3	4	7.8 "

Note: All samples etched in Argon except where noted.
Etching conditions for illustrations taken from reference (8) are listed in Table II.

TABLE II. COMPARISON OF CONDITIONS FOR CATHODIC BOMBARDMENT ETCHING
Note--Reproduced from Armstrong, Madden, and Sykes (9) with modifications and additions.

Investigation	Armstrong, Madden and Sykes	Newkirk and Martin	Holland	Padden and Cain	Carlson, Williams, Rogers and Manthos	Other Investigators					
Bombarding gas	argon	krypton	argon	argon	argon						
Gas pressure	30 μ	8 to 15 μ	5 to 60 min	50 μ	15 μ						
Induction period	3 min	few seconds		½ to 2 min	few minutes						
Etching times	2 to 10 min	1 to 25 min		1 to 3 min	3 min						
Time for operation	10 min			20 min	≈ 45 min						
Voltage	2 to 5 kv	3½ to 4½ kv	2 kv	5 kv	5 kv						
Current density	0.5 to 0.80 ma per sq cm	1½ to 2 ma per sq cm	1 to 3 ma per sq cm	≈ 4.5 ma per sq cm							
Magnetic field	≈ 100 oersted	≈ 100 oersted									

Materials	kv	ma per sq cm	time, min	kv	ma per sq cm	time, min	Pressure, μ	Time, min		Gas	Pres-sure, μ	kv	ma per sq cm	Time, min	Reference
Aluminum	3 to 4	0.5 to 0.65	3	4.0	2	20	10		X						
Duralumin								20							
Aluminum-7% magnesium		0.5	2½					15							
Bismuth	2 to 2.5	0.5 to 0.55	2												
Cadmium															
30% cadmium-70% silver									rapid etch, satisfactory in air						
Cobalt	4 to 5	0.5 to 0.75	3							A	25	2		155	McCutcheon and Pahl
Vitallium									X						
Graphitic stellite		no success							X						
Chromium		0.5 to 0.80	10						X						
Columbium	4.5														
Copper	4 to 5.5	0.5 to 0.65	3 to 4	3.5	1.5	1	8	10	X	Ne	30 to 200	2 to 25		45 to 60	Spivak et al
α brass															
Copper-nickel								5 to 10	X						
Copper-nickel-iron				3.5	2	1	15								
Copper-2.5% iron										A	25	3.8		15	McCutcheon and Pahl
23% Cu-75% Ag-2% Ni															

PRACTICAL APPLICATIONS OF CATHODIC VACUUM ETCHING

Material																		
Iron	4 to 4.5	0.5 to 0.65	2	4.0	1.5	3	10					X						
Medium C steel											X		A	25	4		60	McCutcheon and Pahl
Low C steel											X							
Pearlitic steel	4 to 4.5	0.5 to 0.65	2	4.5	1.5	3	8			4								
Stainless steel																		
Exhaust valve steel (19% Cr, 8% Ni, 0.40% C)											X		air	15	4		100	McCutcheon and Pahl
Magnesium	2 to 3	0.5 to 0.55	3								X							
Magnesium-zinc	4.5 to 5	0.5 to 0.80	5								X							
Molybdenum	4 to 4.5	0.5 to 0.75	5															
Nickel	4 to 4.5	0.5 to 0.65	3					30 to 60										
Electrodeposited coatings																		
Palladium								10										
Platinum																		
Silicon	4 to 5	5 to 0.8	10 to 20															
Tantalum	4 to 4.5	0.5 to 0.7	7															
Titanium	4 to 4.5	0.5 to 0.75	6					30										
Tungsten	4 to 5	0.5 to 0.8	7															
Thorium	4 to 4.5	0.5 to 0.75	3 to 5										A+Xe	2	4		45	Morris
Thorium-zirconium	4 to 5	0.5 to 0.75	3 to 5															
Thorium-2% iron																		
Thorium-carbon																		
Uranium	4 to 4.5	0.5 to 0.65	3	4.0	2	25	10			1 to 3	X X		O	>15	2	≈16	60	Englander et al Berlein
											X		Kr	75	3	3		
Uranium-cobalt	4 to 4.5	0.5 to 0.65	3															
Uranium-molybdenum	4 to 4.5	0.5 to 0.65	3															
Uranium-columbium	4 to 4.5	0.5 to 0.65	3															
Uranium-columbium-molybdenum	4 to 4.5	0.5 to 0.65	3															
Uranium-silicon	4 to 4.5	0.5 to 0.65	3								X							
Zirconium	4 to 5	0.5 to 0.75	3 to 6															
Zirconium-columbium	4 to 5	0.5 to 0.75	3 to 6															
Uranium dioxide	4 to 4.5	0.5 to 0.65	3										A	100	3		6	Cain
Uranium carbide	4 to 4.5	0.6 to 0.65	3															
83% Cr carbide 15% Ni binder											X X							
70% Cr, 30% alumina																		

metals according to their sputtering rates under various conditions. From these lists the metals most likely to polish-etch could be selected.

EQUIPMENT OPERATION

The Burrell Model 5 CVE instrument is very simple to operate. A total time of less than three minutes is required for sample loading, pump down, flushing, etching and unloading for materials such as tantalum, uranium and zirconium.

A properly mounted and polished sample and a teflon aperture are affixed to the cathode with a threaded plastic hold down cap, Figure 2. The bell jar is lowered and the system is evacuated to approximately 5 microns. The system is then flushed once or twice with argon and again pumped to 5 microns. The desired flow of argon into the system is established and etching commences when the high voltage has been turned on and adjusted to the desired level.

The key variables influencing etching results are gas pressure, voltage and time. Current density considered significant by some investigators (8,9) need not be used as a control factor for the Burrell etcher.

Temperature of the sample is also an important factor, however a water cooled cathode maintains a low sample temperature. Pure tin, melting at approximately 232°C has been etched for several minutes and exhibits no evidence of melting.

The etching conditions used to prepare a variety of materials for this paper and a few other publications are shown in Table I. A compilation of conditions used by other investigators is shown in Table II, reproduced from reference (9) and includes some changes.

APPLICATIONS

Cathodic vacuum etching is a practical means of etching numerous materials such as pure metals, alloys, cermets, ceramics, semiconductors, non conductors, etc. Following are a number of specific applications with illustrations. Since the etching conditions are shown in Table I, only the material identity, etching time and the magnification will be captioned with each illustration.

Pure Metals

Figures 7 through 12 contain a few examples of pure metals that have been cathodic vacuum etched. Note the combination polished-

PRACTICAL APPLICATIONS OF CATHODIC VACUUM ETCHING 217

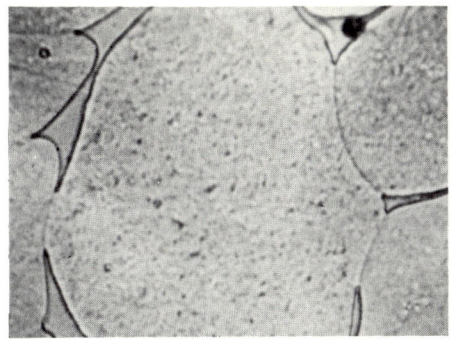
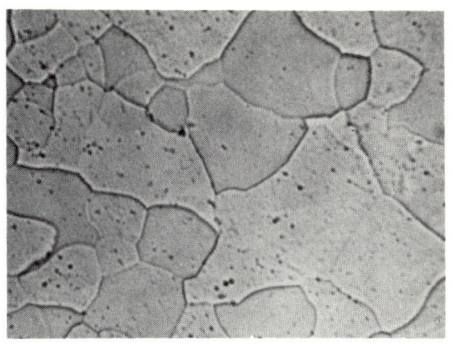

Figure 7. Tungsten, etched 1½ minutes, 500X.

Figure 8. Molybdenum, etched 3 minutes, 500X.

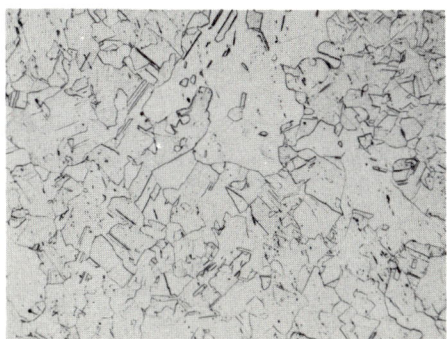
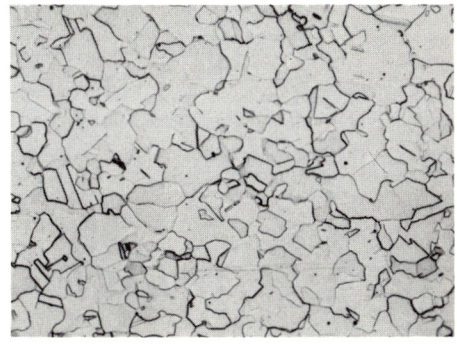

Figure 9. Copper, etched 1 minute, 100X (8).

Figure 10. Nickel, etched 3 minutes, 250X (8).

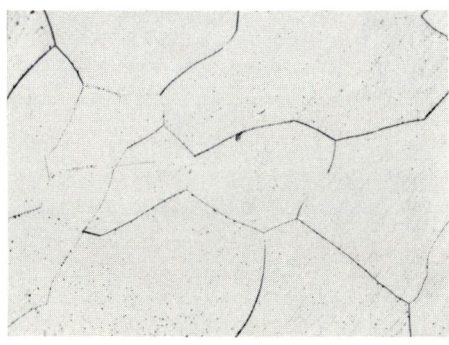
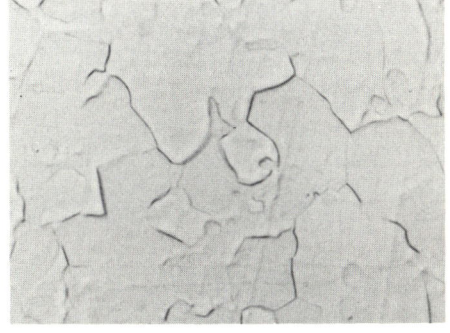

Figure 11. Aluminum, etched 6 minutes, 150X (8).

Figure 12. Tantalum, etched 1 minute, 800X.

(All reduced 40 percent for reproduction.)

etched effect on the tantalum sample which had a 600-grit SiC soft finish before etching, Figure 12.

Alloys

Various alloys that have been etched by the technique are pictured in Figures 13 through 22. Also there is a multitude of various alloys that have been etched by other investigators as can be seen in Tables I and II.

Dissimilar Metal Couples

The satisfactory etching of dissimilar metal couples is usually considered to be a difficult problem by conventional methods. The chemical that is suitable for one side of a couple will generally pit, stain, or obliterate the other side and/or excessively attack the interface. In contrast to this, CVE by its nonspecific nature, tends to develop the structure on both sides of many couples simultaneously (9). This capability is illustrated in Figures 23, 24 and 25 (8). The probability of success with any particular couple can be estimated from the sputtering rates of the materials involved (11).

Complex Materials

A number of publications have demonstrated the application of the technique to complex materials (8,9,11). Graphitic stellite and chromium carbide are shown in Figures 26 and 27, respectively. Figures 28 and 29 are comparisons between tungsten carbide cathodic vacuum etched and chemically etched. The cobalt binder in Figure 28 was over etched in 1½ minutes.

Uranium and Uranium Alloys

Uranium and uranium alloys that require considerable care when etched chemically are readily handled by CVE. A large percentage of the publications on CVE that included uranium and uranium alloys reported good results.

Figures 30, 31 and 32 were reproduced from (5). Figure 31 is the same sample as that shown in Figure 30 but etched for an additional two minutes. Note the planar boundaries and the added detail from the deep etch. Figure 32 is the same sample as that shown in Figure 30, etched for 30 seconds and heat tinted.

PRACTICAL APPLICATIONS OF CATHODIC VACUUM ETCHING

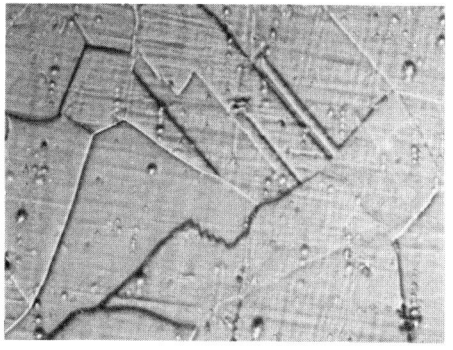

Figure 13. 304 stainless steel, etched 1½ minutes, slightly oblique, 400X.

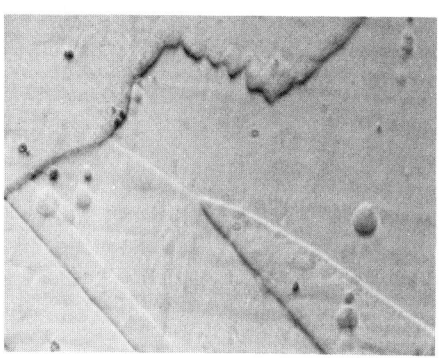

Figure 14. Same sample as Figure 13, 800X.

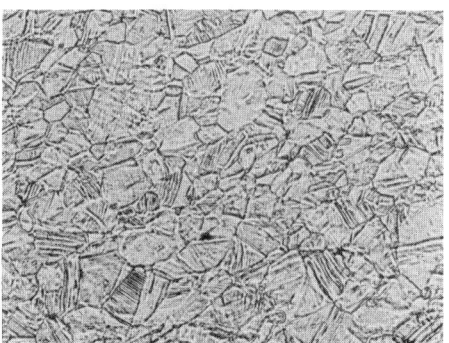

Figure 15. Phosphor-bronze, etched 45 seconds, 130X.

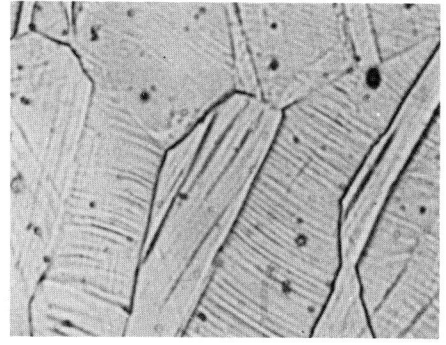

Figure 16. Same sample as Figure 15, 800X.

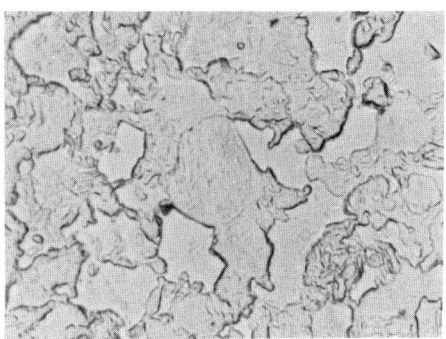

Figure 17. Carbon steel, etched 1 minute.

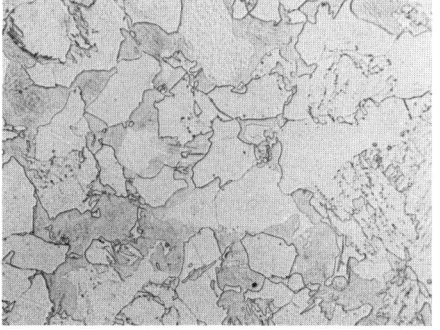

Figure 18. Same sample as in Figure 17 etched in 2% Nital, 800X.

(All reduced 40 percent for reproduction.)

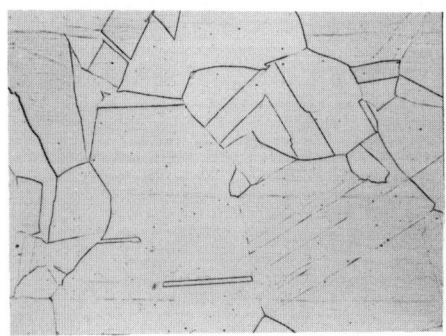 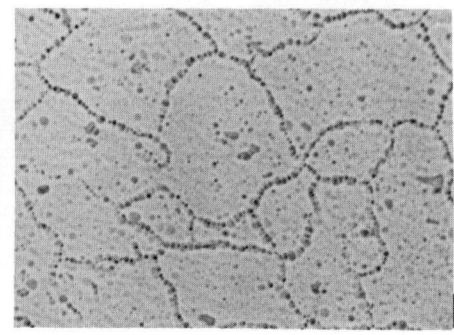

Figure 19. Silver, cadmium, etched 1 minute in air, 75X.

Figure 20. Aluminum alloy, etched 1 minute in 15 second intervals with 15 second cooling between etching periods, 600X.

 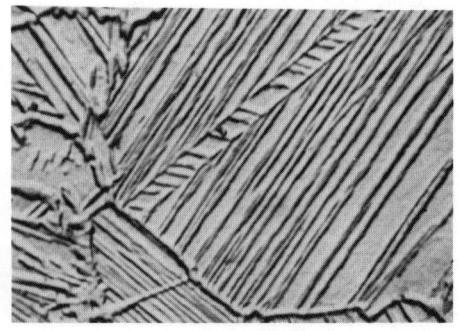

Figure 21. Experimental alloy, etched 2 minutes, 250X (10).

Figure 22. Same sample as Figure 21, 1500X (10).

(All reduced 40 percent for reproduction.)

PRACTICAL APPLICATIONS OF CATHODIC VACUUM ETCHING 221

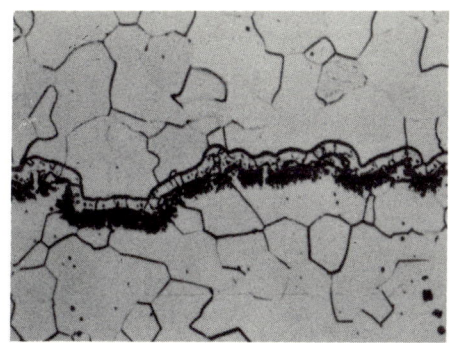

Figure 23. Zirconium (top)-uranium (bottom) couple, etched 3 minutes, 250X (8).

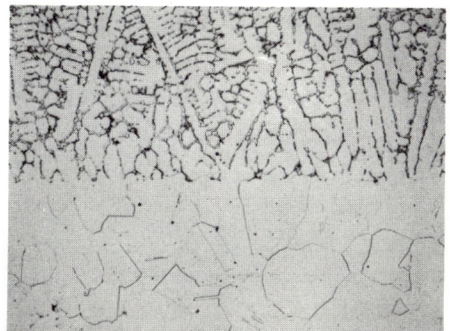

Figure 24. Stellite (top) on type 347 S. S. (bottom) etched 3 minutes, 150X (8).

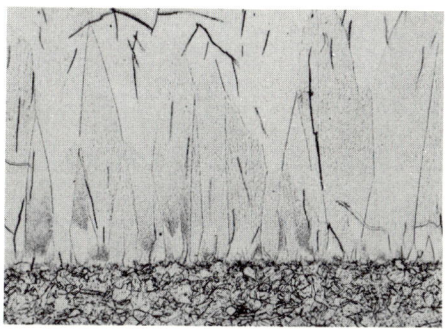

Figure 25. Hard chrome plate (top) on 17-4 PH S. S. (bottom), etched 3 minutes, 400X (8).

(All reduced 40 percent for reproduction.)

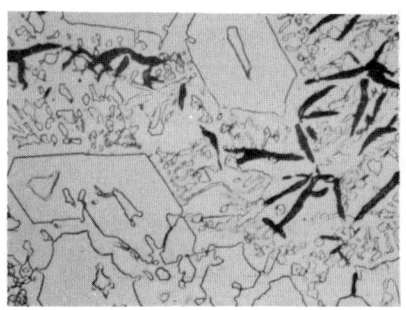
Figure 26. Graphitic stellite, etched 3 minutes, 150X (8).

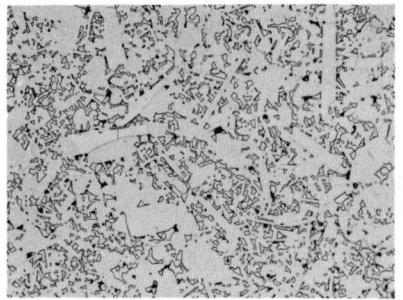
Figure 27. Chromium carbide in 15% Ni binder, etched 3 minutes, 250X (8).

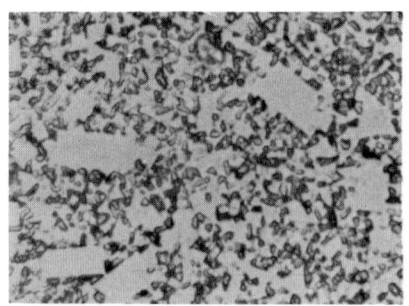
Figure 28. Tungsten carbide, etched 1½ minutes, 1500X.

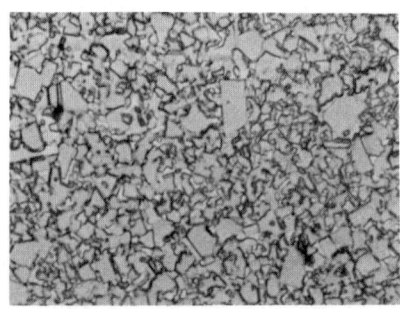
Figure 29. Same sample as Figure 28 etched in KCn-KOH, 1500X.

(All reduced 40 percent for reproduction.)

PRACTICAL APPLICATIONS OF CATHODIC VACUUM ETCHING 223

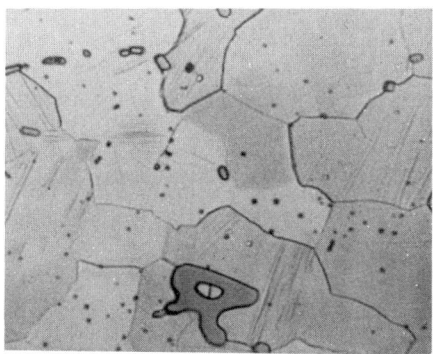

Figure 30. Natural uranium, etched 30 seconds, 500X (5).

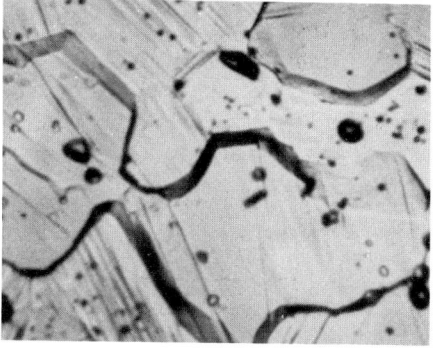

Figure 31. Same sample as Figure 30, etched 2½ minutes, 500X (5).

Figure 32. Same uranium sample as Figure 30, etched 30 seconds and heat tinted, 1000X (5).

(All reduced 40 percent for reproduction.)

Duran and Fisher (1) concluded in their paper that the NUMEC etcher had given very clean etched surfaces on isothermally treated uranium alloys, that were free from interfering oxide films. Their etched samples exhibited sharp, fine details in the microstructure and good definition of grain boundaries.

Ceramics

UO_2, ThO_2, Al_2O_3, PuO_2 and other ceramics have been etched by CVE (5,12,13,14). Figures 33 through 36 are examples of the application to some ceramics. It should be noted that the PuO_2 samples were prepared using a glove box etcher.

Deep Etching for Light and Electron Metallography

A type 304 stainless steel sample, the same field etched a total of 25 minutes in six consecutive steps and photographed after each etch, is shown in Figure 37. The photomicrographs in Figure 37 were reduced from an original magnification of 500X. Additional examples of deep etching for light metallography can be seen in Figures 38 through 40.

A sample of type 322 stainless steel reproduced from reference (8) demonstrates the application of CVE to electron microscopy, Figure 41. Bainbridge (12) reported very favorable results when using CVE for electron metallography.

Rexer and Vogel (2) using a NUMEC etcher stated in their summary: "The structure of practically all materials can be cleanly and reproducibly revealed by ion bombardment. The technique can be used to particular advantage on porous materials and on combinations of materials with dissimilar chemical properties. The clear development of the finest structural details makes this method interesting for use in electron microscopy."

Remote Applications

The handling of chemical solutions for swabbing, dipping or electrolytic etching poses problems in remote metallography. Moreover, the thorough washing and drying of samples is difficult under the best remote conditions. Ultrasonic cleaners are helpful in this regard but will not remove stains or films from the surfaces of highly reactive materials.

CVE greatly simplifies remote etching for many materials as has been demonstrated by Westinghouse, General Electric, Harwell, Hanford and other remote laboratories.

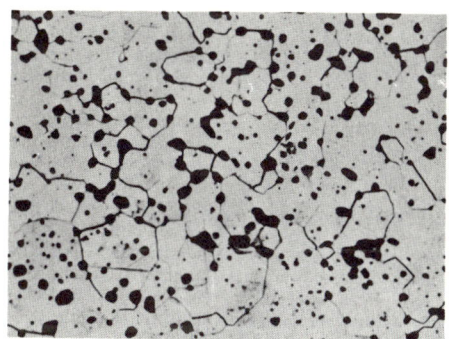

Figure 33. UO_2 compact cold pressed and sintered, CVE 6 minutes, 500X (5). (Reduced 25 percent for reproduction.)

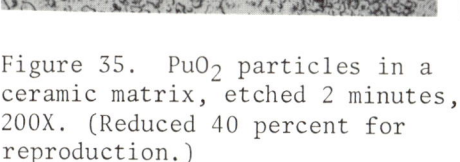

Figure 34. PuO_2 Compact, etched 1½ minutes, 400X. (Reduced 40 percent for reproduction.)

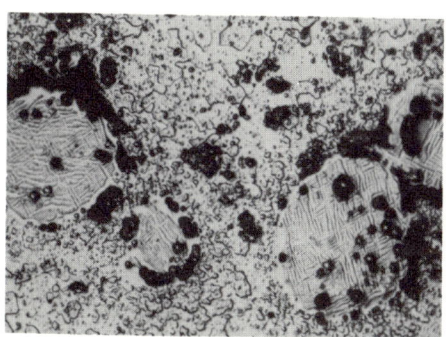

Figure 35. PuO_2 particles in a ceramic matrix, etched 2 minutes, 200X. (Reduced 40 percent for reproduction.)

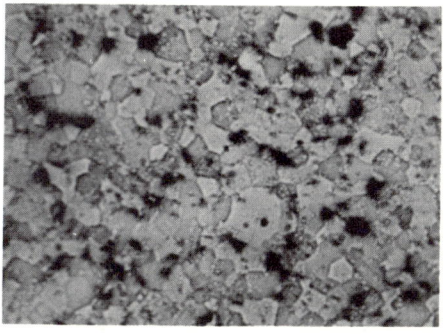

Figure 36. Alumina, etched 1 hour, 200X. (Reduced 40 percent for reproduction.)

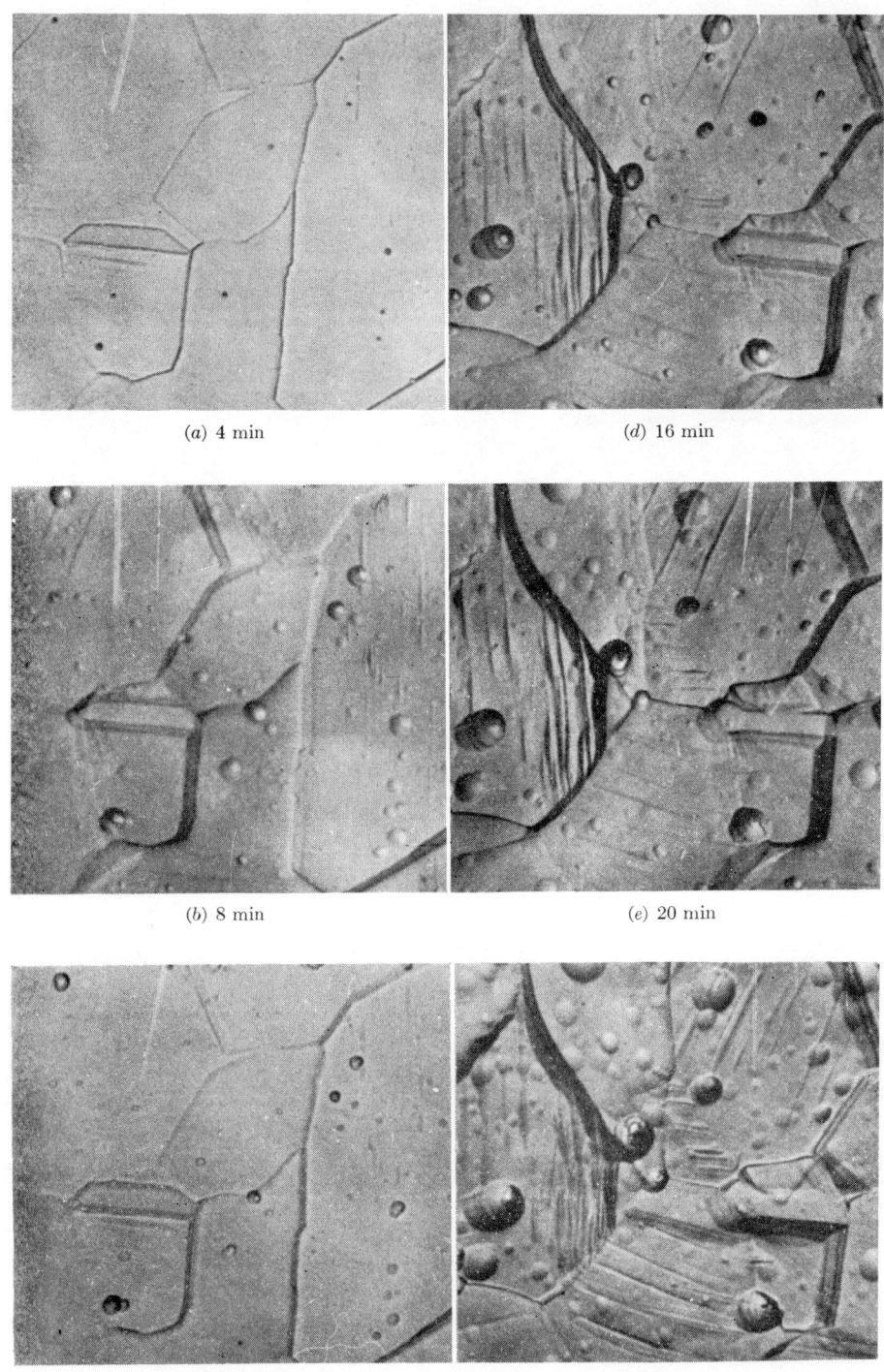

Figure 37. The same field in Type 304 stainless steel photographed after each of 6 consecutive cathodic vacuum etches, 500X (5).

PRACTICAL APPLICATIONS OF CATHODIC VACUUM ETCHING 227

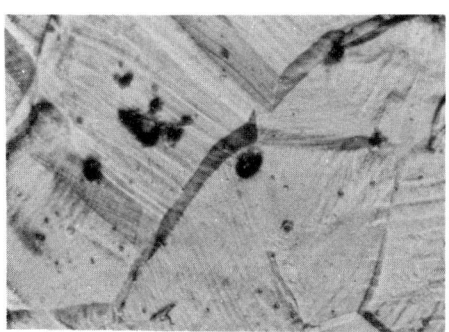

Figure 38. Phosphor bronze, etched 1½ minutes, 800X.

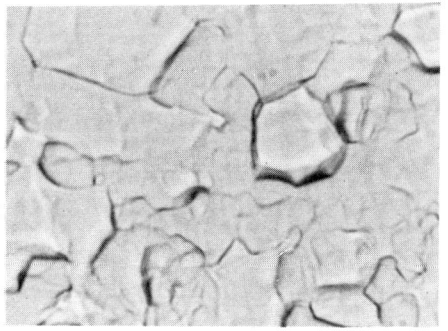

Figure 39. Tantalum, etched 1½ minutes, 800X.

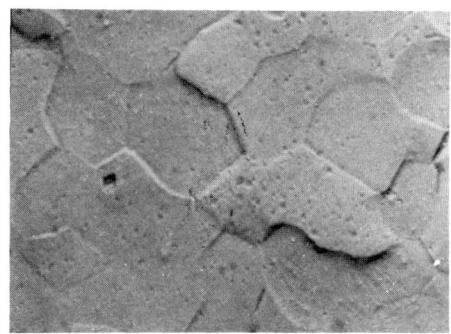

Figure 40. Hot-pressed zirconium, etched 1½ minutes, slightly oblique, 800X.

Figure 41. 322 stainless steel, etched 4 minutes, 23,000 (8).

(All reduced 40 percent for reproduction.)

Bainbridge (12) prepared a wide variety of materials for both remote optical and electron metallography. Bierlein and Mastel (15) state that etching by ion bombardment overcomes many drawbacks of other etching techniques for irradiated uranium.

Ion Thinning and Drilling

The thinning of samples for transmission electron metallography has been performed by several techniques. Among these are chemical and electrochemical milling, mechanical and electrical discharge machining, positive ion bombardment, etc. One of the most advantageous methods as reported by Ward (13) and Holland, Hurley and Laurenson (16) is that of positive ion bombardment.

A piece of zirconium foil was thinned on a Burrell etcher for this report. A total etching time of 20 minutes was applied in 15 second intervals with 15 seconds of cooling time between each etching period. The foil was then examined and found to contain a 20 mil diameter hole surrounded by a deeply etched band approximately 5 mils wide, Figure 42. Figures 43 and 44 represent areas adjacent to the hole and in the center of the deep etched band, respectively. Note the polished-etched effect by comparing the surface finish of the foil to the surfaces shown in Figures 43 and 44. Figure 45 taken from the I.D. of the hole exhibits the outline of grains that have been completely removed.

Figure 46 shows a 5 mil hole that resulted from thinning a piece of zirconium foil for only a total etching time of 3 minutes with 15 seconds of cooling between each 15 seconds of etching. In an attempt to fasten the foil to a flat surface with scotch tape it was accidentally pulled apart. The fracture path was noted to propagate along some grain boundaries, Figure 47.

Other Applications

Positive ion bombardment, in addition to the previously listed applications to metallography, has been found useful for a number of other purposes.

Mullaly (17) was able to form thick films of copper using a concave cathode at low pressure. He obtained a deposition rate of up to 2.2 mils per hour.

The spectrochemical properties of the glow discharge of a metal alloy sample were used to determine the availability of useful intensities for major and minor constituents (18). Trace analysis had not yet been considered.

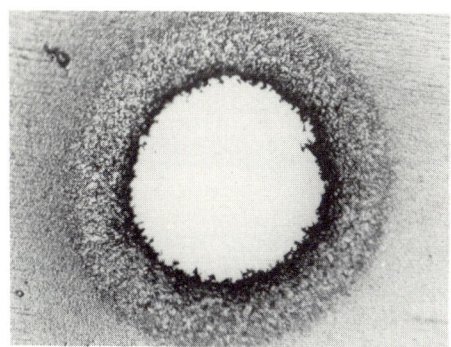 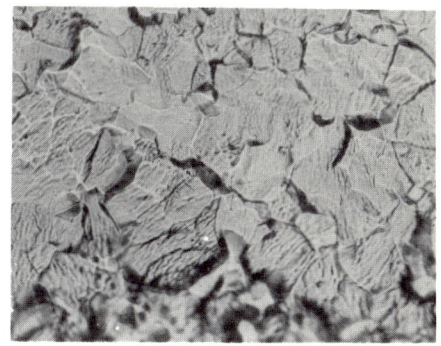

Figure 42. Zirconium foil, 20 mil dia. hold, thinned by 20 minutes of etching time, 75X.

Figure 43. Same sample as Figure 42 at edge of hole, 600X. Note polished-etched effect.

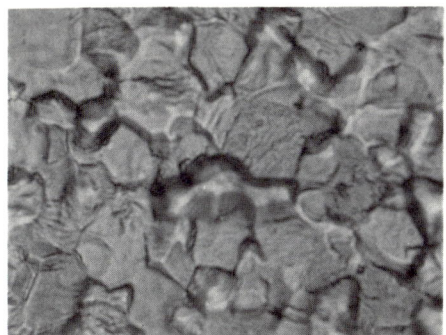 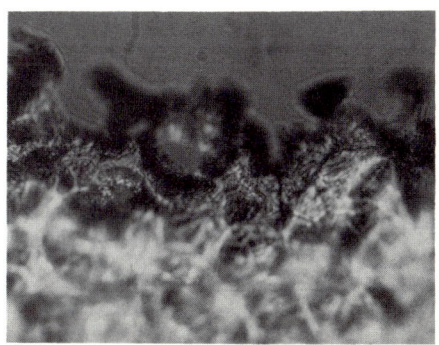

Figure 44. Same sample as Figure 42 at edge of hole, 800X.

Figure 45. Same as Figure 42 focused along inside edge of hole, 800X.

(All reduced 40 percent for reproduction.)

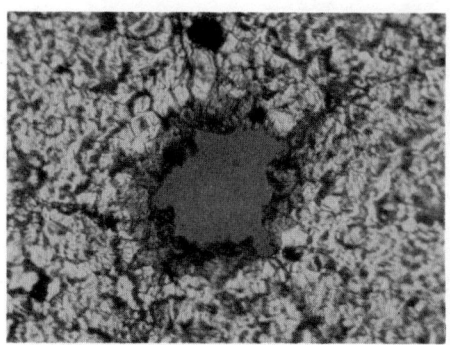 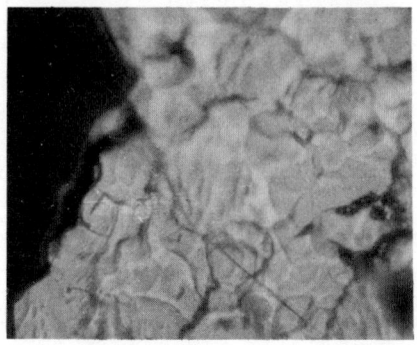

Figure 46. Zirconium foil thinned for 3 minutes, 5 mil hole, 200X. (Reduced 40 percent for reproduction.)

Figure 47. Same sample as Figure 46 after accidental fracture, 1000X. (Reduced 40 percent for reproduction.)

A method which has been developed to allow uniform removal of micro-layers of metal from cylindrical specimens is based on sputtering of the metal from the cathode of a gas glow discharge tube. This technique is directly applicable to diffusion studies and offers many advantages over the other more common methods (19).

Cathodic etching has been used to remove oxide films from samples of U, Th and Fe prior to oxygen determination (20). It was found that results were consistently lower than when sections of the same sample were cleaned by acid treatment or by filing.

The technique has also been considered as a means of preparing standard surfaces for corrosion studies (8).

CONCLUSIONS

Based on the results reported in referenced publications plus the results of laboratory work performed for this paper the following concluding statements can be made:

1. Cathodic Vacuum Etching is a versatile and practical tool for light and electron metallography.
2. It can be used to advantage in conventional, glove box, and remote laboratory operations.
3. The Burrell-NUMEC commercially manufactured equipment has given clean, reproducible results on many materials.

ACKNOWLEDGEMENTS

Thanks are expressed to all authors and investigators whose works are referenced and to my Lord and Savior, Jesus Christ, for making it possible for me to prepare this paper.

REFERENCES

1. J. B. Duran and R. E. Fisher, "Cathodic Vacuum Etching of Uranium Alloys", Microstructures, October-November 1971.
2. Jurgen Rexer and Margrit Vogel, "Experience with the NUMEC Vacuum Cathodic Etching Apparatus for Ion Etching", Praktische Metallographie, July 1966.
3. F.M. Cain, Jr., "A Simplified Procedure for the Metallography of Zirconium and Hafnium and their Alloys", Zirconium and Zirconium Alloys, Am. Soc. Metals, 1953.
4. J.F.R. Ambler, "Attack Polishing of Metals and Ceramics", IMS Proceedings, 1970-1971.
5. F.M. Cain, Jr., "Simplified Metallographic Techniques for Nuclear Reactor Materials", Symposium on Methods of Metallographic Specimen Preparation, ASTM 1960.
6. F. M. Cain, Jr. and F. O. Bingman, "Remote Metallographic Equipment and Practices", Sixth Hot Laboratory Equipment Conference, March 1968 Supplement.
7. J. Strong, Procedures in Experimental Physics, Prentice-Hall, Inc., New York, N.Y. (1938).
8. T. R. Padden and F.M. Cain, Jr., "Cathodic Vacuum Etching", Metals Progress, 1954.
9. D. Armstrong, P.E. Madsen and E.C. Sykes, "Cathodic Bombardment Etching of Nuclear Materials", Journal of Nuclear Materials, Vol 2, 1959.
10. Robert Ball, "Experimental Alloy" Burrell Corp., Pittsburgh, Pa., 1971.
11. ASM Committee on Metallography, Metallography, 1954 Metals Handbook Supplement, p. 174.
12. John E. Bainbridge, "Cathodic Vacuum Etching-a Technique for the Surface Preparation of Irradiated Materials Prior to Electron Metallography", Praktische Metallographie, March 1971.
13. John W. Ward, "High-Vacuum Glow-Discharge Etcher and Ion-Thinner", Microstructures, April/May 1971.
14. F.M. Cain, Jr., USAED Document TID 7532 (Part I) 1955.
15. T.K. Bierlien and B. Mastel, "Metallographic Studies of Uranium", Second United Nations International Conference on the Peaceful Uses of Atomic Energy, June 1958.
16. I. Holland, R.E. Hurley and L. Laurenson, "The Operation of a Glow-Discharge Ion Gun used for Specimen Thinning", J. Physics E (Sci. Instruments) March 1971.
17. James R. Mullaly, "Crossed Field Discharge Device for High Rate Sputtering", Research/Development, Feb. 1971.

18. P.W.J.M. Boumans, "Studies of Sputtering in a Glow Discharge for Spectrochemical Analysis", Analytical Chem., 44, No. 7, June 1972.
19. Thomas F. Fisher & C.E. Weber, "Cathodic Sputtering for Micro-Diffusion Studies", Journal of Applied Physics, 23, Feb. 1952.
20. John H. Hill, C.J. Morris and Jack W. Frazer, "Cathodic Etching to Prepare Metals for Oxygen Determination", Analyst, Feb., 1970.
21. L.B. Loeb, Fundamental Processes of Electrical Discharge in Gases, John Wiley and Sons, Inc., New York, N.Y. 1939.
22. J.D. Cobine, Gaseous Conductors, Theory and Engineering Applications, McGraw-Hill Book Co., Inc., New York, N.Y. 1941.
23. C.H. Townes, "Theory of Cathode Sputtering in Low Voltage Gaseous Discharge", Physical Review, 65, 1944.
24. J.B. Newkirk and W.G. Martin, "A Method for the Etching of Metals by Gas Ion Bombardment", Transactions, Am. Soc. Metals, Vol. L, p. 572 (1958).
25. T.K. Bierlien & B. Mastel, "Improved Method of Etching by Ion Bombardment", Review of Scientific Instruments, 30, No. 9, (1959) pp. 832-833.
26. M. Kaminsky, Atomic and Ionic Impact Phenomena on Metal Surfaces, Springer-Veriag, Berlin-Heidelberg-New York, (1965).
27. W. Feitknecht, "Uber den Angriff von Krystalled durch Kanalstrahlin", Helvetica Chemical Acta, 7, 1924, pp. 825-848.
28. C.S. Smith, "Note on Cathodic Disintegration as Method of Etching Specimens for Metallography", Journal of the Institute of Metals 38, (1927) p. 133.
29. D.M. McCutcheon, "Cathodic Vacuum Etching of Metals", Journal of Applied Physics, 20, April, 1949.
30. D.M. McCutcheon and W. Pahl, "Cathodic Vacuum Etching of Metals", Metal Progress, Nov. 1949.
31. L. Holland, Vacuum Deposition and Thin Films Chapman Hall, 1956, 439-443.
32. G.V. Spivak, V.E. Iurosova, I.N. Prilizhaeva and E.K. Pravdina, Bull. Akad. Sci. U.S.S.R. (Physics Series) 20, (Columbia Technical Translations) p. 1073.
33. Gottfried Wehner, Sputtering, Science & Technology, September, 1968.
34. F. Llewelyn-Jones, The Glow Discharge, Methuen, Wiley, London, New York, 1966.
35. G. Carter and J.S. Colligon, Ion Bombardment of Solids Heineman, London, 1968.
36. G.F. Weston, Cold Cathode Glow Discharge Tubes, Iliffe, London, 1968.
37. A. von Engel, Ionized Gases, Clarendon Press, Oxford, 1965.

METALLOGRAPHIC METHODS APPLIED TO ULTRATHINNING LUNAR ROCKS, METEORITES, FOSSILS, AND OTHER BRITTLE MATERIALS FOR OPTICAL MICROSCOPY

R.H. Beauchamp and J. F. Williford

Battelle - Pacific Northwest Laboratories

Richland, Washington

BACKGROUND

Petrography is an old method, and thin sections of thirty micron thickness have been used for well over 100 years. Indeed, early petrographers, such as H. C. Sorby, actually started metallography on its way. Since Sorby's day, metallography and the preparation of metallic specimens have come a long way. During the same period, optical petrography has remained very static and conservative.

Optical petrography, as traditionally practiced, suffers from two deficiencies:

-- The unpolished surfaces are not usable in reflection (limiting optical methods).
-- The relatively great specimen thickness of 30 microns leads to opacity and poor resolution, particularly with fine-grained dark materials.

Kennedy (1) produced and reported polished thin sections of conventional thickness in 1945, approximately 100 years after the beginning of petrography. Other advances have been made in the preparation of brittle materials, either in the form of polished ceramographic specimens for reflected light or in the form of a less-than-conventional thickness section (e.g., 10-15 microns). It is significant that many of these advances have come from materials sciences people.

---

*The lunar rock work was sponsored by NASA Manned Spacecraft Center, Houston, Texas, under contract NAS 9-11993.

Those of us who work mainly with man-made materials have not had the same license to dismiss invisible portions of microstructure as "ground-mass"--a convenient practice still available to the geologist. We must see phase segregation, precipitates, porosity and microcracking in order to understand material characteristics and to assure the quality of manufacturing processes.

It is therefore not surprising that highly significant improvements in the art of preparing and examining petrographic sections should come from a metallographer. We have been working on procedures for ultra-thinning for about seven years, and will describe what we now know rather than review how we came to know it.

PREPARATION OF SECTIONS

There are a number of key developments that have all become available fairly recently and make ultrathin (1 to 5 microns thickness), doubly-polished sections possible. These are:

-- Metal-bonded-diamond wheels
-- High-strength, thermosetting polymers
-- Closely graded micron and submicron diamond slurries

Mounting

Our sections are first mounted in 655 Maraset epoxy*, using 555 catalyst. Vacuum impregnation is always used. Polydimethylsiloxane (Silastic RTV) molds are used to make the mounts. The molds are only partially filled during vacuum impregnation. While the specimen is being impregnated, a portion of the same batch of epoxy is filled with 0.3 micron alumina powder by stirring. When the epoxy in the mount has stopped outgassing, the mold is taken from the bell jar and glass beads are carefully placed in around the perimeter of the specimen. The alumina-loaded epoxy is then used to finish filling the mold. The casting is oven cured at 80°C, generally overnight.

This molding technique has a number of advantages. The glass balls and alumina make it easier to keep flatness than does a simple plastic mount. Also, the alumina, which sinks to the bottom (hence the face of the mount), later serves as a reference material for control of thickness and wedging during thinning.

---

*Marblette Corp.

First Side Preparation

The first side is initially ground flat against a 120-grit silicon carbide belt sander or on a 45-micron diamond wheel (depending on hardness). It is then sequentially lapped through 30-, 15- and 9-microns on sintered diamond wheels. Water cooling is used for all grinding steps.

There are two important things to remember during grinding. First, loose abrasive grinding--particularly using grains larger than about 5 microns--must be avoided. The depth of subsurface fracturing is much greater with a rolling abrasive grain than with a fixed grain. The second thing is that exceptional flatness must be maintained when sections only a few microns thick are wanted. The tolerable radius of curvature of each surface is expressed in thousands of meters.

Polishing is done on Syntron vibratory polishing tables using water-dispersed diamond slurries of 1-micron particle size and less. Poromeric laps such as Pellon PAN-W are generally used. The lapped specimen is clamped in a polishing weight of 800 to 1100 grams and allowed to run on the vibratory polisher overnight.

The above procedure generally yields a highly-polished, flat surface with coplanar relationships between all phases. With some materials, an additional polishing step with 1/2-micron diamond provides some improvement.

Gluing to the Slide

After the first side has been completed and has been carefully examined to assure that it is free of relief and is flat, the cast mount is clamped in a diamond saw and a slice is made parallel to and containing the polished surface. The thickness of this slice is important. If it is less than a mm, it will be liable to bend or break. If it is greater than 3 mm, it will be too rigid to conform to slight dimensional changes during curing of the adhesive. Thus, air will be pulled into the glue line.

The polished surface is cemented to the well side of a standard petrographic slide with Maraset 655 epoxy and 555 hardener. It is well to outgas the glue line in the vacuum chamber and carefully work down the glue line thickness by pressing on the section afterwards. The glued joint is oven-cured at 80°C, preferably overnight, and is removed from the oven with some care to avoid breakage by thermal shock.

There are a number of things that are important here, principally clean surfaces on the glass and specimen and care to avoid air bubbles in the glue line. It is important to use the well side of the slide so that the specimen is not torn off by shear stresses at the joint perimeter during lapping.

Second Side Preparation

Second side preparation is the same as the first side sequence. The section is hand-held against the belt or diamond wheel for rough thinning, periodically lifting the section and looking at the alumina surrounding the specimen to assure that wedging is under control. Rough thinning is stopped when light can just be seen through the alumina. Lapping continues through 15 micron on diamond wheels. Depending on the specimen, this process of hand-held thinning is discontinued at about 15 to 20 microns thickness. The alumina is highly transduced at this point. Thinning may be continued on a nine-micron wheel, on the vibratory table using 3-micron diamond slurry or polishing may directly commence using 1-micron diamond. The glass slip is glued to the face of the polishing weight with cellulose cement.

Obviously, enough thickness must be left for complete removal of all subsurface damage, and the lapped surface must be flat. The specimen must be periodically lifted and inspected during polishing, since the difference between an ultrathin section and no section is only a few microns.

RESULTS AND USES

Ultrathin, doubly polished sections of brittle materials present a new level of transparency and resolution to the microscopist. The improvements are most striking with fine-grained and/or optically dense materials.

Figures 1-10 show examples of samples prepared by the above described techniques. Most of the original photomicrographs were in color, but, unfortunately had to be reproduced in black-and-white here.

Figure 1 shows lunar basalts in ultrathin section. The photomicrograph (1a) is a typical unpolished 30-micron section of Columbia River basalt. The poor resolution is caused by multi-stacking of crystals resulting in only about 50% interpretation. The section is used to show the advantages of ultrathinning. Figures 1b, 1c, and 1d are of a double polished 3-micron section of an ophitic lunar basalt. Ultrathinning eliminates stacking of grains. The entire section may then be resolved bringing out such

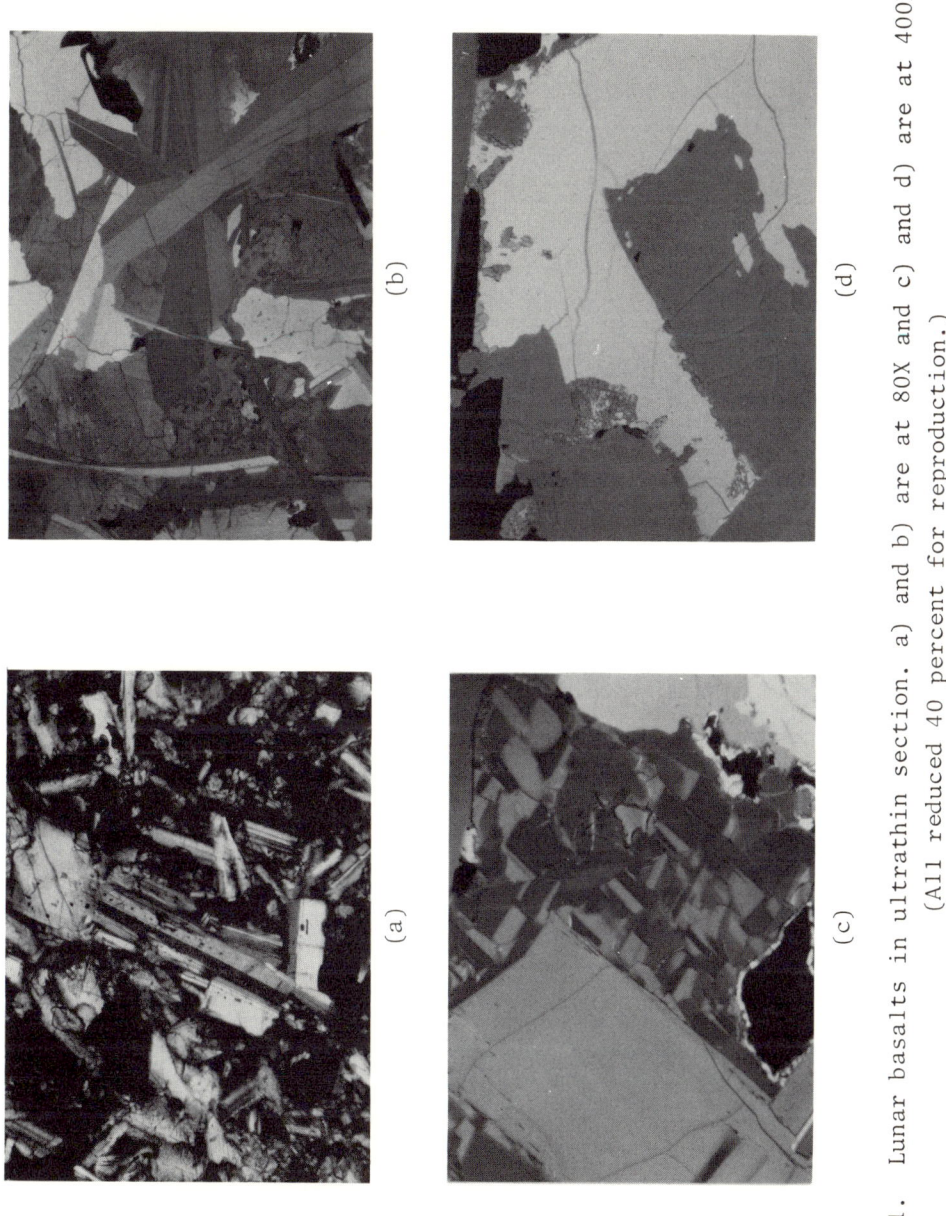

Figure 1. Lunar basalts in ultrathin section. a) and b) are at 80X and c) and d) are at 400X. (All reduced 40 percent for reproduction.)

Figure 2. Lunar basalts in ultrathin section. a) and b) are at 80X and c) is at 400X. (All reduced 40 percent for reproduction.)

METALLOGRAPHIC METHODS APPLIED TO BRITTLE MATERIALS 239

Figure 3. Lunar basalts in ultrathin section. a) is at 80X, b) and c) are at 160X and d) is at 400X. (All reduced 40 percent for reproduction.)

Figure 4. Lunar breccias in ultrathin section. 160X. (All reduced 40 percent for reproduction.)

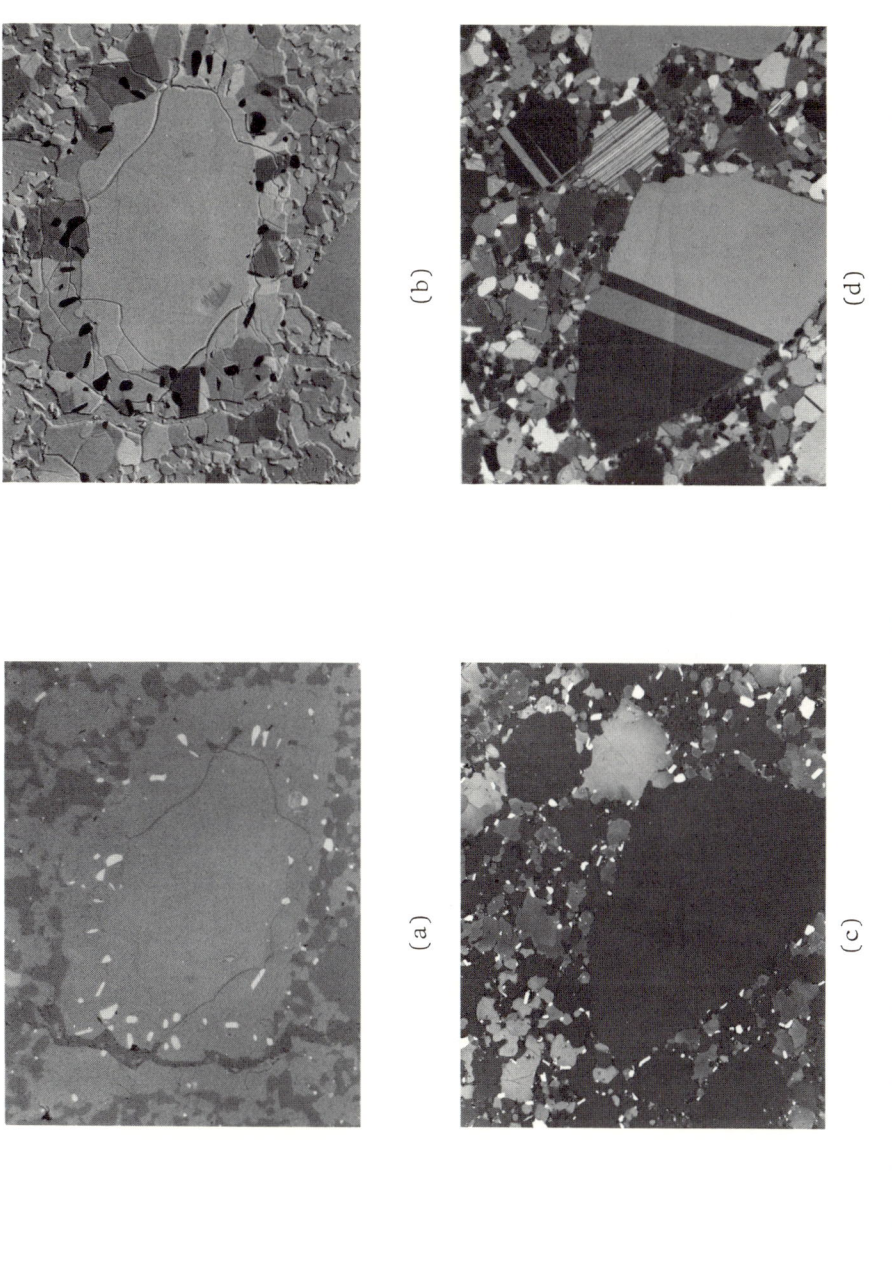

Figure 5. Lunar breccia subjected to post densification heating. a) and b) are at 500X, c) and d) are at 400X. (All reduced 40 percent for reproduction.)

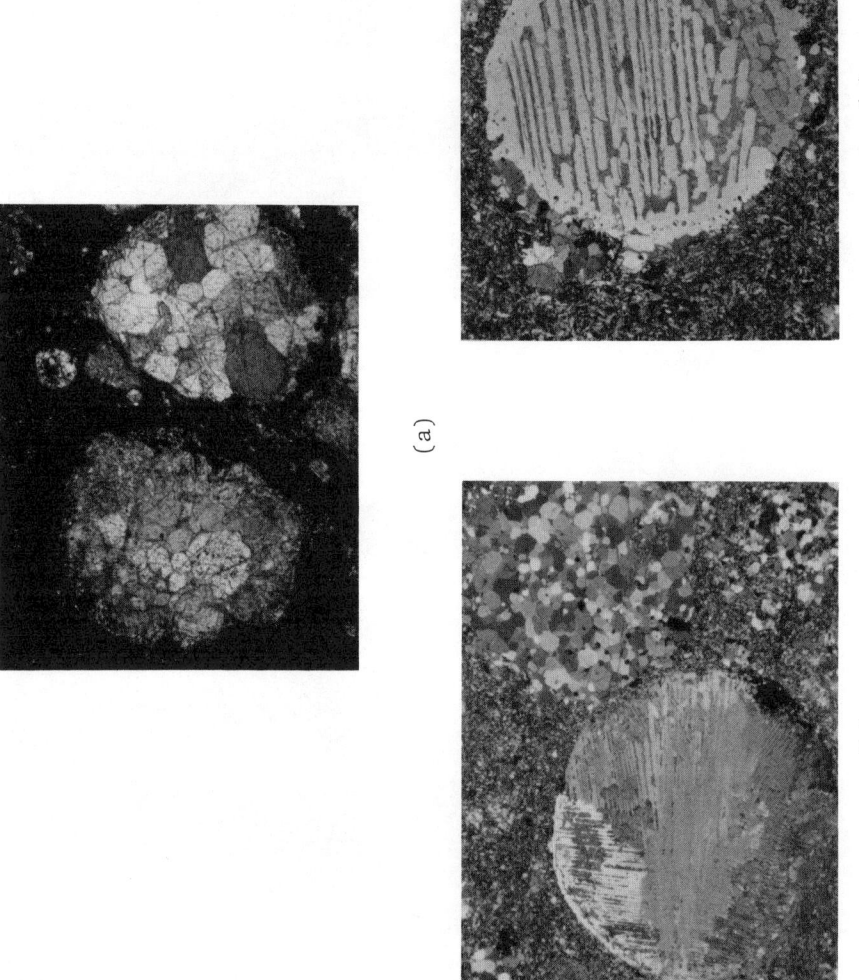

Figure 6. Ultrathin sections of the Allende meteorite. a) and b) are at 80X and c) is at 160X. (All reduced 40 percent for reproduction.)

METALLOGRAPHIC METHODS APPLIED TO BRITTLE MATERIALS 243

Figure 7. Chondrules in groundmass of meteorite. a) is at 100X, b) is at 400X and c) is at 1000X. (All reduced 40 percent for reproduction.)

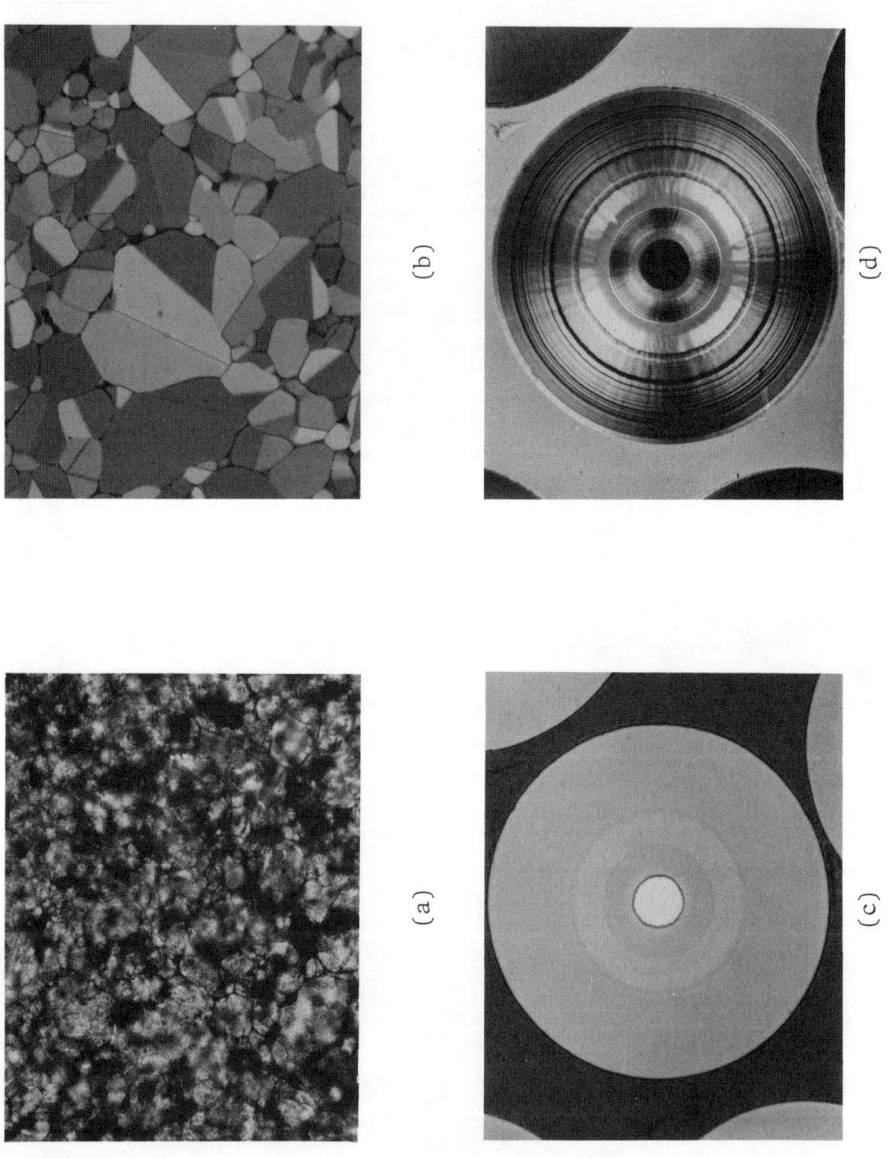

Figure 8. a) and b) are ultrathin sections of tin oxide. 500X. c) and d) are of silicon carbide filaments. 800X. (All reduced 40 percent for reproduction.)

METALLOGRAPHIC METHODS APPLIED TO BRITTLE MATERIALS 245

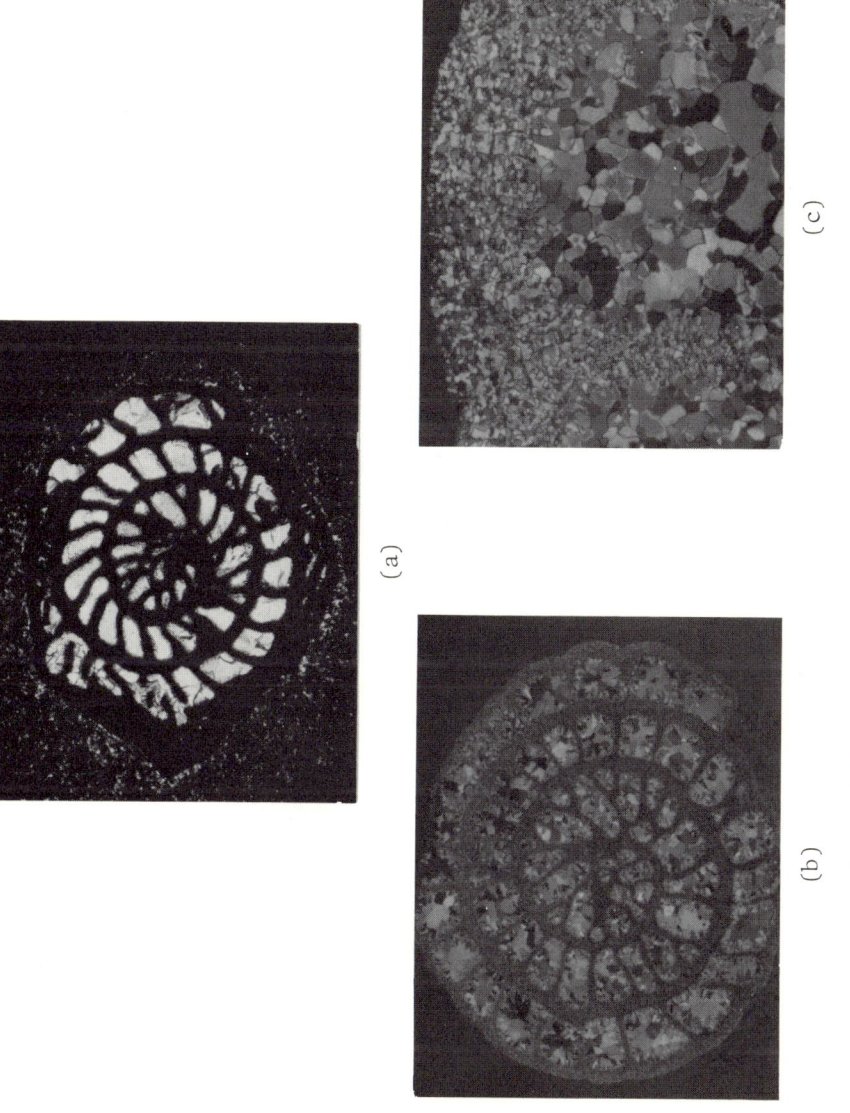

Figure 9. Ultrathin sections of micropaleontology materials. a) and b) are at 50X, c) is at 200X. (All reduced 40 percent for reproduction.)

Figure 10. Ultrathin sections of fossilized conodont in reflected light. a) is at 75X and b) is at 150X. (All reduced 30 percent for reproduction.)

features as microcracking, twinning, reaction rims, and new mineral growth at grain boundaries.

Figure 2 also shows lunar basalts in ultrathin section. Figure 2a shows well-developed undulatory extinction in pyroxene and large inclusions of ilmenite (black), Figure 2b shows large grains of pyroxene, lath-shaped plagioclase and interstitial crystobalite, and Figure 2c is a high magnification photomicrograph showing twinning in crystobalite.

Figure 3 also shows lunar basalts in ultrathin section. Figures 3a and 3b show large amounts of plagioclase (lath-shaped crystals), some pyroxene, and small amounts of ilmenite (black). Figures 3c and 3d show a dense lunar rock from the Apollo 12 mission which resembles terrestrial basalts in composition, but has been altered by reheating some time in the rock's history. The plumose or feathery texture is the result of complex intergrowth of the minerals plagioclase and pyroxene. Ilmenite is present as long, black needles.

Figure 4 also contains photomicrographs of lunar breccias in ultrathin section. The lunar breccia shown in Figure 4 represents another type of problem where the ultrathin section is highly advantageous. The angular grains of this microbreccia, consolidated from lunar soil, vary greatly in size. Much of the rock is made from materials such as ilmenite, troilite, and meteoritic irons, which are opaque even in ultrathin sections. In conventional sections, the fine-grained groundmass is completely opaque. In Figure 4a and 4b the textural details of the rock are sharply resolved. In Figure 4c we see resolution of microcracking, exsolution lamellar displacement and shock lamellar in a pyroxine grain.

Figures 5a and 5b show yet another breccia, subjected to some postdensification heating. The reflected light photo in Figure 5a shows a solid state interaction between troilite opaques and other minerals, leading to grain growth and the formation of new minerals. Figure 5b shows the same area in transmitted Normarski interference contrast. This technique is well adapted to the ultrathin section, but cannot be used with conventional sections. Figure 5c and 5d shows how the double-polished thin section of a lunar breccia may be used in reflected light (Figure 5c) and in transmitted polarized light in the same area of interest.

Figure 6 is of the Allende meteorite and presents much the same kind of problem as the lunar breccia. Figure 6a is a 30-micron section that reveals the large round structures (called chondrules), but does not resolve the very fine grained groundmass that surrounds them. Figure 6b and 6c show the various type of chrondrules as well as the fine grained groundmass.

Figure 7a shows yet other types of chondrules in the fine grained groundmass. Figure 7b shows the interior of a chondrule in which many details such as interphase reactions and glassy devitrification can be seen with great clarity. As shown in 7c, the individual grains in the groundmass can be seen in an ultrathin section approaching maximum resolution of the optical microscope.

A direct comparison is shown in Figure 8a (30 microns) and 8b (3 microns) of tin oxide, an electrode material. Materials such as beta silicon carbide filaments are shown in Figure 8c (reflected light). Because of the double-polished thin section, the same filament may be viewed in transmitted orthoscopic light, (Figure 8d).

Micropaleontology materials, such as the Protozoa (Foraminifer) can be readily resolved in ultrathin section. Figure 9a shows the fossil 30 microns in thickness. Figure 9b and 9c show clear delineation of wall structures and calcite replacement.

Figure 10a shows a fossilized conodont in reflected light. Ultrathinning techniques, as shown in Figure 10b, bring out very clearly details of wall structure and storage capacity in the conodont. This was photographed using transmitted Nomarski interference contrast.

CONCLUSION

A whole new order of optical access to microstructural detail in brittle substances has been made possible by the application of metallographic techniques, using off-the-shelf equipment and supplies. The new sections are usable with a full range of reflected and transmitted light methods and can be made from a wide range of specimens. Materials as diverse as calcium carbonate and boron nitride have been thinned and polished below 5-microns thickness routinely, and in some cases to thicknesses of one micron or less.

Application of these methods to electrical ceramics, fine-grained single and mixed oxides, hot pressed nitrides and refractories should yield useful results. Although the geologist is rather more conservative in the use of microscopy, it is expected that geologists will increasingly use ultrathin sections to delineate textural details, depending on the electron microprobe for compositional details.

Although the resolution of detail with ultrathin sections presently exceeds that of the electron microprobe, we expect this situation will change, perhaps through the use of the ion microprobe. In such a case, the ultrathin section will be a fine complement to these methods.

ACKNOWLEDGMENT

The authors wish to express their appreciation to Dr. Dave McKay and Dr. Mike Duke, NASA Manned Spacecraft Center, Houston, Texas, for the opportunity to publish the lunar work.

THE METALLOGRAPHIC SAMPLE PREPARATION OF FIBER-REINFORCED COMPOSITES

T. J. Bertone

The Aerospace Corporation

Los Angeles, California 90045

INTRODUCTION

Researchers in the field of fiber-reinforced composites have applied a variety of methods to evaluate not only the strength of the composite but also the interaction of the fibers and matrix at the fiber-matrix interface (1,2). The bond strength which is related to the wetting of the fiber by the matrix material is of great interest in studying the load transfer characteristics of composites (1-3). Furthermore, studies to evaluate thermochemical and mechanical degradation of fiber-reinforced composites in service resulting from physical and thermal treatments have been conducted. These studies depend heavily on metallography.

Although metallographic techniques for the preparation of samples involving many combinations of fibers and matrices are described in several texts (4-6), examination of photomicrographs presented in recent literature (2, 7-11) suggests that major improvements are needed in this area. In this respect it is apparent that the development of composite materials has advanced more rapidly than have metallographic techniques that can be used for the study of these materials.

The purpose of this paper is to describe techniques developed at the Aerospace Corporation for the metallographic preparation of a variety of composite materials containing both hard and soft fibers. The techniques presented here were specifically developed to eliminate mechanical damage to the fibers and matrices during metallographic preparation.

METALLOGRAPHY OF HARD FIBER, SOFT MATRIX MATERIALS

Historical

If only standard metal preparation procedures are used for composites, a great deal of difficulty is encountered. Severe fiber damage results and topographic relief can be observed as a result of 320-grit rotary grinding. Figure 1a shows a tungsten core, boron fiber/aluminum matrix composite after grinding with a 320-grit wheel. Figure 1b is a photomicrograph of this material after further grinding, ending with 600 grit followed by polishing with 1.0 micron alumina on Microcloth*. Pieces of the boron filaments have broken off and imbedded into the aluminum matrix. The surface of the composite is heavily worked making analysis of the structural characteristics virtually impossible. The final polishing operation has created a large topographic variation between filament and matrix.

A metallographic technique developed specifically for the preparation of hard-fiber, soft-matrix composites was described by Snyde et al. (4) They suggested that the sample be lapped with a figure 8 motion using diamond compound on an Oxford No. 50 index card to remove the topographic variations between the fibers and matrices. The author has applied Snyde's technique to undamaged tungsten core, boron fibers/aluminum matrix specimens and has observed structures with little or no topographic variations between the fibers and matrices, Figure 2a. While this technique achieved the desired microstructural characteristics when applied to the undamaged composite, it was not suitable for composite materials suspected of being fractured or in a damaged condition, Figure 2b. When using the paper card, the soft cellulose fibers snag the filaments in the composite extending the fracture damage which is present at the surface into the initially undamaged areas of the fibers within the matrix metal. This may possibly lead to an erroneous interpretation when a study of damage mechanisms or failure analysis is attempted.

Present Procedure

Since the two above described techniques produced fiber damage, relief between the fibers and matrices, and embedding of particles of the fibers in the matrix material, a new grinding and polishing technique was developed to minimize these effects. This method combines standard metallographic procedures with those used in optical

---

*Buehler Ltd., Evanston, Illinois.

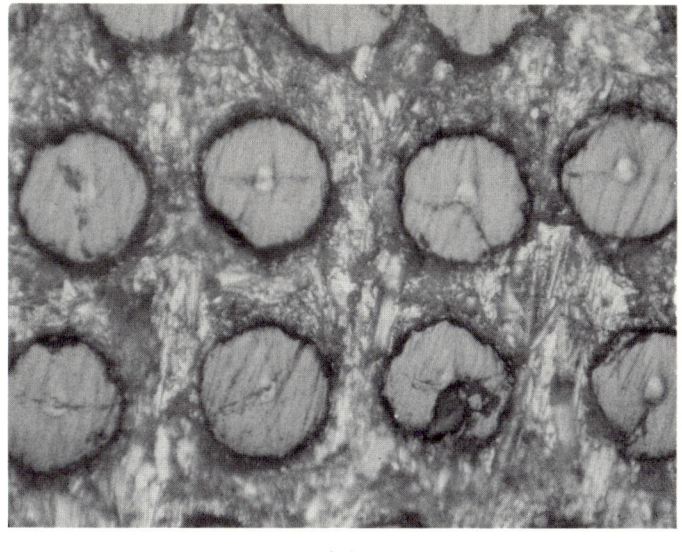

(a)

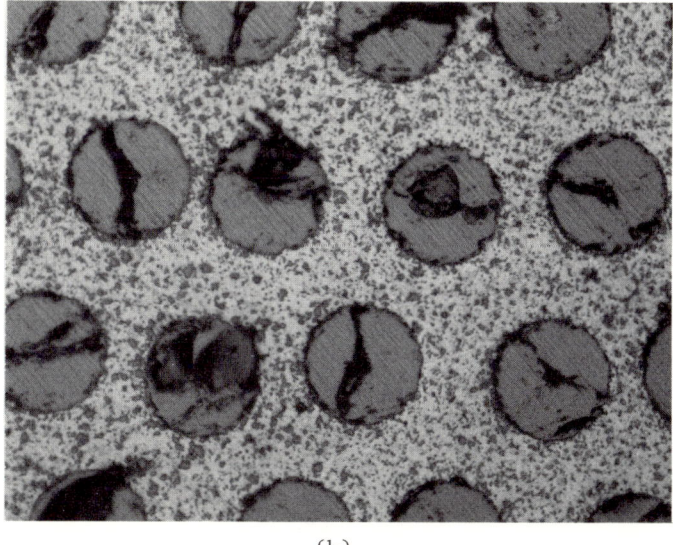

(b)

Figure 1. (a) Damage created while wet rotary grinding with 320-grit silicon carbide paper. 250X.

(b) Embedded particles of grinding grit and resulting fiber damage after 600-grit silicon carbide rotary grinding. 250X.

(Reduced 20 percent for reproduction.)

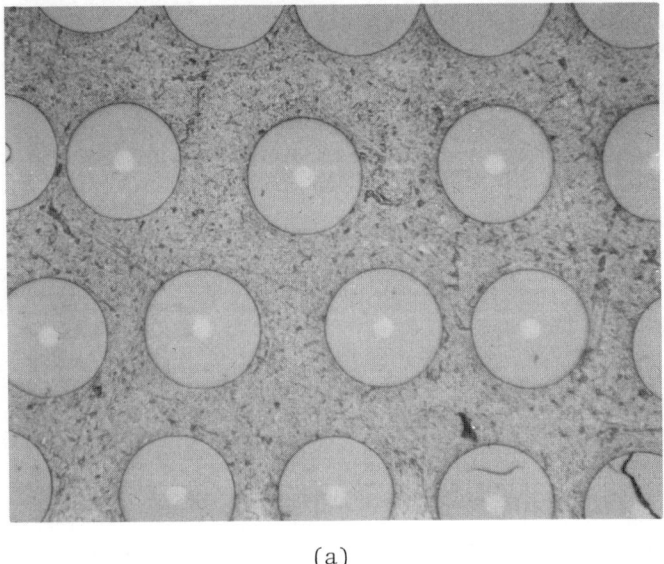

Figure 2. (a) An undamaged tungsten core, boron/aluminum matrix composite processed using Snyde's technique. 200X.

(b) The same composite as Figure 2(a) bend tested to damage fibers prior to using Snyde's technique. 200X.

(Reduced 20 percent for reproduction.)

component fabrication. The factors considered important for this technique are:

1. Minimal damage to the specimen during cutting; for this purpose, a wire saw is recommended (12).
2. The mounting media should have hardness and abrasion resistance characteristics that closely match those of the composite matrix material. This aids in edge retention.
3. The procedure chosen should be designed to progressively reduce the mechanical damage induced in the composite from the cutting and subsequent grinding operations in order to obtain an undisturbed surface condition after completion of final polishing.
4. The grinding and polishing operations must not damage the filaments by chipping or bending, and must not induce or propagate existing cracks in the filament.
5. Topographic variations should be held to a minimum, 5 micron or less, in order to obtain a metallographically acceptable sample that can be examined at high magnification with relative ease.

Tungsten core, boron fiber/aluminum matrix specimens were selected for developing the procedure. Monolayer and multilayer composite systems were prepared in identical fashion.

The initial grinding operation is done on 600-grit wet or dry silicon carbide grinding paper with water as a lubricant. The grinding paper is held stationary while the sample is guided over the grit with figure-eight lapping motions as shown in Figure 3. This technique requires a heavy flow of water to prevent large particles of sample material and grit from galling the composite sample. Slight rotation of the paper helps to keep the sample on fresh cutting grit. As shown in the figure, the sample should be grasped firmly with the thumb and middle finger on the edge and the index finger on the top approximately at the center. The best results are obtained by applying pressure, i.e., the grinding force, with the index finger only, using the other fingers to hold and slide the sample over the grinding paper with figure-eight lapping motions. If a large amount of mounting medium or sample must be removed, the grinding paper should be changed frequently.

The pressure should be as heavy as possible while holding the sample as described above so that the cutting action will be rather fast. The mounting medium will be ground more rapidly than the composite sample. Approximately 1 minute of grinding on 600 grit will cause 0.005 to 0.010 in. of sample relief. Sample relief is desired after 600 grit grinding, thus assuring a nondamaged filament at the matrix interface as shown in Figure 4. The matrix material is heavily abraded in comparison with the filaments, which already show signs of polishing. This is shown schematically in Figure 5.

Figure 3. Lapping technique used with grinding papers.

Figure 4. Sample produced by grinding on 600-grit grinding paper. 125 X. (Reduced 20 percent for reproduction.)

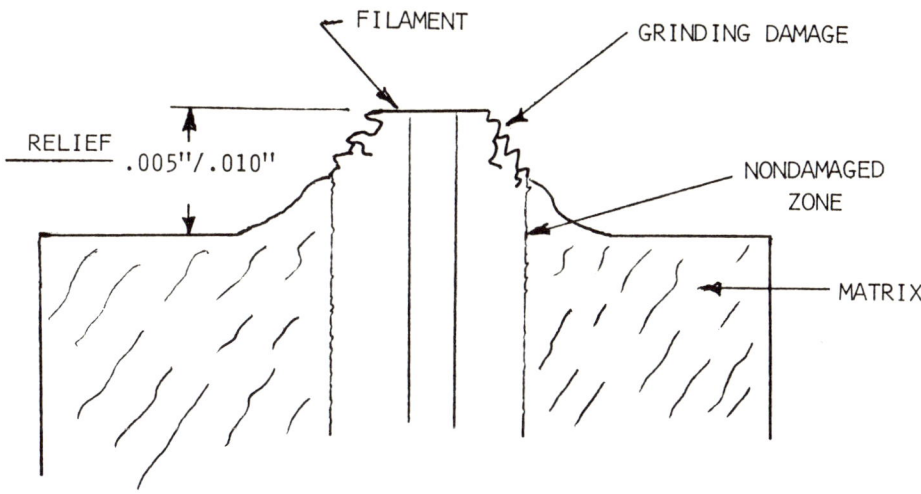

Figure 5. Schematic representation of filament extension from matrix after 600-grit figure-eight motion grinding.

Following a thorough ultrasonic cleaning, the sample is ready for optical grinding. This grinding is accomplished on a three-inch diameter, half-inch thick glass plate. The grinding surface is optically polished and flat within about 4 waves concavity. The first cutting compound should be 3-micron diamond paste. A very small amount of the diamond abrasive is brought to a thin fluid consistency by adding about two drops of Metadi* fluid. The sample should be hand-lapped on the optical flat, applying pressure with figure-eight lapping motion until the fibers are ground topographically to the same level as the matrix material, Figure 6. The time required for this operation depends on the size of the composite sample. The sample and the optical flat should then be washed and ultrasonically cleaned. One-micron diamond paste should then be added to the optical flat along with Metadi fluid, and the optical grinding operation repeated. An equal amount of time should be spent grinding with 1-micron diamond paste. The finely abraded texture produced by these steps is shown in Figure 7.

---

*Buehler Ltd., Evanston, Illinois.

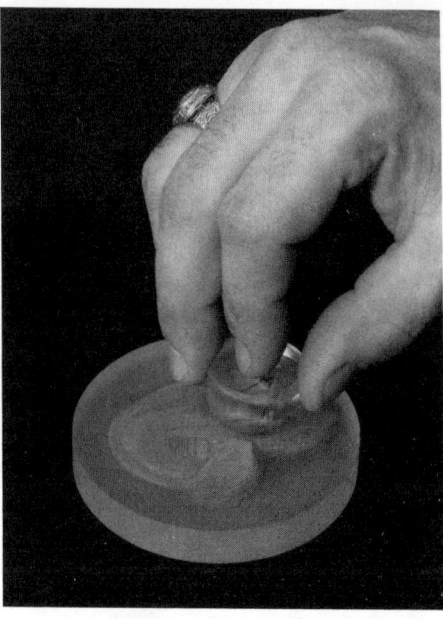

Figure 6. The proper technique of optical grinding used in this procedure for the preparation of composite samples is shown in the above photograph. The figure-eight path of grinding can be seen.

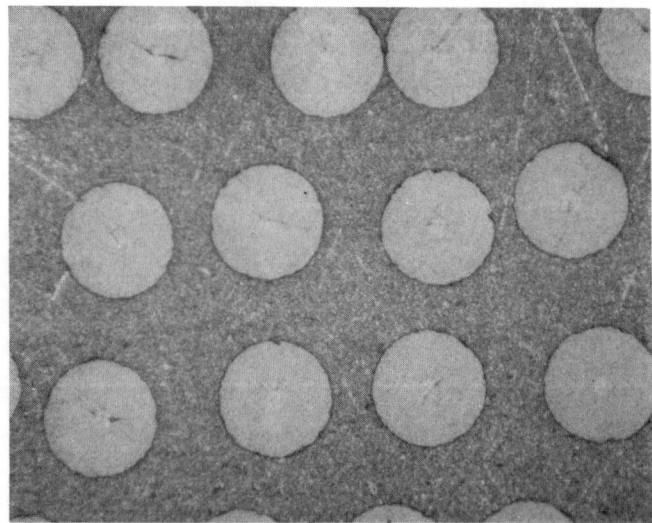

Figure 7. Sample produced by cutting action of diamond paste and optical flat grinding. 200X. (Reduced 20 percent for reproduction.)

SAMPLE PREPARATION OF FIBER-REINFORCED COMPOSITES

The sample should then be washed and ultrasonically cleaned in preparation for final polishing. A Syntron* automatic polishing machine equipped with a silk polishing cloth, and 0.3-micron alumina polishing compound should be used. The ratio of compound to water should be 200 c.c. compound to 600 ml distilled water. The sample is fastened into the holder and allowed to rotate on the Syntron at approximately one revolution every 15 to 20 sec. Care should be taken to ensure that the mount holding the sample rotates and does not remain fixed in one position, in which case the sample and polishing cloth would be damaged. To avoid this problem, an empty sample holder may be placed in the polishing tank along with the composite sample holder, see Figure 8. It is recommended that only one sample be polished on the Syntron during the final polishing operation.

Although polishing time will vary with sample size, excellent results can be obtained with a minimum of 2 hours of final polishing. Figures 9-11 show samples, prepared with this procedure, at progressively higher magnifications. The lack of topographic relief between the fibers and the matrix material is evident even at high magnifications.

Figure 8. The Syntron automatic polishing machine is shown fitted with a tightly stretched silk polishing cloth. A sample mounted in a holder is ready to be placed into the polishing tank along with an empty pusher holder. The weight of the sample holder is approximately 300 grams.

---
*Syntron Co., Homer City, Pennsylvania.

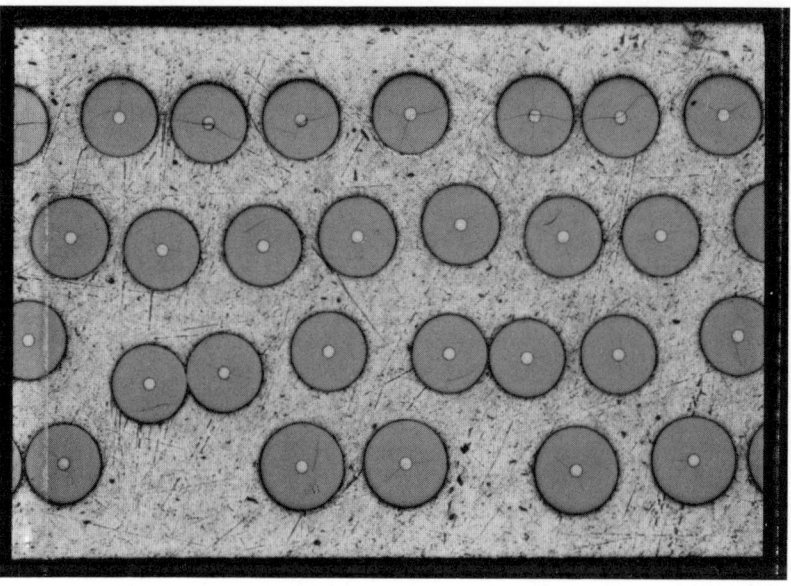

Figure 9. Tungsten core, boron/aluminum composite after two hours polishing on a Syntron automatic polishing machine using 0.3-micron alumina compound. 125X. (Reduced 15 percent for reproduction.)

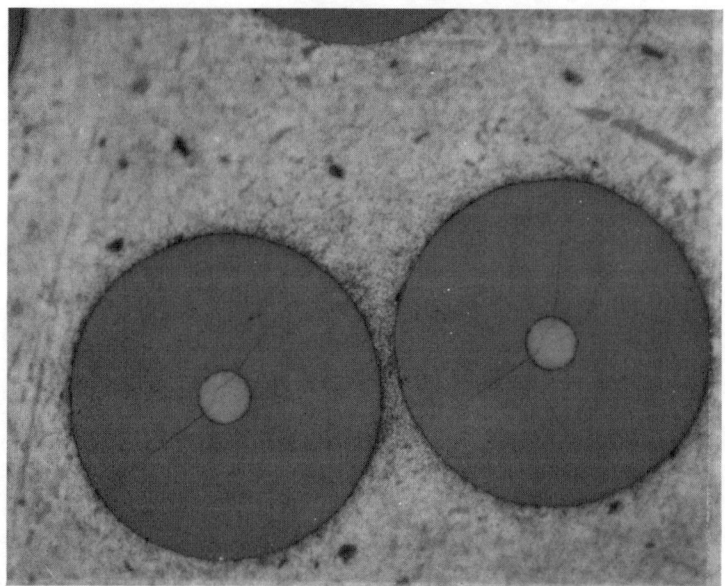

Figure 10. The same composite as Figure 9 showing minimal topographic relief between fiber and matrix. 500X. (Reduced 15 percent for reproduction.)

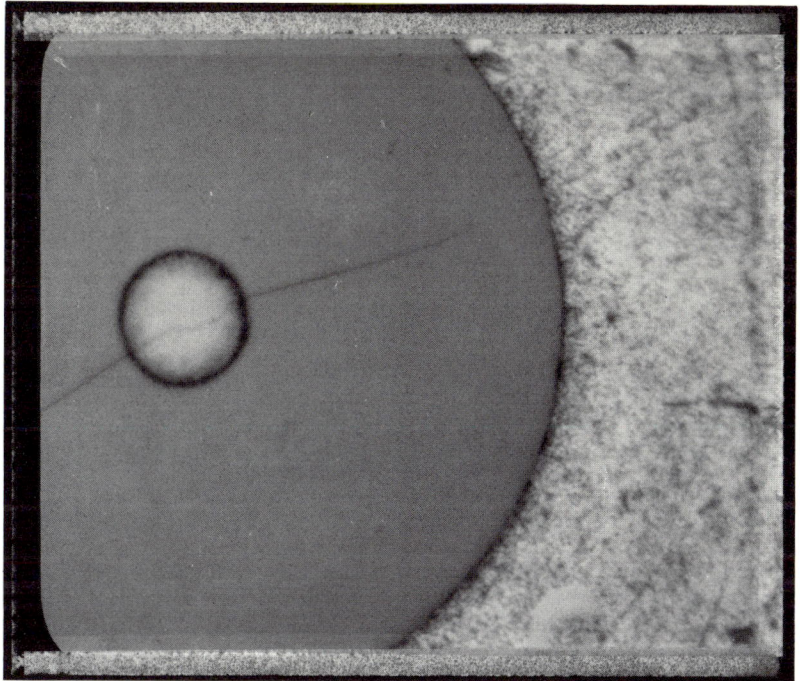

Figure 11. Micrograph of a fiber such as is shown in Figures 9 and 10 illustrating the excellent bond characteristics between fiber and matrix. 1400X. (Reduced 15 percent for reproduction.)

The aluminum matrix however, requires a further final polishing step to achieve a work-free surface. A Microcloth polishing cloth is prepared for this purpose by boiling the cloth in distilled water for approximately 1/2 hour prior to placing the cloth on an aluminum polishing wheel. The purpose of boiling the Microcloth is to soften the stiff fibers which have a tendency to scratch the soft aluminum matrix. Residual water is removed from the wheel by operating it briefly at a high rpm. One-quarter-micron diamond paste is then applied liberally to the central area of the cloth. The wheel is then rotated during this operation at 125 rpm. While adding liberal amounts of distilled water, the specimen is rotated in a clockwise direction near the center of the cloth with extremely light pressure. The polishing time will be very short, on the order of a few minutes. The results of matrix polishing can be seen in Figure 12.

Discussion

The techniques described in this section eliminate many of the difficulties associated with the metallographic preparation of hard filament-reinforced composites, primarily the damage induced in the filaments.

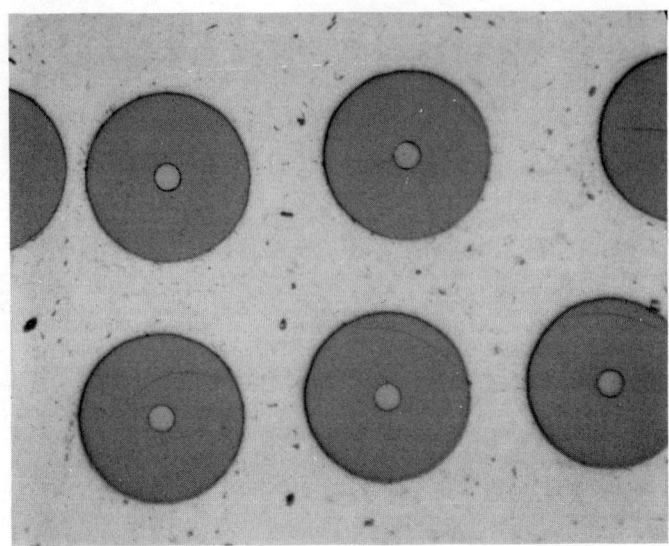

Figure 12. Tungsten core boron fibers in a polished 6061 aluminum matrix. 300X. (Reduced 20 percent for reproduction.)

Following the 600-grit grinding phase, the filaments protrude from the matrix. When these filaments are polished on a cloth, as in standard metal polishing procedures, they are snagged by the nap of the cloth, bent and fractured. Also, the matrix is preferentially removed at a higher rate than the fibers; this exposes new fiber and perpetuates the fracturing. Thus, the longer the sample is polished on cloth, the greater the damage to the fibers.

The optical flat eliminates filament snagging; filaments glide over the smooth surface of the flat and are abraded only by the diamond compound. By alternating microscopic examination with optical flat grinding, this operation can be properly controlled.

The concave curvature of the optical flat aids in the final polishing operation by producing a sample that is slightly convex. Some of the softer mounting materials have a tendency to resist the cutting action of the figure-eight lapping on the optical flat. When such samples are placed on the Syntron, since they lack a convex curvature, a cavity is created between the sample and the polishing cloth. It is suggested that this cavity traps more compound than is desired for proper cutting action and cavitation of the composite matrix material can occur. With a slightly convex sample surface, this problem is eliminated.

Although these procedures require a greater control of technique and are somewhat time consuming, the effort is justified by the excellent metallography that can be obtained with a variety of composite systems, Figures 13-16.

SAMPLE PREPARATION OF FIBER-REINFORCED COMPOSITES 263

(a)

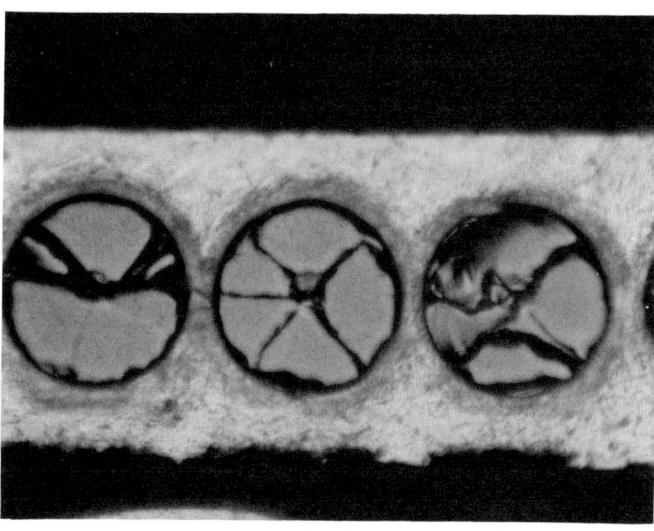

(b)

Figure 13. (a) Microstructure of thermally damaged composite (tungsten core, boron/Al matrix) as prepared by the present technique 350X.
(b) Same sample as above; standard metallographic technique 350X.
(Reduced 20 percent for reproduction.)

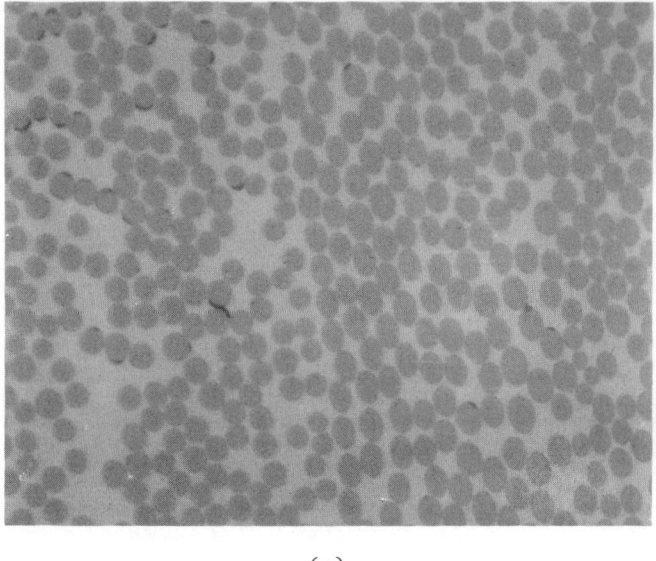

(a)

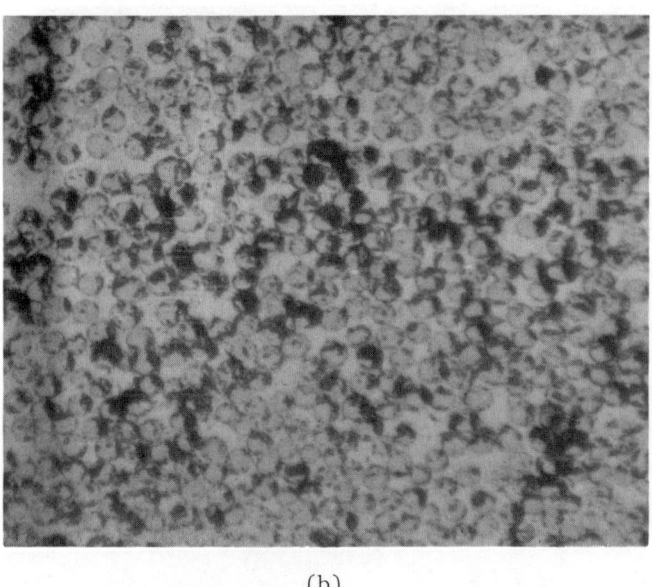

(b)

Figure 14. (a) Microstructure of glass fibers in an epoxy matrix as prepared by the present technique. 500X.

(b) Same sample as above; standard metallographic technique. 500X. (Reduced 20 percent for reproduction.)

SAMPLE PREPARATION OF FIBER-REINFORCED COMPOSITES 265

(a)

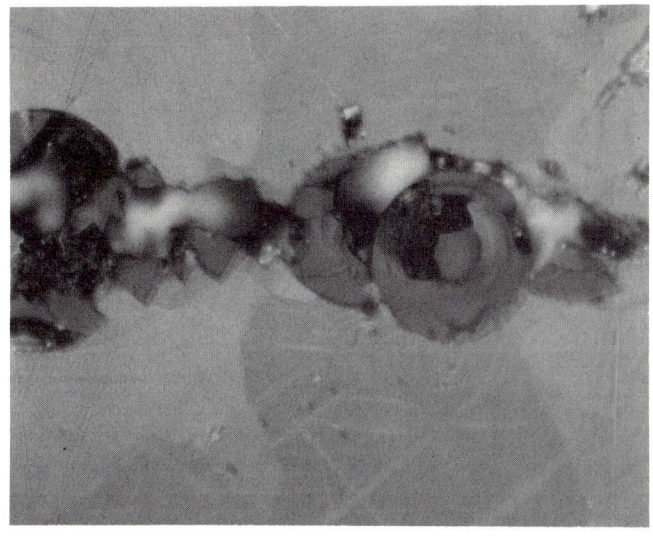

(b)

Figure 15. (a) Structure of carbon core boron fiber in a Be matrix as prepared by the present technique. Polarized light. 350X.

(b) Same sample as above; standard metallographic technique. 350X.

(Reduced 20 percent for reproduction.)

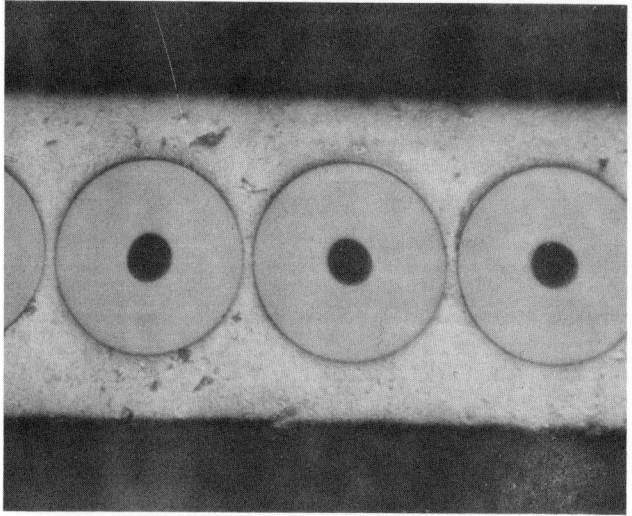

(a)

(b)

Figure 16. (a) Structure of silicon core boron fibers in an Al matrix as prepared by the present technique. 350X.

(b) Same sample as above; standard metallographic technique. 350X.
(Reduced 20 percent for reproduction.)

PROCEDURE FOR SOFT-FIBERS, SOFT-MATRIX COMPOSITES

The metallography of soft-fiber, soft-matrix composites illustrated in a number of papers (13-16) is excellent, however the methods of sample preparation are not usually described. Therefore, the methods employed at Aerospace Corporation are described here.

The two types of reinforcing filaments most commonly encountered in aerospace composites research programs are carbon or graphite fiber. The abrasion resistance of the graphite fibers is somewhat greater than that of the carbon fibers. The abrasion resistance of the carbon fibers more nearly matches that of the typical matrix. The composites based on these two types of fiber require somewhat different polishing procedures.

These composites often have porosity within the matrix material with the result that special potting methods are required to hold unsupported fibers during sample preparation. Vacuum potting is used to minimize this problem (17).

The grinding operation should begin with 600-grit silicon carbide paper. A liberal flow of water is necessary during grinding to prevent excessive buildup on the paper of material which will gouge the composite causing damage to the fibers. Larger grit sizes used in grinding have a tendency to chip the graphite filaments causing severe damage which may not be removed during subsequent polishing operations. A moderate pressure may be applied to the specimen during low-speed rotary grinding until the excess potting material has been removed and the surface of the specimen begins to be cut.

Light pressure on a new grinding paper is recommended for the remaining grinding operation. The specimen should be oriented 90 degrees from the last grinding scratch direction and held stationary while grinding on the new paper for a short period of time. The grinding scratch direction will rapidly change across the specimen surface, as both composite and mounting media are easily cut. Slow grinding speed and light pressure is very important during this operation to prevent deep scratches from damaging the composite material. These deep scratches are not readily visible to the unaided eye.

Polishing for the more abrasion-resistant fibers is accomplished as follows. An aluminum polishing wheel is cleaned and tightly fitted with a silk polishing cloth. The cloth is liberally impregnated with 9-micron diamond paste as shown in Figure 17. During polishing, the cloth is kept wet with a small amount of distilled water. The wheel is rotated at approximately 200 rpm. The specimen is counter-rotated at about 1 rps on the cloth using heavy pressure. Care should be taken not to let the cloth dry out while polishing. Scratches from the 600-grit grinding will be removed easily within the first minute of polishing.

The advantage in using a silk polishing cloth impregnated with diamond paste is that the harder fibers will be cut at the same rate as the matrix material. There is no nap in the silk cloth, its function is only to hold the diamond paste during polishing. Topographic variation between fiber and matrix should not be evident after this operation. The specimen should be examined microscopically after cleaning to verify this before proceeding to the next step.

A boiled Microcloth is prepared on an aluminum wheel in the same fashion as the silk cloth. The cloth is impregnated with 1-micron diamond paste and rotated at approximately 150 rpm. The specimen is rotated as previously described, with light pressure and lubricated with a small amount of distilled water until all the scratches from the 9-micron polishing are removed. This polishing step will take somewhat longer than the 9-micron polishing. The specimen should be examined microscopically to assure complete removal of the former polishing scratches. Figure 18 represents a

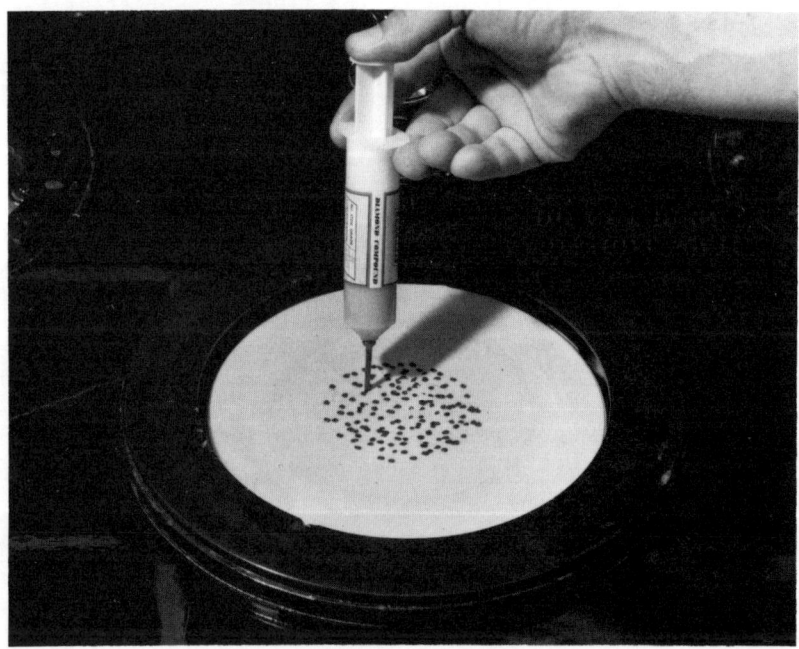

Figure 17. 9-micron diamond paste impregnated in a silk polishing cloth.

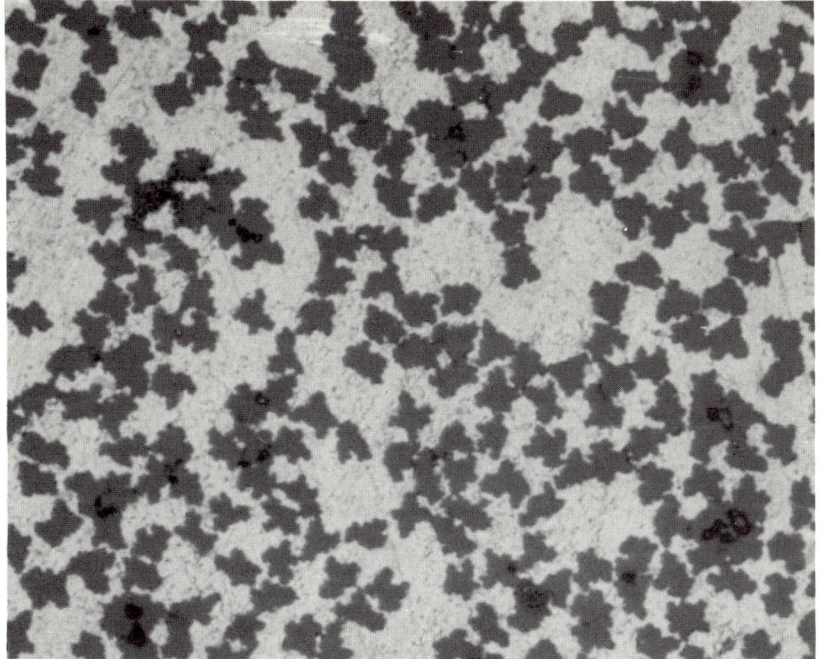

Figure 18. Specimen processed through 1-micron diamond polishing. 1000X.

typical composite microstructure as processed through 1-micron diamond polishing.

The final step in polishing will produce a specimen free from scratches and surface topography. This step requires another boiled Microcloth prepared in the same fashion as the previous step. The cloth is heavily impregnated with 1/4-micron diamond paste and the wheel rotated at approximately 200 rpm. Distilled water is used as the lubricant for this operation. The specimen is rotated as previously described with very light pressure, virtually floating the specimen on the wet polishing medium for about one minute. This step may be repeated more than once to remove the fine microscratches. Care should be taken not to touch the specimen after the final polishing operation to avoid staining the surface. The final cleaning of the specimen should be with distilled water and the specimen should be carefully air blown to dryness.

The polishing of those composites in which the abrasion resistance of the fibers more nearly matches that of the matrix is now described. These composites are easily prepared through grinding and polishing operations with little topographic difference occurring between fiber and matrix material.

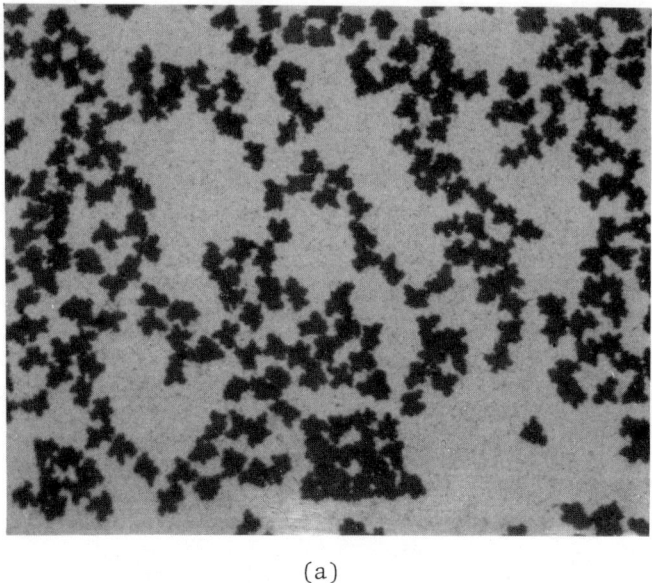

(a)

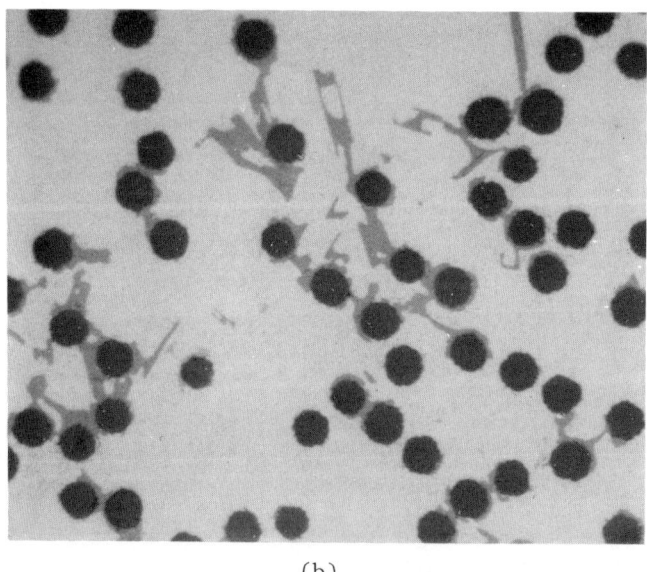

(b)

Figure 19. (a) Typical cross section of Thornel, HMG and GLCC graphite fibers. 1000X.
  (b) Typical cross-section of Courtaulds graphite fibers. 1000X.
    (Reduced 20 percent for reproduction.)

SAMPLE PREPARATION OF FIBER-REINFORCED COMPOSITES 271

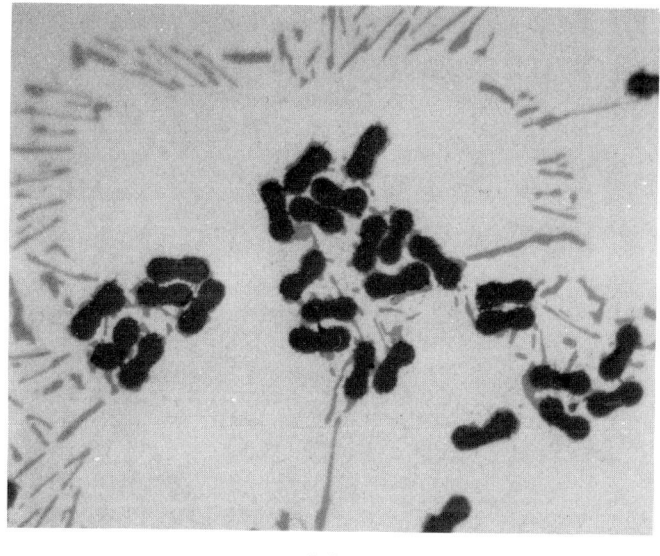

(a)

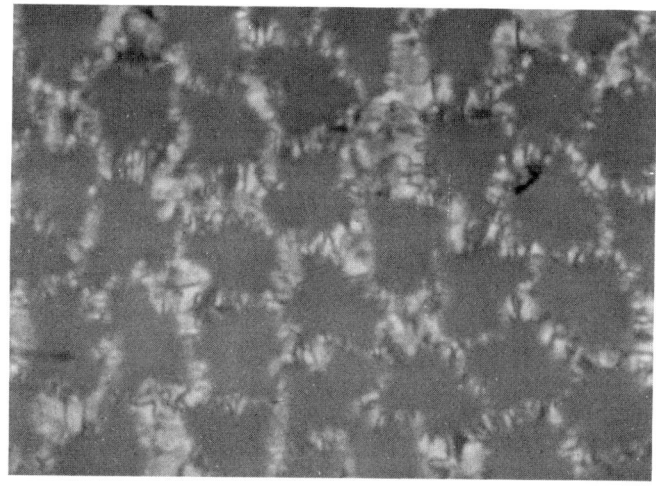

(b)

Figure 20. (a) GLCC's "dog bone" shaped fibers in a silicon aluminum matrix. 1000X.

(b) Carbon-Carbon composite weave. Polarized light. 2500X.

(Reduced 20 percent for reproduction.)

A kitten-ear polishing cloth is prepared on an aluminum wheel and rotated at approximately 150 rpm. A 1.0-micron alumina slurry in distilled water is liberally applied to the cloth. The specimen is polished with light pressure for a short time using the same rotation technique previously described until all scratches from 600-grit grinding are removed. The cloth should be kept wet at all times with the alumina slurry to assure proper cutting action. The specimen should be examined microscopically to verify scratch removal. This polishing operation may be repeated if necessary.

A boiled Microcloth is prepared on an aluminum wheel in the same fashion as the kitten-ear cloth and rotated at approximately 200 rpm. A slurry of 0.3-micron alumina and distilled water is liberally added to the polishing cloth. The specimen is slowly rotated with very light pressurefor about one minute to remove the scratches created by the 1.0-micron alumina abrasive.

The operation described above is repeated on another boiled Microcloth using a 0.05-micron alumina distilled water slurry. Extremely light pressure is recommended for this final polishing step. The specimen should be frequently examined with a microscope during polishing to assure complete removal of all micro scratches. The final cleaning of the specimen should be with distilled water and the specimen should be carefully air blown to dryness.

This method may produce a slightly rounded edge, however the microstructure will be scratch free and excellent in detail.

The following series of photomicrographs shows a variety of different soft fiber reinforced composite materials which is typical of the high quality microstructures obtained by the metallographic sample preparation techniques described herein, Figures 19-20.

REFERENCES

1. J. R. Hancock, J. Composite Materials, 1 (1967), p. 136-142.
2. J. L. Camahort, J. Composite Materials, 2, No. 1 (Jan. 1968).
3. A. G. Metcalfe, J. Composite Materials, 1 (1967), p. 356-365.
4. J. A. Snide, L. G. Bates, F. A. Ashdown and J. R. Myers, J. Composite Materials, 2, No. 4 (Oct. 1968), p. 509-512.
5. G. H. Kehl, The Principles of Metallographic Laboratory Practice, McGraw-Hill (1949).
6. B. Chalmer and A. G. Quarrell, Physical Examination of Metals, 2nd Edition Arnold, London (1961).
7. D. M. Schuster and R. P. Reed, J. Composite Materials, 5, (July 1969), p. 562.
8. D. M. Schuster and R. P. Reed, J. Composite Materials, 4, (Oct. 1970), p. 514.

9. H. R. Thornton, J. Composite Materials, $\underline{2}$, No. 1 (Jan. 1968), p. 32.
10. I. Ahmad, J. P. Greco and J. M. Barranco, J. Composite Materials, $\underline{1}$ (1967), p. 18.
11. R. K. Robinson, Bureau of Naval Weapons, Laboratory Report BNWL-SA-550 (1966).
12. H. McLaughlin, "Use of Wire Saws for Metallographic Sectioning", Specimen Preparation Techniques for Optical and Electron Metallography, edited by J. L. McCall and W. M. Mueller, Plenum Press, New York, 1974.
13. R. T. Pepper et al. The Aerospace Corp., Report No. TR-0066 (9250-03)-3, April 1970.
14. D. M. Goddard et al. The Aerospace Corp., Report No. TR-0172 (9250-03)-1, August 1971.
15. C. E. Browning and J. A. Marshall, II, J. of Composites $\underline{4}$, (July 1970), pp. 390-398.
16. R. J. Dauksys and J. D. Ray, J. of Composites $\underline{3}$, (October 1969) pp. 684-688.
17. J. H. Richardson, Optical Microscopy for the Material Sciences, Marcel Decker, Inc., New York (1971), pp. 251-265.

PLUTONIUM METALLOGRAPHY*

E. M. Cramer

University of California, Los Alamos Scientific Laboratory

Los Alamos, New Mexico 87544

   Perhaps no other materials in history have had the concentrated attention, the dedication of financial resources, and the limited application than have been shared by plutonium and uranium (1,2). The groundwork for their elevation to such prominent positions was laid in 1919 by Rutherford's work on nuclear disintegration. The twenty years following his revelation that atomic nucleii are not indestructible saw first a relatively slow advance in physics and then rapid progress in the development of machines to accelerate charged particles and in understanding the reactions thus produced. Atom splitting became commonplace, fissioning of uranium-235 was observed and verified, and the magnitude of the energy released in the process became known. With regard to plutonium, the period culminated in the concept of radiative capture by Frisch and Meitner.

   By 1940 the concept of the element that was to be called plutonium had been proposed by L.A. Turner and others. They conceived of the formation, by radiative capture, of a nucleus having an atomic number of 94 and an atomic weight of 239 and which would probably be fissionable in the manner of uranium-235. Moreover, it appeared that the element could be synthesized in quantity and, therefore, would not be limited in amount by its natural occurrence. Nor would recovery of the new element be complicated by the formidable isotopic separation of uranium-235 from 238, in which uranium-235 occurs only to the extent of 0.07 percent. Of course, military interest in the prospect of releasing energy through fission was well established and growing.

---
*Work performed under the auspices of the U. S. Atomic Energy Commission.

Governmental involvement in the field of nuclear energy, other than in the academic world, first took place in the United States in the fall of 1939 when the Advisory Committee on Uranium was formed by the president in the interest of national security. Governmental funding began in the spring of 1940 with the allocation of $6,000 for the purchase of graphite and uranium oxide to determine the absorption cross section of carbon. By August the need had risen to $40,000, and the first contract was awarded to Columbia University.

Early in 1941 other contracts were let to Princeton University and the University of Chicago to coordinate with Columbia in achieving a sustained chain reaction, and to the University of California to investigate the production of element 94 in particle accelerators. The first chain reaction was achieved in December and a source of plutonium, its position now assured, came into being. The first product of the pile, however, was the isotope 238. The first fissionable isotope, plutonium-239, came from the bombardment of uranium solutions in cyclotrons at the University of California and Washington University. By the end of 1942, microgram quantities of plutonium-238 had been produced, and plutonium was considered to be an element whose behavior was as well known as those of some other relatively obscure elements in the periodic table.

Now the phenomenal effort that was to be devoted to the new element as a source of nuclear energy for military use became apparent. In January, 1943, the decision to build the Hanford Engineer Works had been made, and construction began in April. Project Y to produce a nuclear device began at Los Alamos in March. A pile to produce plutonium and a pilot plant for its chemical separation were being built at Clinton, Tennessee. The Clinton pile was started in November. The pilot plant was started in December, and its first plutonium product, consisting of 190 milligrams, was obtained in February, 1944. Of more specific interest here, the first metallographic examination of plutonium was made late in 1943 by R. S. Rosenfels at the University of Chicago with cyclotron-produced material (3). This achievement took place less than four years after the concept of plutonium had first been proposed.

The Hanford Engineer Works began operation in 1944. Reduction of the metal in gram quantities began at Los Alamos in February, 1945, when the first shipment of the product of the Hanford reactors and separation plant was received. The Chemistry and Metallurgy Division was formed that year with the responsibility of fabricating a military device in the shortest possible time. Included in this Division was a Physical Metallurgy Group, under the leadership of G. L. Kehl, in which a systematic study of plutonium was to take place. The first observations, at Los Alamos, however appear to have been made in May, 1944, by A. U. Seybolt and his associates. It is not clear whether the material examined was a product of cyclotron irradiation or the Clinton pile. Examples of the early work with both alpha and stabilized delta phases are included here in Figures 1 and 2. The microstructures show, in retrospect, that a great deal of skill and patience went into their preparation.

Metal in sufficient quantity for metallurgical investigation became available in England in 1951 and in France in 1956. The first account of work on plutonium in Russia was presented in 1955, the presumption being that the metal became available for investigation several years earlier. In all instances micrography played an early and important role in the burgeoning studies of plutonium. Numerous articles describing, or alluding to, metallographic practices appeared in the open literature after 1955, and presentations of phase diagrams were richly illustrated with photomicrographs. Metallography has continued to play an essential role in the accumulation of knowledge about this unusual element.

Expansion of laboratory facilities for studying plutonium continued through the 1950's, but not with the single-mindedness of purpose nor the intensity that had prevailed in the immediately preceding decade. At Los Alamos the military applications effort continued, and was joined by similar efforts at the Hanford Laboratories and the Rocky Flats Plant. Battelle Memorial Institute became involved on a contract basis. The Lawrence Livermore Laboratory, a late comer, committed its metallographic facility to plutonium in 1962.

In peaceful uses, Los Alamos began a study of applicable binary alloy systems in 1948. The Argonne National Laboratory, which supplanted the Metallurgical Laboratory of the University of Chicago, began work with plutonium in 1954 in connection with reactor development. The Mound Laboratory joined in this effort in 1956. The Savannah River Plant, the most recent, has been engaged primarily in the production of nuclear fuels.

With this prestigious background, what have we metallographers accomplished with plutonium? Do the results of our efforts compare favorably with those of our contemporaries who have worked in less restricted fields? There seems little doubt that, with respect to the state of the art, our workmanship and instrumentation compares favorably with those prevailing in the ferrous and nonferrous fields. The technical elegance of our microstructures, however, is not comparable. They are not as crisply defined. Photomicrographs of plutonium and its stabilized phases continue to exhibit characteristics of electropolished and weakly etched specimens. Methods of specimen preparation that will consistently produce comparable results are yet to come, and they will come only with greater effort than was initially expended in preparing the metals.

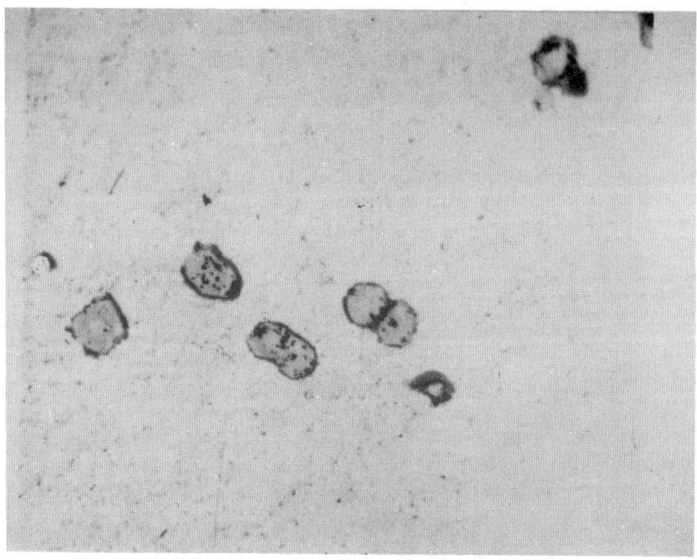

Figure 1. Plutonium metal, mechanically polished with tin oxide and ethyl alcohol. 500X. (A. Gerds, July 27, 1944.) The inclusions are plutonium hydride.

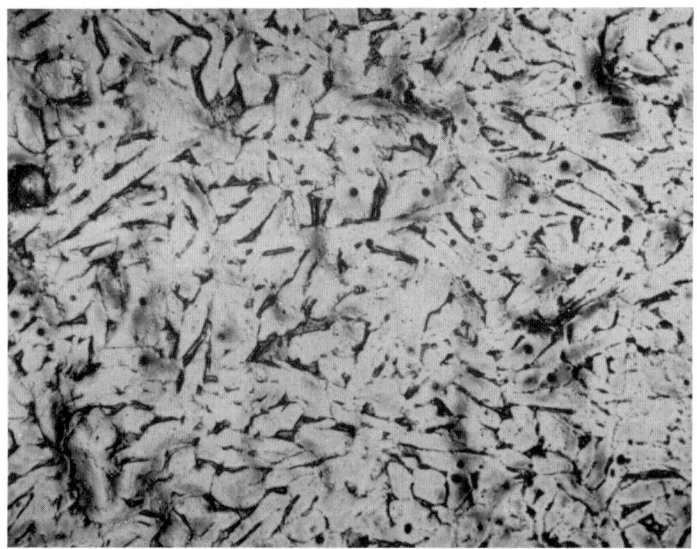

Figure 2. Plutonium-1 wt. percent gallium alloy, cast. 500X. (A. Gerds, August 28, 1945.) This cored specimen was electroetched in a solution of 6 parts orthophosphoric acid, 5 parts glycol, and 9 parts water.

The difficulties occur almost entirely in the electrochemical stage of preparation. Plutonium, in all of its commonly encountered phases, responds well to mechanical grinding and polishing. Conventional abrasives and practices work well if nonaqueous lubricants are used. Carbon tetrachloride has had the widest acceptance but kerosene, glycol, trichloroethylene, and methyl chloroform have also been used. The alpha phase, being the hardest, is also the easiest to polish, and a polished surface suitable for photography readily achievable. The delta phase presents a greater problem. Being relatively soft, it is subject to galling; however, the careful use of clean, sharp abrasive papers and clean laps makes the preparation of delta surfaces suitable for etching not a difficult task.

It happens infrequently that a mechanically prepared delta surface will be photographed. It is quite reactive, as are all the plutonium phases, and the rate of metal removal in the polishing process must be rapid, otherwise the rate of oxidation and corrosion by moisture-bearing lubricants will overtake polishing. Being softer, the surface of polished delta has a higher energy than does polished alpha and initial oxidation occurs extremely rapidly. Consequently, fine scratches that usually can be overcome optically stand out in a distressing manner. The disturbed surface layer, however, may be readily removed electrolytically.

It is in this final stage of preparation that problems arise. Plutonium's activity and tendency to form insoluble compounds with strong acids have precluded chemical etching. The same characteristics have also limited the time it may be exposed to aqueous electrolytes. Thus the selection of suitable electrolytes consist essentially of nonaqueous solutions in which ethylene glycol, glycerine, ethyl and methyl alcohols, and 2-ethoxy ethanol are the common diluents in combination with tetraphosphoric, orthophosphoric, citric, nitric, acetic, or lactic acids. Achieving distinct polishing or etching conditions with these electrolytes is difficult. Their polishing curves of potential versus current do not exhibit the desired plateau which indicates optimum conditions for polishing, but are generally only slightly inflected. The result appears to be an overlapping of the two processes wherein one or the other can only be emphasized.

The metal is inherently in a state of stress as a consequence of perpetual damage from self irradiation. In addition, phase transformation may have occurred at a relatively low temperature. This state contributes to masking normal stress concentrations at structural discontinuities that are the basis for the selective solution that makes grain boundaries visible in reflected light.

The response of plutonium to etching does not seem to be particularly sensitive to isotopic composition, but mechanical and thermal histories and impurities are important in this respect. This characteristic may account to some extent for our inability to achieve consistent results, but it falls short of explaining an exasperating unpredictability that occurs from day to day or from environment to environment.

Current practices appear to be the result of continued probing by individuals who are unsatisfied with their own methods and who are disappointed when they try to adopt procedures that are used at other laboratories. This may be due to personal preferences, although having been unsuccessful in transferring procedures from Los Alamos to the Livermore Laboratory, the author has found this explanation inadequate. Phosphate electrolyte recipes that were part of well-established routines at Los Alamos failed to give comparable results at Livermore. For example, the tetraphosphate solutions were discarded completely. The orthophosphate solutions continued to be used with both alpha and delta plutonium although they gave satisfactory results only under widely different conditions. Instead of a short etch time at 6 to 8 volts in the familiar 8-5-5 formulation of orthphosphoric acid, glycerine, and alcohol, it was necessary to limit the potential to the range of 2-1/2 to 4 volts and to extend the time manyfold, sometimes to as much as hours. The microstructures, however, were of better quality and satisfied a more exacting demand. One of the tetraphosphate solutions was replaced with the citric acid-based etchant introduced as Battelle-Northwest Laboratories (4) as a routine reagent for delta.

Relative humidity has been mentioned as a factor contributing to the success or failure of preparation techniques, but this, too, seems inadequate since humidity is often comparable at Los Alamos and Livermore. Perhaps air pollution is a factor, although no correlation was observed with the widely variable degree of pollution at Livermore. At any rate, it seems that the unpredictability of plutonium is a somewhat baffling reality and each metallographer must develop his own way of coping with it.

Vibratory polishing and electropolishing techniques have been adopted to replace, or supplement, the conventional methods of mechanical preparation. Most of the effort, however, has been expended in researching the electroetching processes. At Los Alamos in 1964 an alternating current technique was introduced that gave a somewhat wider selection of electrolytes and improved the quality of alpha surfaces.

Early efforts in France (5) were directed toward developing an epitaxial oxide on alpha plutonium to permit examination under polarized light in the manner of uranium. (For many years this was the only mode in which the structure of alpha could be observed.)

Electrolytes consisting of phospheric, sulfuric, or nitric acids in combination with nonaqueous diluents were explored. One of these, a solution of nitric acid in glycol, frequently modified, has been used extensively as a bright field etchant in the United States.

A method of electroetching alpha in a nitrate solution was introduced at the Battelle Memorial Institute (6). Their method involved the use of nitric acid, alcohol, and water and successfully produced structures for examination under bright field illumination. A similar etchant was adopted at the Battelle-Northwest Laboratory and at Los Alamos where it is used in the alternating current mode.

At the Battelle-Northwest Laboratory a visual method of observing and controlling the etching process was introduced in 1966 (4) and applied to both alpha and delta plutonium. This adaptation seems to offer one of the best solutions to the problems encountered in the use of low conductivity electrolytes. Phosphoric, citric, and nitric acid-based electrolytes are recommended for the method.

The effectiveness of low conductivity electrolytes was improved at the Rocky Flats Plant (7) by swabbing with heated electrolytes containing varying proportions of orthophosphoric acid in glycerine and alcohol. The heated solution is applied to the specimen with a hand held swab through which the solution is forced by an external pump. The swab also acts as the anode in the cell.

An ion etcher was frequently used at the Livermore Laboratory to prepare samples for bright field examination in preference to etching in nitrate solutions. Adequate cooling for small specimens of alpha was accomplished with a conductive mount containing iron powder. Following etching, a cleaning step to remove loose particles of plutonium was required and noticeably degraded the etched surface. An ion etcher, however, is a desirable tool. In this regard, one that is presently used at Los Alamos (8) to etch thoria appears capable of alleviating the heating problem. In principle, the ion beam is drawn from a glow discharge that is maintained within a hollow-anode ion gun. The energy of the beam is continuously controllable and it can be focused, or restricted, to minimize the energy imparted to the specimen. The effectiveness of the design with regard to plutonium has not yet been determined.

These innovative techniques have helped to overcome some of the inadequacies that have been apparent in plutonium metallography during the past two decades. Since we are aware of the inadequacies in our method, more answers will be sought and found by those who have chosen this restricted but rewarding field. The descriptions of techniques and processes included here have deliberately been made very brief since each would be considered a complete article in the technical literature (9,10).

REFERENCES

(1) H. D. Smyth, "A General Account of the Development of Methods of Using Atomic Energy for Military Purposes under the Auspices of the United States Government". U.S. Gov't Printing Office, (1945).

(2) A. S. Coffinberry, M. B. Waldon, Progress in Nuclear Energy, Chapt. 4. VI, Series 5, Pergamon Press, Ltd., London, (1956).

(3) E. M. Cramer, F. W. Schonfeld, "Techniques for the Metallography of Plutonium", Proceedings of the Second International Conference on the Peaceful Uses of Atomic Energy, Geneva, 17, (1958).

(4) D. D. Hays, "A Bright-Field Grain Structure Etch for Delta-Phase Stabilized Plutonium", BNWL-188, J. Nucl. Materials, 19, (1966).

(5) A. S. Coffinberry, W. N. Miner, The Metal Plutonium, University of Chicago Press, (1961).

(6) V. Storhok, R. Snider, M. Farkas, "A Bright Field Etch for Alpha Plutonium", Proceedings of the Eighteenth AEC Metallographic Group, NMI-5025, (1964).

(7) R. Greeson, W. Johns, R. Jackson, "A Special Metallographic Technique for Plutonium-Electrolytic Swab Etching", RFP-543, (1965).

(8) J. W. Ward, "High-Vacuum Glow-Discharge Etcher and Ion Thinner", Microstructures, 2, #3, (Apr./May, 1971).

(9) K. Johnson, "The Alternating Current Electroetching of Plutonium", USAEC Report LA-3173-MS, (1964).

(10) F. Cochran, "Preparing Plutonium for Optical Analysis: A Precis", Microstructures, 1, No. 1, (Aug./Sept., 1970); 1, No. 2, (Oct./Nov., 1970); 2, No. 3, (Apr./May, 1971).

ON THE USE OF ELECTRON BEAMS TO CHARACTERIZE THE MICROSTRUCTURE
OF RADIOACTIVE MATERIALS

H. S. Rosenbaum

General Electric Company, Vallecitos Nuclear Center

Pleasanton, California 94566

INTRODUCTION

When we consider solid materials that have been neutron irradiated, we are confronted with a myriad of microstructural effects. Irradiated materials have very complex microstructures. The characterization of such microstructures is an important part of the nuclear industry, which depends on the utilization of materials as fuels and structural materials in neutron environments. Let us consider some of the microstructural effects of neutron irradiation. First, there is the physical displacement of atoms from their normal positions. This effect leads to an immediate disturbance and relaxation event that results in the formation of small vacancy-rich zones. With some thermal annealing these zones become dislocation loops and tangles. Under conditions of high neutron flux at elevated temperatures most metals can form voids by the precipitation of vacancies. Second, there are chemical effects. Transmutations change the chemical composition resulting in the formation either of solid solutions or of precipitated phases. In a neutron flux the kinetics of phase changes are affected by altered diffusivities and by the creation of metastable phases. Third, we have severe temperature effects. There are such steep temperature gradients that high thermal stresses are imposed. Mass transport occurs by diffusion, effusion and convection along steep temperature gradients. Sudden temperature changes produce thermal shock, and of course, phase changes can occur as the temperature is changed.

Practically all of these effects are known to occur in nuclear fuels of urania or of urania-plutonia mixtures. There are atom displacements by neutrons and energetic fission fragments; there are chemical changes due to fission and other nuclear transmutations;

there can be very high temperatures (>2000°C) and steep temperature gradients (~10^3-10^4 °C/cm) as well as sudden temperature changes. The temperature effects by themselves are sufficient to produce a gradient of microstructural situations including: recrystallization, grain growth, and cracks. Such a microstructure is illustrated in Figure 1. Prior to irradiation the microstructure was relatively uniform, that of a sintered ceramic pellet. The columnar grains and the central void were formed as the pores in the sintered structure moved up the temperature gradient. Cracks were produced by the thermally induced stresses. Superposed on this already complex microstructure are the complexities previously mentioned, the physical and chemical effects of irradiation.

In this paper we will discuss some experimental aspects in the characterization of irradiated materials by electron beam methods: electron probe microanalyzer, scanning electron microscopy (SEM), and transmission electron microscopy (SEM). Discussed will be those experimental aspects that are affected by the sample and sample preparation. The radioactivity of the specimen (and therefore the specimen size and configuration) can affect the data. Because ceramic oxide fuels have complex microstructures and present great difficulties in specimen preparation, they will be taken as prime examples. However, much of the discussion will be applicable to other radioactive materials.

ELECTRON PROBE MICROANALYSIS

Electron probe microanalysis is a technique that is well known to materials scientists and engineers. It is the analysis of the characteristic x-ray spectrum generated by the interaction of a focussed electron beam with the atoms of the specimen. In the context of this discussion the reader might wish to refer to any of the treatments of the subject that is available in the literature.

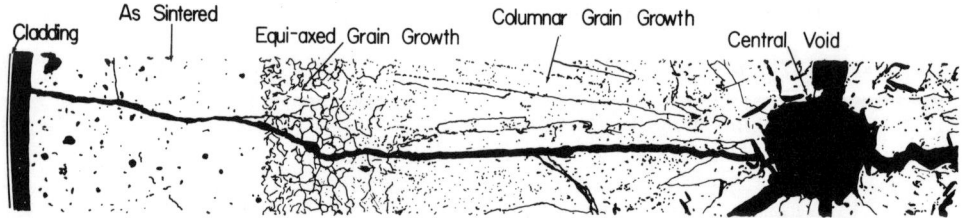

Figure 1. Microstructural changes caused by irradiation in urania fuel (Schematic).

USE OF ELECTRON BEAMS TO CHARACTERIZE MICROSTRUCTURE

The recent review by Beaman and Isasi (1) is quite complete and current.

The importance of microprobe techniques to study the chemical effects of neutron irradiation on nuclear fuels was recognized early, and some excellent work was reported using very small specimens. Larger specimens could be analyzed if the x-ray detectors were shielded; thus, several laboratories enhanced their capabilities for analysis of irradiated nuclear fuels by modifying commercial instruments by adding shielding to the spectrometers. Much of this work was reviewed by Rosenbaum, et al (2) and by Cummings, et al (3). By making a fully shielded microprobe (3) full cross-sections of highly irradiated nuclear fuels could be safely analyzed with very acceptable analytical results.

Of course, the specimen preparation of such highly radioactive materials must be done remotely, usually using well established remote metallographic techniques. A description of a remote metallography set-up was described by Gray, et al (4). In Figure 2 the specimen configuration is illustrated. The fuel is first impregnated with an epoxy resin. Then a section, including the metal fuel sheath or cladding is cut. Even with a fully shielded instrument the specimen thickness must be kept to a minimum to prevent unnecessary background signals in the x-ray detection equipment. The copper

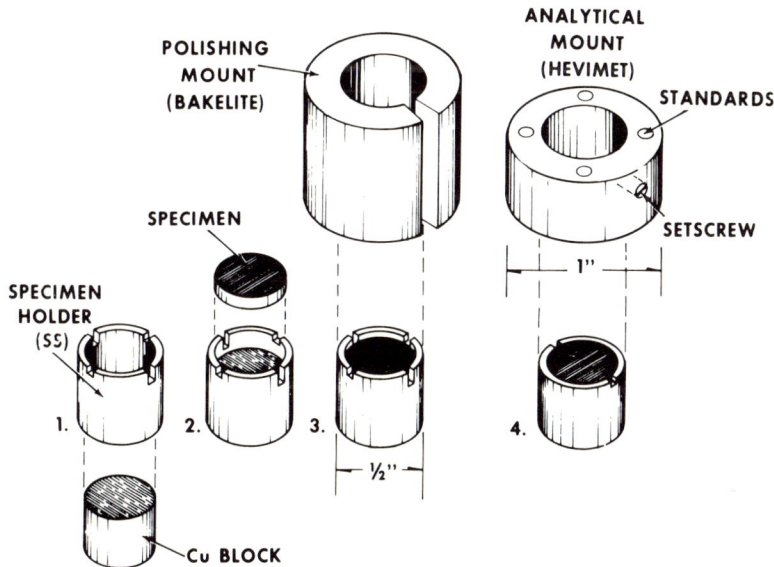

Figure 2. Arrangement for polishing irradiated fuel specimens for microprobe examination.

block is used as a heat sink, and the stainless steel holder with notches provides a guide to the metallographer in grinding the specimen to the required depth. The fuel specimen is held in a heavy metal (usually a tungsten alloy) ring that also holds the standards needed for the analysis.

The proper metallographic procedure for specimen preparation prior to microprobe analysis depends on the specific information sought in the analysis. If we are interested only in the stable components of the fuel microstructure (e.g. the fuel matrix and the stable fission product phases) then normal metallographic procedures will suffice, leaving the specimen in the "as-polished" condition. However, sometimes we wish to retain water soluble fission products (e.g. CsI, the alkali metals, etc.). In such a case the entire metallographic procedure must be done in non-aqueous media. Kerosene or a similar liquid (e.g. Monsanto HB-40) can be used as lubricant in cutting and grinding. Commercial diamond pastes can be used with kerosene or xylene. All operations must be done without water or alcohol; we use xylene, and we accept a certain amount of staining. After metallographic preparation and photomicroscopy the specimen is coated with vapor deposited carbon to provide electrical conductivity and to fix loose radioactive particles. It is a good idea to make a trial run, including microprobe analysis, with the particular materials used. We have found that some mounting materials, diamond pastes, cutting fluids, etc. have chemical elements that might confuse the analyses. If SiC abrasives are used in grinding, it is not surprising to find Si contamination on the specimen. We have found that some commercial diamond pastes contain Si, Al and/or Cl. We cannot anticipate all possibilities in the scope of this paper. However, the general caution against chemical contamination of microprobe samples is valid, and workers must be cognizant of the problem. The problem is especially severe in remote metallography and is accentuated by the restrictions on use of water or alcohol in cleaning the specimen surfaces.

In our laboratory the specimen is transferred from the remote handling cells to the microprobe via a shielded transfer cask (3). Natesh, et al (5) have described an interesting pneumatic specimen transfer arrangement for use with a shielded microprobe.

In the electron probe microanalysis of highly radioactive materials we must contend with higher background signal than one normally encounters. Shielding makes this problem manageable, but cannot completely eliminate it. The energetic gamma rays in nuclear fuels penetrate the shielding, and although the intensities can be kept within manageable levels, sensitive x-ray detection equipment is affected. Electronic methods (e.g. pulse height selection) can be used to further ameliorate the background signal, but ultimately we have to contend with a measurable (and in many cases a sizable) background signal that is higher than would be the case for samples that have no gamma activity.

The background invariably is deleterious to the analytical results. Both sensitivity and precision are affected. Consider the sensitivity to trace elements. The "detectability threshold", C_{DT}, is the smallest concentration of a particular element that can be unequivocally detected, and can be estimated from the relation

$$C_{DT} \cong \frac{3\,\sigma_B}{P_o - B} \geq \frac{3\sqrt{B}}{P_o - B} = \frac{3\sqrt{Bt}}{(\dot{P_o} - \dot{B})t} \quad (1)$$

where P_o is the peak number of x-ray counts as measured on a pure elemental standard,
B is the background as measured on the specimen,
t is the total time counted,
σ_B is the standard deviation of the background,
$\dot{P_o}$ and $\dot{B}$ are the corresponding peak and background counting rates or intensities.

Since counting statistics follow a Poisson distribution, the standard deviation is expected to equal the square root of that number of counts. Notice that for a given peak x-ray count P, for the element in question C_{DT} is raised (i.e. sensitivity lowered) as B or $\dot{B}$ increases. Low C_{DT} is favored by low B and high t. Therefore, the effects of high background can be somewhat ameliorated by long counting times and by use of an intense electron beam. The peak intensities, $\dot{P}$, are proportional to the electron beam current that impinges on the specimen.

The effect of background on the precision of a quantitative analysis can be seen by the expression representing the relative standard counting error

$$\frac{\sigma_k}{k} \geq \left[\frac{P+B}{(P-B)^2} + \frac{P_o + B_o}{(P_o - B_o)^2} \right]^{\frac{1}{2}}$$

where $k = (P-B)/(P_o - B_o)$, the first approximation of the concentra- according to Castaing (1),
σ_k = standard counting error (expected standard deviation estimated from counting statistics),
P,B = peak and background counts, respectively for the element sought as measured on the sample, and
P_o, B_o = peak and background counts as determined on a pure elemental standard.

Notice the error increases with B and B_o. One of the most important aspects of controlling the level of B and enhancing the precision of the analysis is making the specimen as thin as practical.

Wave-Length Dispersive X-ray Analysis

A wave-length dispersive x-ray spectrometer of a fully shielded microprobe is illustrated in Figure 3. For a given level of beta plus gamma activity the background is controlled by the shielding (item 1 in Figure 3) and by collimation (item 6 in Figure 3). Collimation limits the intensity of beta and gamma which can interact with, and produce fluorescence by, the crystal and hardware components of the spectrometer. The shielding eliminates the beta completely from line-of-sight interaction with the x-ray detector and reduces the gamma-ray intensity. Collimation is ineffective for a point source of radioactivity; it is most effective with a fairly large sample of low specific activity.

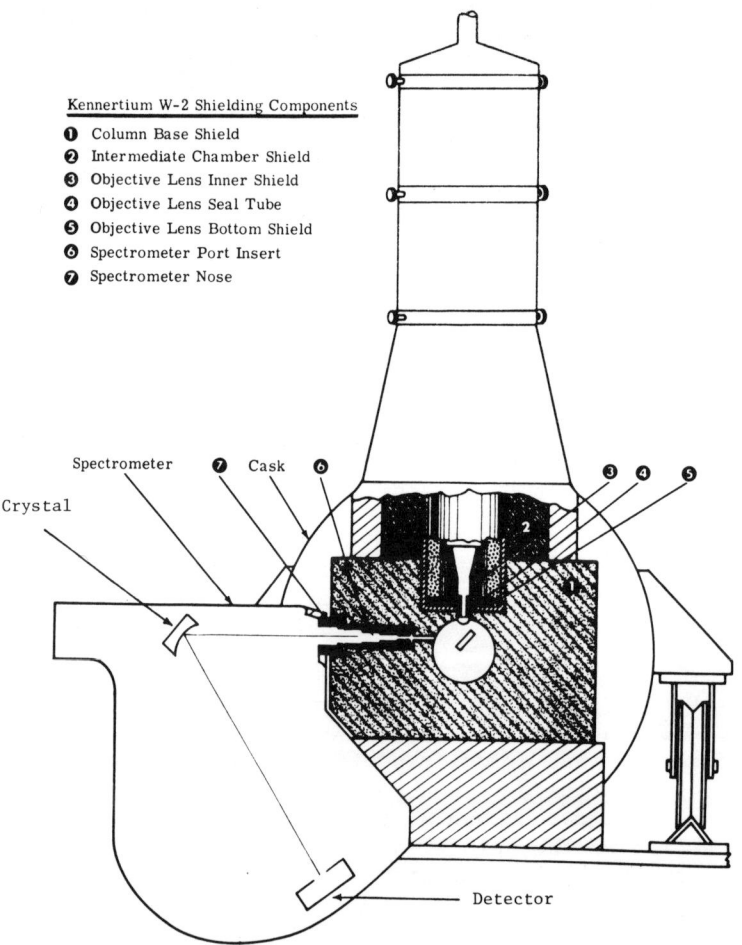

Figure 3. Shielded microprobe with wave-length dispersive spectrometer [adapted from Cummings et al (3).]

Energy Dispersive X-ray Analysis

In energy dispersive x-ray analysis the usual situation is to place the detector (Li drifted Si) as close as possible to the specimen with only a thin beryllium window to stop the secondary electrons. However, with a highly beta plus gamma active specimen, the background activity can easily swamp the solid state x-ray detector. Since the energy dispersive detector must have direct line-of-sight access to the specimen, shielding is impossible; collimation is the only recourse. Placing the energy dispersive x-ray detector in the spectrometer (in place of the crystal in Figure 3) takes advantage of collimation and provides a useful way to do energy dispersive x-ray analyses on radioactive materials.

It is interesting to note that with such an arrangement one can observe the characteristic x-ray spectrum from the specimen <u>without</u> the electron beam. The x-rays are excited by secondary fluorescence induced by the radioactivity within the specimen. When secondary fluorescence from the radioactivity is a significant contribution to the observed x-ray intensities, then the spatial resolution of the microprobe is very much degraded. Only the collimation limits the area from which x-ray signals are received, typically several mm diameter instead of the several μm that we normally expect from microprobe analyses.

SCANNING ELECTRON MICROSCOPY (SEM)

Scanning electron microscopy has become an invaluable tool for the examination of radioactive specimens. Used in the secondary electron image mode, the image quality is somewhat affected by the beta-rays and the photoelectrons emitted from the specimen. These have the effect of a "noise" or background signal that tends to wash out contrast and thereby degrade the resolution. Nonetheless, samples of irradiated urania fuel whose beta plus gamma activities were approximately 1 R/h at 1 ft. have been examined at magnifications up to 5000X with image quality that is quite acceptable.

The most impressive work to date on irradiated fuel was reported by Baker, <u>et al</u> (6). They prepared discs of the fuel by mechanical cutting and grinding, followed by cathodic etching (7). Then they either fractured the discs or separated them on pre-existing cracks. The resulting shards were mounted on the SEM specimen stubs and the fractured surfaces coated with vapor deposited aluminum. With this technique they successfully analyzed the irradiated urania for bubbles of precipitated gas on grain boundaries and on grain edges.

Examination of fuel brings the danger of radioactive contamination. In our laboratory (8) specimens with loose contamination are

treated with a drop of amyl acetate prior to coating with metal
(usually gold) by vapor deposition. In this way loose particles of
oxide and fission products tend to remain stationary. Even loose
urania powder can be examined by fixing with anyl acetate and coating
with gold.

Perhaps the most severe contamination problem in our business
is encountered in examination of <u>unirradiated</u> powders containing
plutonia. The isotopes of plutonium are so alpha-active that they
constitute a serious hazard, requiring special handling procedures.
The alpha-activity is difficult to detect because the alpha particles have a very short range in practically any material. Ironi-
cally, the contamination problem is less severe if the material has
been neutron irradiated to give it some gamma activity, which then
makes it easier to detect and control. For SEM studies of such
toxic materials a special transfer chamber has been devised (8) to
take the specimens from glove boxes to the SEM. The chamber is
shown in Figure 4, and it mates to the entrance port of our SEM.
Specimens are held on mounts (one shown in foreground) which can be
transferred through the large gate valve into the specimen chamber
of the SEM. For plutonia, the chamber is evacuated during transfer,
and the specimens remain in vacuo during the entire examination in
the SEM. Such an arrangement would be useful for any toxic material
or for any material that might react with air (e.g. alkali metals).

In conjunction with SEM work it has become very popular to use
energy dispersive x-ray analyzers to gain chemical information
about the microstructures. With radioactive specimens, however,
the standard commercial energy dispersive x-ray set-ups are practically useless; the high background inundates the detector, as was

Figure 4. SEM vacuum transfer chamber for alpha-active materials.
[courtesy W. Oh]

discussed in the section on electron probe microanalysis. Since the energy dispersive x-ray detector is in direct line-of-sight with the specimen, shielding is not feasible; collimation is our only option. We approached the problem by arranging the detector so that the solid state Si(Li) x-ray detector was placed further from the specimen than is normal. We then had a heavy metal collimator designed and fabricated (9). The collimated detector is illustrated in Figure 5. Its essential feature is a conical aperture such that the entire detector surface receives radiation from a small solid angle originating where the electron beam impinges on the specimen. Actually, there is a penumbra shadow effect, so the area of the specimen that has line-of-sight access to the detector is several mm in diameter. Using such a device we have had good qualitative analyses on specimens whose beta plus gamma activities were approximately 4 R/h at 2 inches.

TRANSMISSION ELECTRON MICROSCOPY (TEM)

The complex microstructures of oxide nuclear fuels (see Figure 1 and the Introduction) require the ability to sample the material at known and predetermined locations. Metallographic information is required to select the areas of interest. Therefore, it is practical to take TEM specimens from a metallographic section of the fuel. Michels and Dragel (10) used a vibratory scriber mounted on a micromanipulator which had mechanical drives in three orthogonal directions. They used the scriber to produce small craters in the fuel and they then studied the bubble distribution by replication methods. Katz (11) used a microhardness indentor to locally break the fuel and then used a replica technique to pick up chips and slivers of fuel that were thin enough at the edges for direct TEM examinations. Rosenbaum, et al (12) described the use of a methacrylic methylester (Technovit) to simultaneously replicate the fuel and pull out fuel particles, whose locations can be identified on the replica. The fuel and the replica with fuel particles on it are shown in Figure 6. By applying a vapor deposited carbon film through a mask with a ½ mm diameter hold, fuel slivers can be obtained for direct TEM examination whose positions in the fuel are known within the area of the aperture of the mask. The primary replica (with or without the fuel particles) can be examined directly in a SEM.

Manley (13) successfully obtained urania fuel specimens for direct TEM by cutting and trepanning a fuel section, mechanically grinding and then pre-thinning it with an air-borne jet of abrasive. Final thinning was done with a chemical polish. The apparatus for pre-thinning irradiated urania with a jet of abrasive particles is shown in Figure 7, from the work of Marlowe (14). Chemical polishing can be useful, but the response of the urania to the polish depends

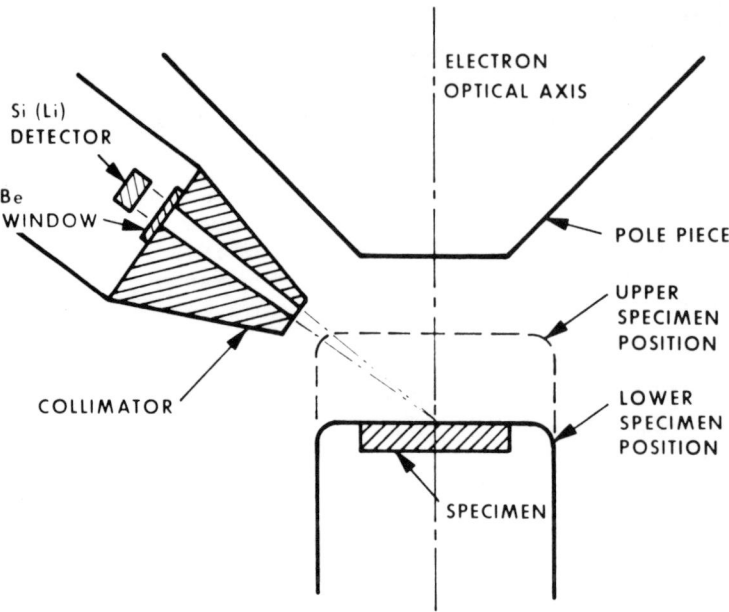

Figure 5. Collimator for energy dispersive x-ray analysis of radioactive materials in an SEM. [courtesy U.E. Wolff]

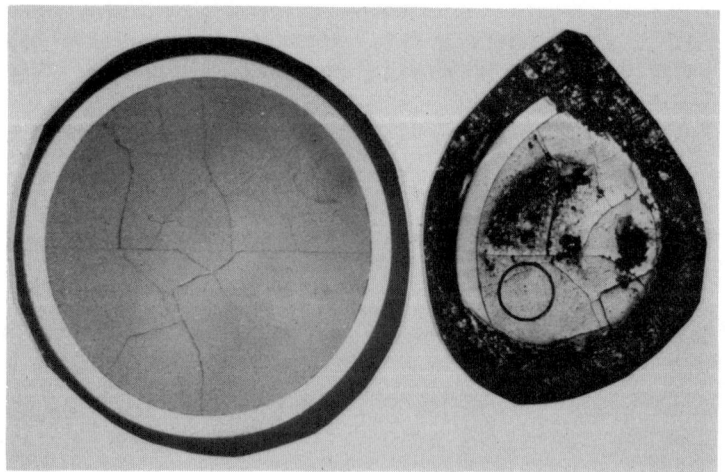

Figure 6. Fuel cross-section (left) and Technovit replica plug (right) with fuel particles.

Circle indicates position of a mask aperture used for coating with vapor deposited carbon. Slivers on the carbon film was examined by direct TEM.

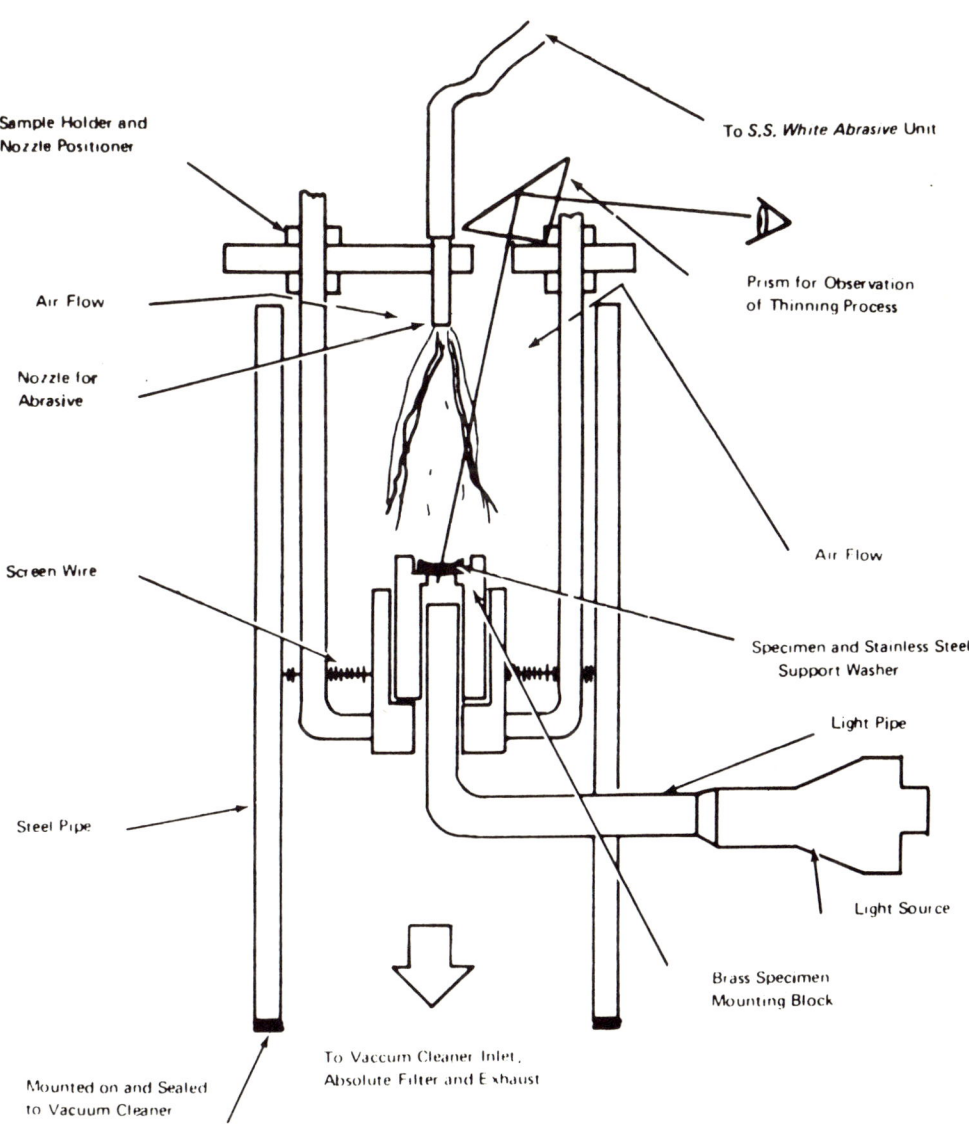

Figure 7. Pre-thinning of irradiated UO_2 wafers by use of airborne abrasive. [after M.O. Marlowe (14)]

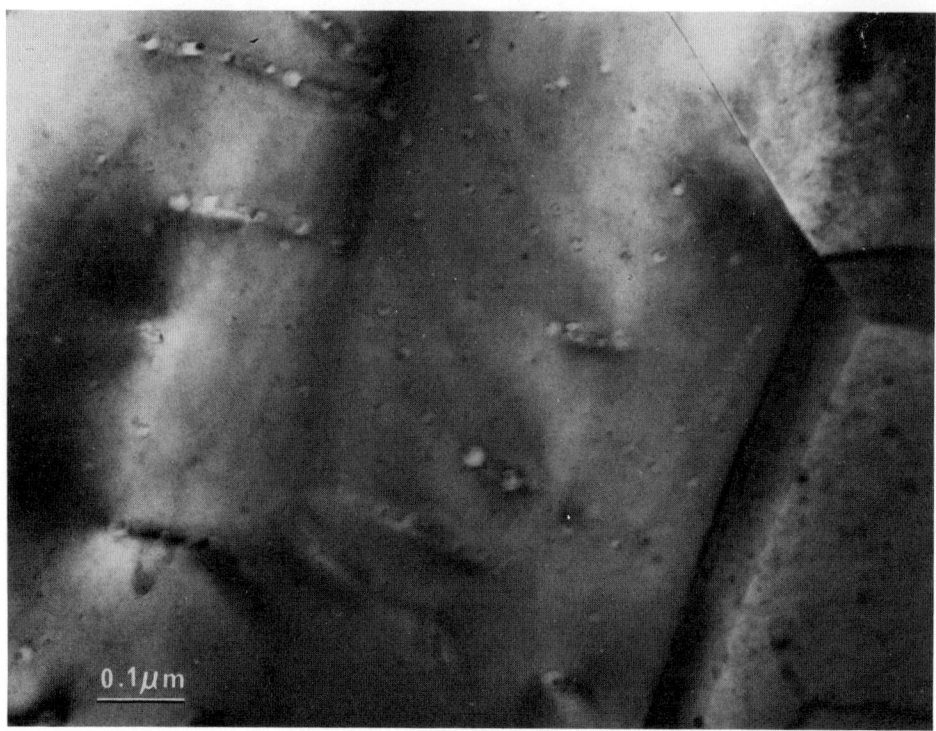

Figure 8. Transmission electron micrograph of urania with precipitated fission gas bubbles. The urania was irradiated to 2.3×10^{20} f/cm^3 at less than 600°C and then annealed at 1600°C for 1 hour to precipitate the fission gas. [courtesy M.O. Marlowe and U.E. Wolff]

on the fuel chemistry. Further, soluble impurity or fission product phases can be leached. We have replaced the chemical polish with ion milling (12). An example of irradiated urania that had been prepared by ion milling is shown in Figure 8. In this instance the urania did not come from a fuel rod, but was part of an experiment (14) where irradiation was done isothermally at less than 600°C with a minimal temperature gradient, and then the specimen was annealed at 1600°C for 1 hour to precipitate the fission gas.

It appears that the combination of prethinning with a jet of abrasive particles followed by final thinning by ion milling is the most generally applicable method for TEM of irradiated oxide fuel materials. Notice the good definition of grain boundaries in Figure 8. When the replica method is used to mechanically sample the fuel, fracture tends to occur at grain boundaries, so good images of grain boundaries are unlikely.

We have concentrated on the TEM specimen handling problems of ceramic nuclear fuels because those are the most challenging specimens to prepare. The metallic fuel sheaths and reactor structural components are easier by comparison. Preparation of highly radioactive austenitic alloys was described recently by Fulton and Cawthorne (15). Any metallic fuel sheath can be handled in similar fashion. The great advantage of metallic specimens is the availability of reliable electropolishing methods.

The general procedure for preparing TEM specimens from cylindrical fuel sheaths is: (a) with remote handling equipment cut a segment of the fuel approximately 1 cm long; (b) mechanically remove the fuel from the ring of cladding; (c) decontaminate the ring; (d) remove the ring from the remote handling cell to a shielded fume hood and spark cut (by electro-discharge machining) disc samples sized to fit the electron microscope specimen holders; (e) grind discs flat to convenient thickness (approximately 0.01 inches); (f) electropolish. The electropolishing conditions will obviously depend on the material. With radioactive metallic materials a jet with a light and photocell arrangement to signal the first penetration is very helpful. Such devices are available commercially.

EPILOG

In the microstructural characterization of radioactive materials specimen preparation requires special techniques. Not only must we contend with the hazards of radioactivity to personnel, but the radioactivity can directly affect the information that is obtainable. X-ray spectra are especially sensitive to radioactivity. Difficulties can be ameliorated by proper specimen preparation along with shielding and collimation. Electron micrographs by SEM or TEM are relatively insensitive to radioactivity.

ACKNOWLEDGEMENTS

It is a pleasure to acknowledge the help of many of my colleagues at the General Electric Vallecitos Nuclear Center. They have demonstrated ingenuity and perseverence in coping with the difficult problems of characterizing the microstructures of radioactive materials. In particular I thank: W. Oh, U. E. Wolff, R. E. Smith, T. E. Lannin, E. S. Darlin, and M. O. Marlowe.

REFERENCES

1. D. R. Beaman and J. A. Isasi, <u>Electron Beam Microanalysis</u>, ASTM STP 506 (1972).
2. H. S. Rosenbaum, T. A. Lauritzen, W. V. Cummings, and R. F. Peterson, "Electron Probe Microanalysis of Radioactive Materials" G. E. Report GEAP-5344 (June 1967).
3. W. V. Cummings, H. S. Rosenbaum, and R. C. Nelson, "Analysis of Irradiated Materials Using a Shielded Electron Probe Microanalyzer" In <u>Proceedings of 18th Conf. on Remote Systems Technology</u>, Amer. Nucl. Soc. (1970) pp. 179-187.
4. R. J. Gray, E. L. Long, Jr., and A. E. Richt, "Metallography of Radioactive Materials at Oak Ridge National Laboratory". In <u>Applications of Modern Metallographic Techniques</u>, ASTM STP 480 (1970) pp. 67-96.
5. R. Natesh, B. J. Koprowski, E. M. Butler, and D. A. Donahue, "Transfer and Analysis of Highly Radioactive Materials in a Shielded Electron Microprobe", In <u>Proc. of 16th Conf. on Remote Systems Technology</u>, Amer. Nucl. Soc., (1969) pp. 243-252.
6. C. Baker, G. Reynolds, and G. H. Bannister, "A Technique for Preparing Electron Microscope Specimens from Reactor Fuel Elements", <u>Post-irradiation Examination Techniques</u>, Inst. of Civil Engineers (1972) pp. 191-193.
7. R. A. Holm, V. J. Haddrell, and G. F. Hines, "The Preparation of Magnox Fuel Element Components for Quantitative Metallography", <u>Post-irradiation Examination Techniques</u>, Inst. Civil Engineers, (1972) pp. 183-187.
8. U. E. Wolff and W. Oh, unpublished work.
9. U. E. Wolff and R. C. Wolf, to be published.
10. L. C. Michels and G. M. Dragel, "Preparation of Replicas for Electron Microscopy of Irradiated Ceramic Fuels", Argonne National Lab Report ANL-7790 (April 1971).
11. O. M. Katz, "Fission Gas Bubbles in Fractured UO_2 Chips", J. Nucl. Mater., <u>31</u> (1969) pp. 323-326.
12. H. S. Rosenbaum, U.E. Wolff, and T. E. Lannin, "Characterization of Irradiated Ceramic Oxide Nuclear Fuels", In <u>Electron Microscopy and Structure of Materials</u> (G. Thomas, ed.) University of California Press (1972) pp. 1017-1026.
13. A. J. Manley, "Transmission Electron Microscopy of Irradiated UO_2 Fuel Pellets", J. Nucl. Mater., <u>27</u> (1968) pp. 216-224.
14. M. O. Marlowe, "Fission Gas Re-solution Rates", G. E. Report GEAP-12217 (July 1971).
15. E. J. Fulton and C. Cawthorne, "Examination of Irradiated Reactor Cladding and Structural Materials by Electron Microscopy", Post-irradiation Examination Techniques, Inst. Civil Engineers (1972) pp. 201-204.

SPECIMEN PREPARATION TECHNIQUES FOR SCANNING ELECTRON MICROSCOPY

N. M. Hodgkin

Micrographics   Univ. of California
&
Newport Beach, Calif.   Irvine, California

INTRODUCTION

It is well known that the Scanning Electron Microscope, hereafter to be known as the SEM, provides an apparent 3-dimensional view of the subject. The contrast, or shading, in the picture is a function of the surface topography, and the apparent overall sharpness of the picture is because of the large depth of focus. The reasons for these are not necessary for this paper, but it is very important to constantly keep in mind that the finished picture is an artifact that either looks like the subject in reality or not, depending upon the viewer's prior knowledge of the subject. If the viewer has no prior knowledge, he must prepare the specimen in such a manner that he can manipulate it in the SEM so as to be able to build up a mental picture of the subject. It is the very fact that the SEM picture so often looks like something we recognize from the real world that causes us to make some serious misinterpretations about the actual nature of the object in the scanner. For a practical solution to the above problem, it is possible to make stereo pairs in the SEM or, more crudely, to rapidly change the angle of viewing or the rotation and thus come to a mental conclusion about what is under the electron beam.

The requirements for the material that can be put in the SEM are:

1. It must be physically small enough to go into the specimen stage and still allow the degrees of freedom for rotation and tilt that are necessary to characterize the specimen. Larger and larger specimens can be accommodated as long as fewer stage motions are needed.

2. It must be compatible with a vacuum environment, which usually approaches the pressure of 1×10^{-5} Torr. Certain minerals have vapor pressures higher than this, and the physical structure will change at these low pressures. Other materials have components or contaminations that will out-gas and will prevent the attainment of the necessary pressure. Plant and animal tissue, or other hydrated material, will dry up and change shape in the vacuum.

3. It must be an electrical conductor or be a substance that can be made conductive by coating with a metal that can be electrically connected to the stage.

4. It must be a substance that has the ability to emit secondary electrons in fairly large numbers or be, as in No. 3 above, a substance that can be coated with a metal which has a high coefficient of secondary electron emission.

The specific types of material that can easily be accommodated in the SEM are powders, particles, fibers, wires, plastics, paper, woven fabrics, and large pieces of any kind of material. Powders and particles can be prepared by dispersing in a slurry of some liquid that will not alter the physical structure of the material or fill in the fine structure of the particles. Such liquids are alcohol, distilled water, petroleum ether, and sodium pyrophosphate diluted to about 0.001 percent with distilled water.

Small particles also can be dispersed in a liquid and filtered out with a Swinny filter attached to a hypodermic syringe. A very few particles in a fluid such as oil, for example, can be concentrated this way because as much liquid as necessary can be put through the filter and all the particles will be caught on a 13 mm. diameter filter disk which can be glued down to a specimen stage.

Fibers and wires are usually best mounted vertically in holes drilled in the stage so that the microscopist will have the option of looking down on the end from any azimuth angle and of looking at the sides from 360 degrees. Modification of the hole angle will have to be made to conform with the tilt angle limits of the particular SEM in use.

Paper is usually viewed on the flat side, but if it is to be viewed on the edge, a slot cut in the stage will help to support it. It is important to use a glue that will not soak through the paper and alter the surface structure. Alternately, the paper may be stuck down at just two or three points. (See section on glues.)

Plastics are handled like any other bulk material, except that the user must be aware of the fact that some plastics are unstable under the electron beam. If you can tolerate the reduced surface contrast that you get with a higher accelerating voltage, raise the voltage. Another method is to focus on one area of the plastic and take the picture on a fresh area that has never had the beam on it. Replication of the plastic is another solution if the replicating materials are compatible with the plastic. Woven fabric is very difficult to handle. Both sides must be coated with a conductive material in order to electrically connect each exposed piece of fiber with its neighbor and ultimately to ground. If the weave is very tight, even this does not help. You simply have to work with a charging sample. If the fabric is open weave, it must be mounted across two supports or on a stage that has had a slot filed in it and has been painted with a graphite paint to prevent "seeing" the background through the mesh. Another way would be to mount the fabric vertically in a slot in the stage and the stage could be tilted over to an angle of 90 degrees.

With larger pieces of material, it is impossible to give any specific instructions, except to say that the most information is usually gained when the material is broken and the fractured surface examined.

CLEANING

The specimen must be as clean as possible when it is put in the SEM. Grease and oil will degrade the image, and small adherent articles will charge up and spoil the contrast in nearby areas. If a specimen is to be coated, it must be cleaned prior to coating. The general theory of cleaning is to use a solvent or solution that will dislodge, dissolve, or otherwise remove the unwanted material. Then, remove the cleaning solution with a succession of solvents, each one cleaner than the previous one. The following is a list of analytical reagent grade solvents with the "residue after evaporation."

Benzene and Petroleum Ether	10 p.p.m.
Acetone and Methanol	5 p.p.m.
Trichlorethylene	2 p.p.m.
2-Propanol	1 p.p.m.
Bottled Distilled Water	0.3 p.p.m.
Ultra-Pure Water	0.1 p.p.m.

Because alcohol and acetone are miscible in both hydrocarbons and water, it is necessary to use one of these solvents as an intermediate fluid in going from a hydrocarbon-based material such as benzene or petroleum ether to water. A typical regime for oil might be petroleum ether followed by 2-propanol. It is usually not necessary to go to distilled water. If you are using an ultrasonic cleaner with a laboratory detergent, follow that with two changes of distilled water, and then 2-propanol for quick drying if water will harm the material. With these few solvents, you can clean up just about anything. While 2-propanol is miscible with benzene and petroleum ether and will rinse these materials and their dissolved impurities away, it may not redissolve those impurities after they have been left by evaporation of the solvent. Therefore, it is most necessary not to allow the solvent to dry as the specimen is carried from one liquid to the next.

FASTENING THE SPECIMEN TO THE STAGE

The requirements for glues are that they must be fast drying, be strong enough to hold the specimen and yet not so strong that the specimen cannot be removed after it is dried, and they will show little out-gassing after a reasonably short drying time. Some glues that are useful are:

1. Silverprint, manufactured by G.C. Electronics Company; available at any electronics store. This material is fine for small parts, but it is not very strong, and its high capillary attraction makes it flow up over things if it is put on in excess.

2. Silverpaste; available from the Ted Pella Company.* This material is conductive and is quite strong, but it is easily removed after the session. It is soluble in methyl ethyl ketone and methyl isobutyl ketone.

3. Wilhold White Glue. This is a fine general purpose glue. It is excellent for paper and plastic parts. If a specimen is not going to be coated, a conductive stripe must be painted from the conductive specimen to the stage.

4. Double-stick Scotch Tape. This material is useful for small parts and particles, but very small particles in the range of 25 microns will sink in and be lost and the material will flow up the side of the specimen to some extent. The heat of the metallization process will cause cracks which are unpleasing aesthetically.

5. Gum Tragacanth. This is a water based material. Things do not sink in it, but it cracks like Scotch tape; and when it gets old, it flakes off . . . carrying the specimens with it.

6. Mikrostik. Available from the Ted Pella Company. This material is not a strong glue, but it is just about the only thing to use for very small particles such as wood dust and fine powders. The material dries in about ten

---
* P.O. Box 510, Tustin, California, 92680

seconds. It is then tacky and ready to use for about ten days. It does not crack under the beam, and the particles do not sink in it, nor are they wetted by it. This material must be used on a glass or mica-covered specimen stage because it is the smoothness of the substrate that makes it appealing. (See Figure 1.)

7. Colloidian Diluted with Amyl Acetate. This special purpose material, useful for X-ray work, is used to secure small particles that already are on a filter paper or other surface and cannot be removed or handled in any other way. A tiny amount is allowed to run over the surface of the filter, thus effectively sticking down the particles. This will obscure some surface detail, but if X-ray data is to be taken, it will add nothing new to the spectrum as long as your system will not detect carbon.

PREPARING THE SPECIMEN

Often, nothing need be done to the specimen other than mounting. The charging effect can be used to advantage to better see the oxidation product on metals, or to distinguish between conductors and non-conductors and see their spatial relationship as in graphite-epoxy composite materials.

Figure 1. A marine dinoflagellate hystricokolpoma cyst fastened to a stage with mikrostik.

Objects that are conductive need not be coated, but the resolution is not quite as good as when they are coated with a metal. (See section on coatings.)

POLISHING

Polishing is advantageous in SEM work when it is desired to get an X-ray dot pattern of a cross section of a substance that is composed of different materials. Polishing is also the best way to prepare a sample when the backscattered mode is used to delineate different materials in a cross section such as in a braze.

ETCHING

Surface material can be successively etched away by repeatedly dipping and raising the piece in an etchant. The lower part will be etched the most, the upper part the least. X-ray analysis can be used to show the location of the layers. Polished metal surfaces can be etched in the same manner as for standard metallographic analysis and then put in the SEM.

Ion etching is done inside the vacuum chamber of the SEM, and the progress can easily be checked.

COATING

Coating, in SEM terminology, means vapor deposition of a metal onto the surface to be looked at in the microscope.

Coating provides two basic effects. For a non-conductor, it provides conductivity and better resolution. For the already conductive material it only provides better resolution.

Considering only conductivity, it is necessary that all the elements in the structure be electrically connected to each other and to the ground. This is not always possible, especially with composite material such as sandstone; which looks compact, but in actuality is made up of a lot of individual units suspended in a matrix. The vapor comes to the substrate in a straight line, and in a high vacuum some elements in the subject will shadow other parts. For this reason, the specimen must be rotated and tilted so as to be "seen" on all sides by the source of the vapor. To help this stringent requirement, it is possible to allow the metal vapor to come in from different angles by raising the pressure in the vacuum chamber of the coater to about 1×10^{-3} Torr. with a heavy inert gas such as argon. Argon atoms deflect and diffuse the metal vapor somewhat. Of course, the mean free path is reduced as the pressure is raised; and so there is a trade-off between no vapor and a lot of diffusion and much vapor and no diffusion. Rotating at all times and tilting to two distinct angles for two different sources is a practical way of coating over almost the entire hemisphere of the specimen.

Another way to handle the conductivity problem in materials such as sandstone is to soak the rock in a silver nitrate solution, then put it in a reducing agent and allow the silver to be reduced over the inside as well as the outside of the sandstone to effectively connect the grains together even where metal vapor will not penetrate. The precipitation must not be allowed to go on too long because the silver will obscure the surface detail on the outside of the sandstone.

Osmium tetroxide vapor, either directly from a crystal of osmium tetroxide or from a water solution, will also be reduced to osmium metal on the surface of the material, making it conductive.

Considering coating as a way to enhance resolution, the highest resolution requires the smallest electron beam; but the small beam will produce only a few secondary electrons, which are the ones that produce the picture. Thus, in order to get as many secondary electrons as possible, we coat the sample, whether it is conductive or not, with a metal that has a high secondary electron yield. Some of these metals, with their secondary electron efficiencies, are listed below:

Metal	Efficiencies
Platinum	1.8
Silver	1.5
Gold	1.4
Palladium	Over 1.3

Density of the coating material must be high in order to keep the area of secondary emission small. Based on the above list, all these materials should be good coating metals, but the resolution also depends upon how the film is formed. Pure gold goes down in large clumps (1,2). Palladium and gold-palladium alloy form a very fine grain film (2).

REPLICATION

Replication is necessary in SEM work when the object itself cannot be brought into the microscope. Some of the reasons might be that it is too large or valuable to cut up, the specimen is magnetic and cannot be demagnetized, it must be retained as legal evidence, or it is inconvenient to cut the specimen down to size.

The two general types of replicas are negative and positive.

Negative Replication

The negative replica is simple, direct, and is used to look at and measure depth of cracks and holes in materials. The standard replication tape is used as in transmission electron microscopy, but the replica itself is coated and examined in the SEM as though it were a specimen.

Positive Replication

The positive replica is the most interesting, because it reproduces the original surface. Three techniques will be presented. Silicone rubber-epoxy (3), cellulose acetate-epoxy, and electroforming.

Silicone-epoxy - This method uses silicone rubber to make the negative mold. Epoxy is poured into it; and when hard, it is coated with a metal in the conventional manner. Silicone rubber limits the resolution to about 2,000 useful magnification, but the silicone is excellent for deep and intricate parts, areas that are always wet such as inside the human mouth, and specimens that cannot tolerate the cellulose acetate-methyl acetate of the next two methods to be described.

Cellulose Acetate-Epoxy - This method was developed by the author and is described for the first time here. Cellulose acetate is used with methyl acetate in the standard manner, after which palladium is vapor deposited onto the replica, followed by a large vapor deposition of aluminum. The aluminum is used simply because it is cheaper than palladium. Five-minute epoxy is then used to back up the aluminum and the original replicating tape is dissolved away in methyl acetate. This method is fast, but it is not consistent and sometimes the epoxy distorts upon curing and destroys the delicate metal film. These limitations are being investigated. See Figure 2a,b.

Electro-Forming

This method has been very consistent and reproducible. It also was developed by the author and is presented here for the first time. A standard cellulose acetate negative replica is vapor coated with about twice as much palladium as would be used in normal coating. The stage and adherent replica is put into a nickel sulfamate

SPECIMEN PREPARATION FOR SCANNING ELECTRON MICROSCOPY 305

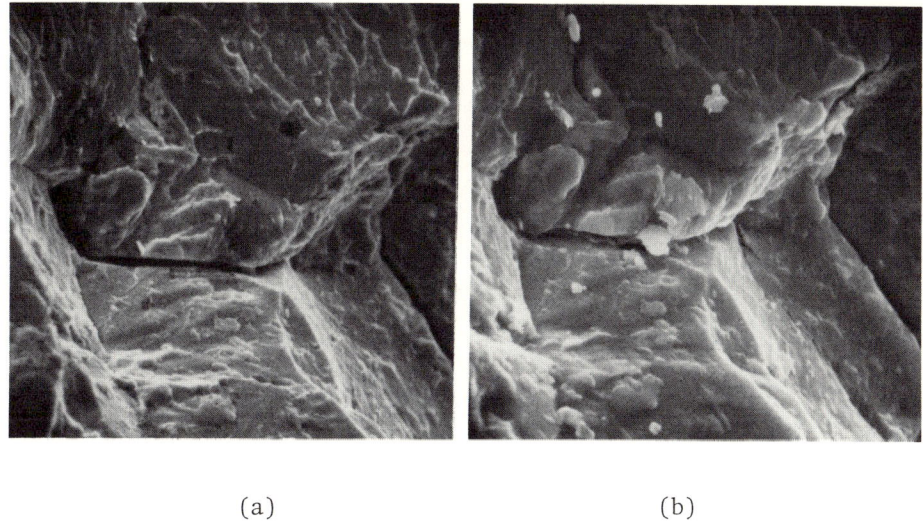

(a) (b)

Figure 2. a.) Original fracture in 4340 steel. 5,000 X. b.) Replica of fracture made by the cellulose-epoxy technique. 5,000X. (Reduced 32 percent for reproduction.)

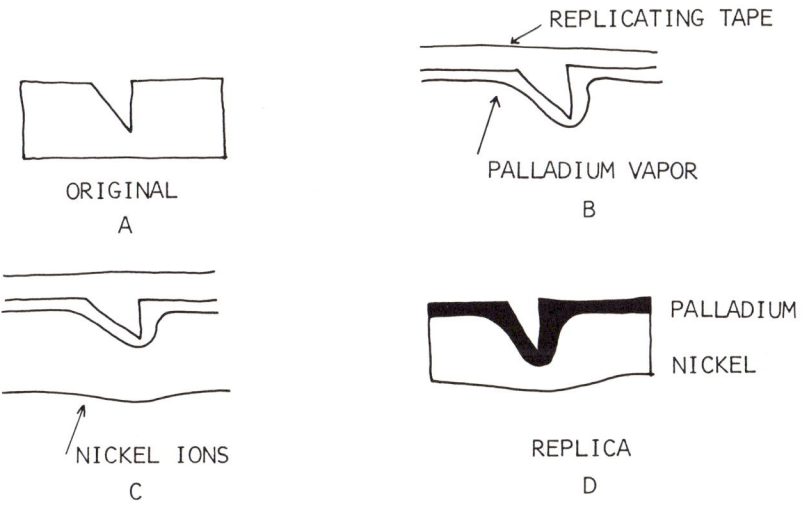

Figure 3. a.) Original surface. b.) Cellulose acetate tape with palladium vapor deposited on the surface. c.) Palladium surface being backed up by a nickel electroplate. d.) Finished replica.

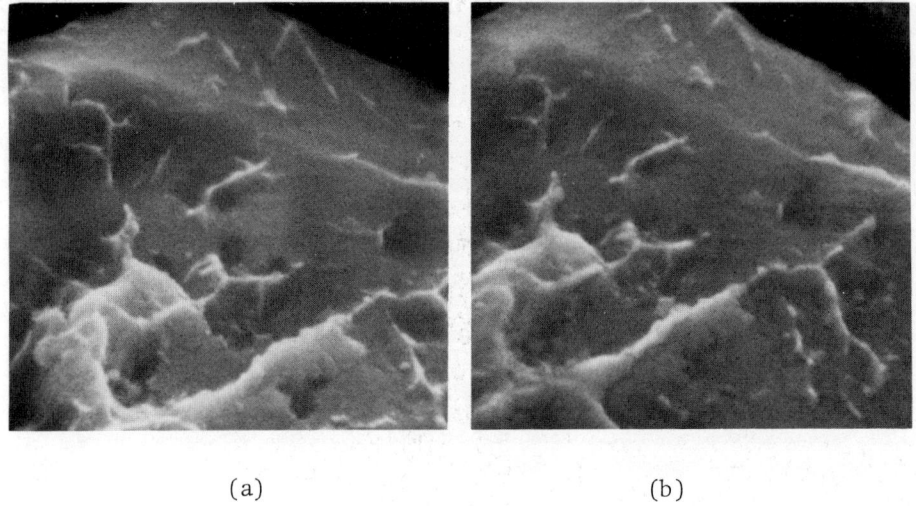

(a) (b)

Figure 4. a.) Original fracture in 4340 steel. 20,000 X. b.) Replica of fracture made by the nickel electroforming technique. 20,000 X.

plating bath, and nickel is electroplated onto the palladium until the tape feels rigid. The plating current is started at 2 amperes per square foot, which is equivalent to 2.7 milliamperes per 1/2 inch round specimen stage. The tape is then scraped off the stage and soaked in acetone until the original cellulose acetate is completely dissolved away. See Figure 3 a, b, c, d. This may take two or three hours and three or four changes of acetone, along with gentle brushing with a camel's hair brush. The resulting sample should be a replica with detail as good as the original, but coated completely with palladium and ready for the SEM. The resolution should be as good as the replicating tape, which is known from transmission EM work to have a resolution higher than the SEM can resolve. See Figure 4 a,b.

REFERENCES

1. Poulson, G.G., and R.W. Pierce, "Texture and Artefacts in SEM Observation Due to Vapor Coating," 30th Annual EMSA Meeting, (1972), p. 406.
2. Hodgkin, N.M., and L.E. Murr, "Quantitative Study of Vapor-Deposited Metal Coatings for Scanning Electron Microscopy," Proceedings of 6th Annual IMS Technical Conference, (1973).
3. Sognnaes, R.F., "The Use of Microreplication for Scanning Electron Microscopy of Dental Lesions," Caries Research, 6 (3) (1972), pp. 87-88.

ELECTROPOLISHING OF THIN METAL FOILS

Charles Hays

Texas A & M University

College Station, Texas 77843

It has been almost a quarter of a century since Heidenreich (1) successfully thinned aluminum metal for transmission electron microscopy via electropolishing. His original apparatus is depicted in Figure 1 and some of his early advice on electro-thinning is also given as Figure 1. Since 1948, much has been done in this area using varied techniques with different polishing appliances. In recent years so much has been done that there now exists much confusion for the young researcher. Oftentimes the following questions are raised:

-- Which foil apparatus should I use?

-- Why all this bother about electron transparencies?

-- Why am I having problems making a foil?

-- What improvements can I offer?

These are not just idle curiosities of the younger set. They are, in fact, real questions of considerable importance to each of us who deal with transmission electron metallography. Because these are real issues and because the subject area has become unnecessarily confusing, it is appropriate that the state of the art should now be reviewed. This paper deals with 25 years of progress in foil making by electropolishing methods. The intent is to describe what is known about these foil making processes with the hope that certain confusing issues can be somewhat better understood.

(A)
HEIDENREICH'S SAMPLE HOLDER FOR ELECTROTHINNING

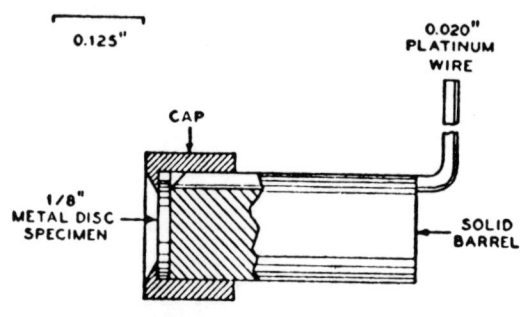

(B)
HEIDENREICH'S VIEWS ON SEQUENCES IN THINNING

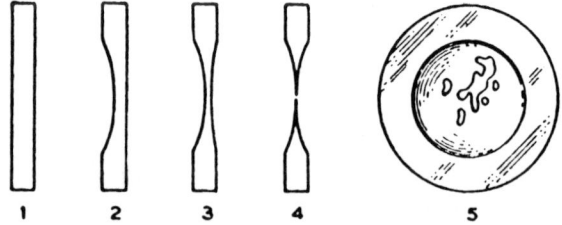

Figure 1. Photographs showing some of Heidenreich's classical work on the successful thinning of aluminum (taken from Reference 1).

CLASSIFICATION OF APPLIANCES

It is quite confusing to the novice that so many different foil-making devices are currently available. This situation is a natural evolution caused by the tremendous amount of metallurgical research on electron transparent foils within the last few years. This information avalanche can be classified owing to the fact that cathode shape or type has been a consistent factor between the different investigations (2). Figure 2 is given to show that all of the literature developments for electropolishing appliances can be classified into just seven categories and these are as follows:

-- Immobile Pointed Cathodes
-- Immobile, Vertical, Parallel Sheet Cathodes
-- Immobile, Non-vertical, Parallel Sheet Cathodes
-- Immobile, Non-vertical, Non-parallel Sheet Cathodes
-- Immobile Complex Cathodes
-- Immobile Pump Cathodes
-- Mobile Sheet Cathodes

ELECTROPOLISHING OF THIN METAL FOILS 309

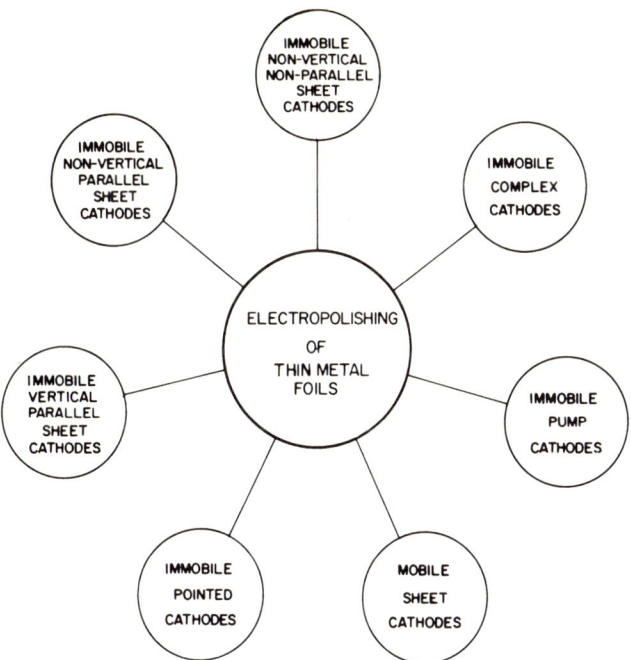

Figure 2. Classification of appliances used to produce thin metal foils by electropolishing.

Thus, it is proper to state that if electropolishing appliances are viewed with respect to cathode size or shape, their variabilities are then less numerous.

Because so much development work has been done using the different types of polishing appliances, it would be virtually pointless to review each of these accomplishments. Instead, it is wise to call attention to the prototype discoveries within each classification. Figure 3 offers information on the classical appliances for thinning by electropolishing (3-10). These important contributions are to be ranked with the earlier exploratory work by Heidenreich (1). Newcomers to the field of foil making by electropolishing would be well advised to become familiar with the different methods depicted in Figure 3.

To this point, attention has been given to the question, "Which foil apparatus should I use?" The best answer here is to use whatever device you have or can expect to have. If a metallographer understands electropolishing and knows the operating details of his own apparatus, foil making is not to be feared. If these very basic fundamentals are not known, the only recourse is to review selected literature topics (11,12).

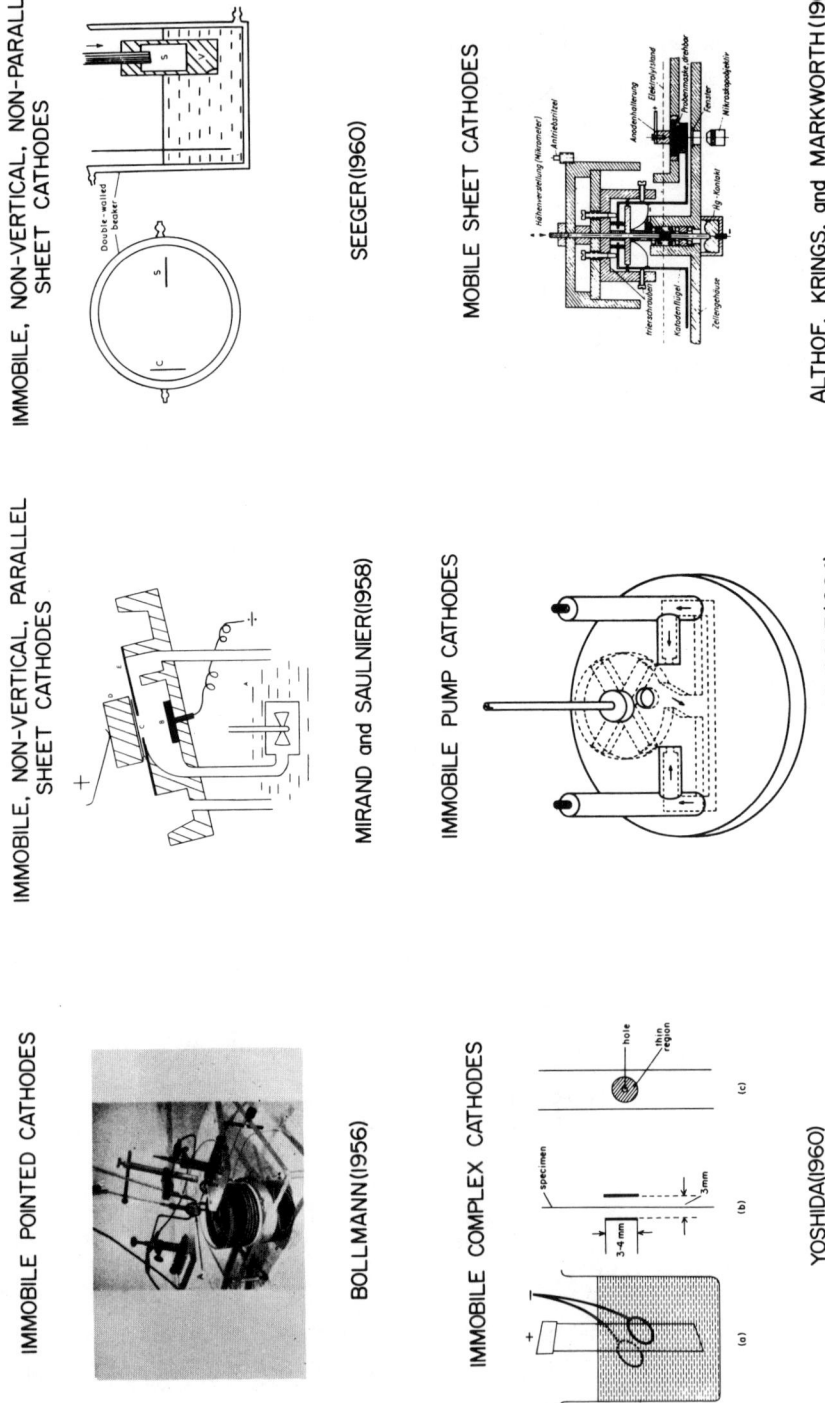

Figure 3. Important contributions for thinning of metals by electropolishing.

SELECTED FOIL APPLICATIONS

What then of the question, "Why all this bother about electron transparencies?" If the past 25 years are evaluated in some detail, it is surprising to learn that so much more is now known about the subject. This situation of enlightened empiricism has developed, to a great extent, because of transmission electron metallography. It is important then that some of these contributions should now be recognized. For example, consider the problem that electron microscopists had during the early years of 1956-1962 when every electron transmission micrograph was attacked viscously as to its being "truly representative of the bulk." Each of us were forced to defend our choice of selected-area micrographs. Figure 4 is an excellent example of how microscopic reliability can be easily enhanced. In this comparison, Glenn (13) demonstrates a proper correlation between the optical and electron microstructure of pearlitic steel and AISI-304 stainless steel.

Another problem so typical of the early era was that uninformed readers did not clearly understand that what one sees is merely a function of how one looks; i.e., foils exhibit a critical dependence on the direction being viewed. If the aspect angle is not considered, one can easily be mislead as to the appearance of true substructure. Figure 5 (14) shows the importance of aspect angle; in one grain the dislocations are essentially random yet in the other grain they are distinctly cellular. The problem of substructural anisotropy for drastically deformed steel has been examined by Glenn and co-workers (15); their results are given as Figure 6. It is here seen that what one sees depends on how one looks; there is a clear indication that the aspect angle is critical.

Figures 7 and 8 give the now classic work on how dislocation arrangements are changed by effects of strain and test temperature (16). In this work it was demonstrated that, at a specific test temperature, the role of increasing strain is to cause the development of a distinct cell structure. In fact, Keh's work clearly established that the existence of a cell substructure is controlled by both strain and temperature. This discovery did much to enhance the general acceptance of dislocation theory as applied to practical engineering problems. The fact that only electron transparent foils could be used to detect such minute variations in the substructure didn't exactly harm the electron microscopy profession either! Perhaps the greatest contribution by electron microscopists during the last decade was the results on subgrain coalescence by Hu (17); his now classic work is depicted in Figure 9. Hu has shown that recrystallization is dependent on subgrain coalescence through rotation until a critical curvature is reached for migration to proceed. Prior to this study, the kinetics of softening were not well understood. More recent studies on softening have offered

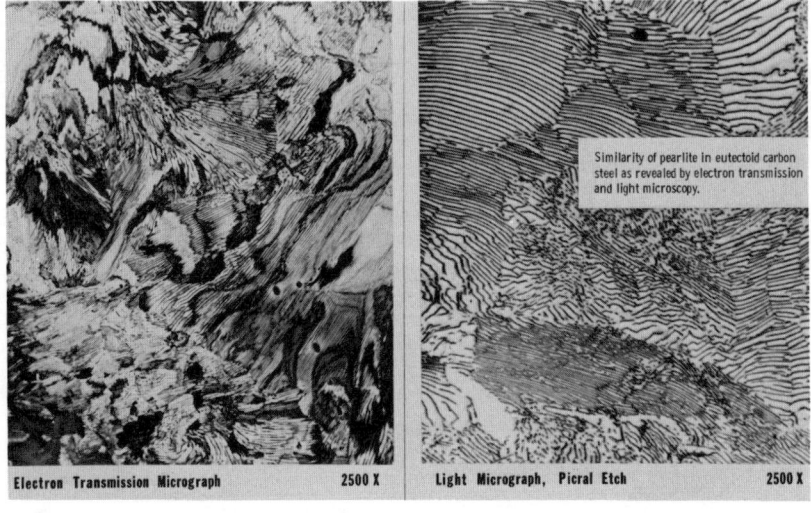

(a)

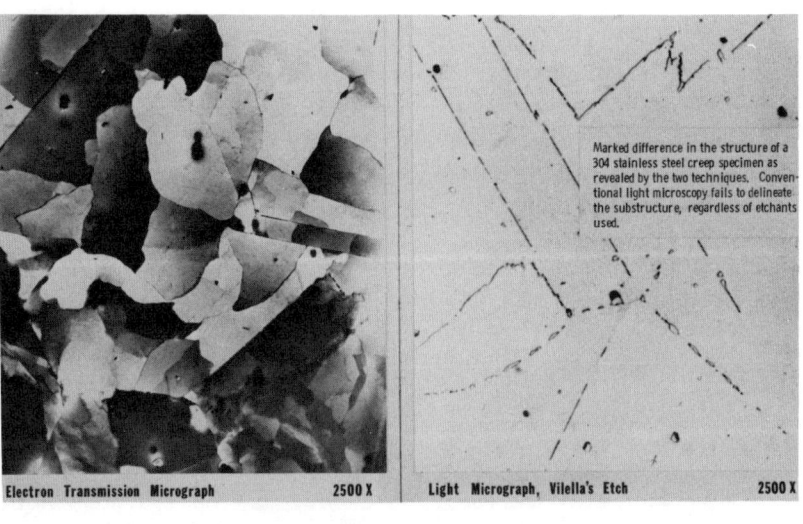

(b)

Figure 4. a) Transmission Electron Micrograph (TEM) of steel which demonstrates that foil samples can be truly representative of the bulk material. A comparison to the optical micrograph shows the foil structure to be real (taken from Reference 13).

b) TEM photograph which shows that the interpretation of a stainless steel microstructure is strongly dependent on a proper consideration of the bulk substructure (taken from Reference 13).

Figure 5. Role of aspect angle in electron metallography (taken from Reference 14.)

the view that recovery rates are perhaps related to defect type (18) as noted in Figure 10. Here again, the use of electron transparent foils was an important laboratory procedure in the investigation.

Currently, it is now common practice to evaluate different aging processes using transmission electron micrographs. Much of this work has followed the early quench aging research by W. C. Leslie (19) and a typical application involving thin foils of strain-aged iron is given as Figure 11 (18). It is here shown that the substructure clearly indicates the success of a mechanical-thermal treatment cycle; the dislocations are randomly grouped and precipitation has occurred. It is also now common practice to evaluate artificial aging treatments using electron transparencies. Figure 12 reveals data taken from Decker (20) on nickel-base superalloys of the type used in gas turbines or reciprocating engines. It is here shown that the details regarding how dislocations move in these alloys is of paramount interest to design engineers. At this point in time, airplane-engine, quality-control technicians are quite familiar with terms such as "dislocation cutting" or "dislocation by-passing". Our field of specialization in metallography has truly come a long way.

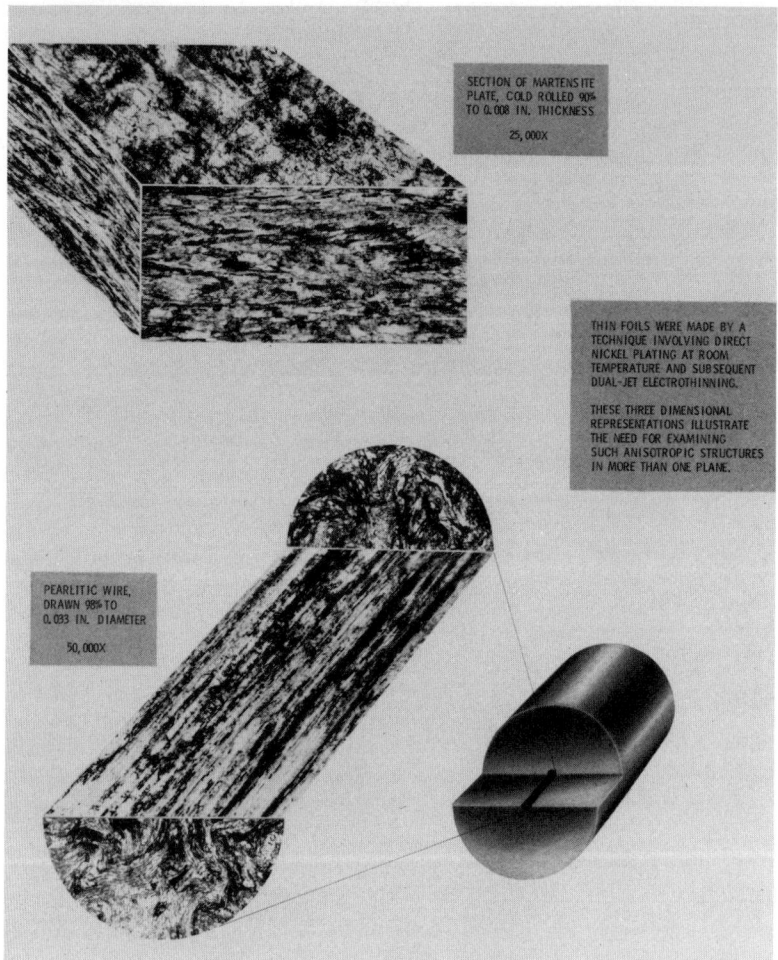

Figure 6. Subgrain anisotropy in swaged iron (taken from Reference 15).

In the same sense that aging temperatures are vital to nickel-base alloys, aging times are equally important. Figure 13 offers evidence (21) to demonstrate that modern metallographers detect over-aging of superalloys through electron microscopic examinations involving ripening. Another classic bit of metallography that comes to mind whenever precipitation may be concerned is the work by Glenn (22) on directional precipitation involving the phase Ni_3Ti. A copy of Glenn's original award-winning micrograph is given as Figure 14. In addition, Glenn (23) has developed electron micrographic methods to investigate magnetic domains in ferro-magnetic materials, note Figure 15. Should it not now be obvious that electron transparencies are of significant importance to product-oriented engineers?

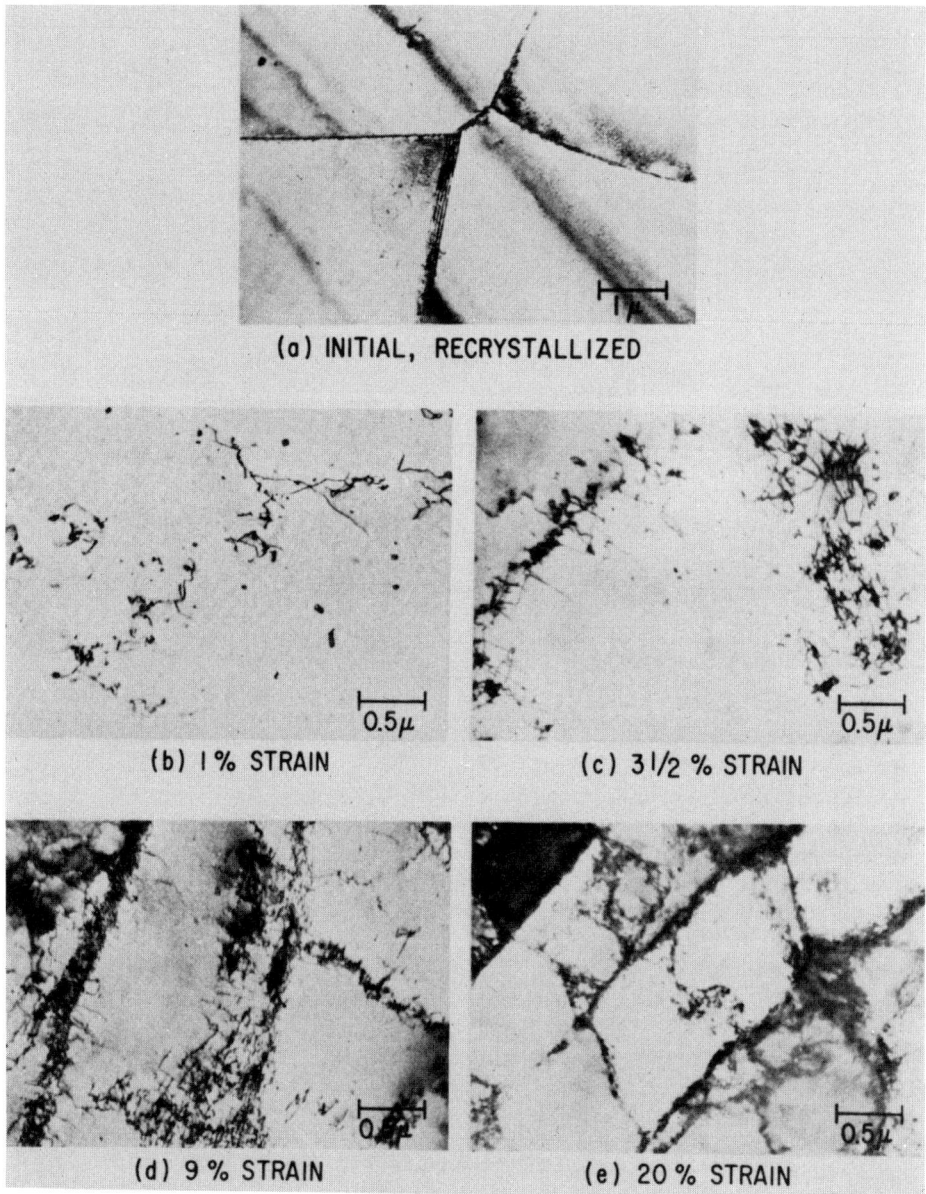

Figure 7. Effect of strain on dislocation arrangements (taken from Reference 16).

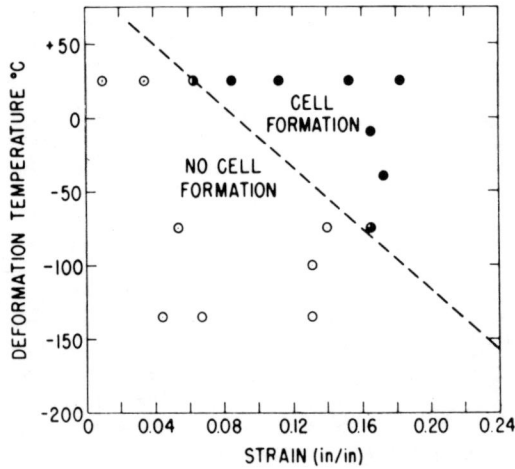

Figure 8. Effect of temperature on dislocation arrangements (taken from Reference 16).

RESEARCH TRENDS

In years past, most of the problems which were encountered in the making of a foil were usually those related to a misunderstanding of procedural details. For that reason, it is appropriate that the critical details of electropolishing thin metal foils should now be briefly reviewed. Figure 16 is a schematic that offers a general guide for the production of an electron transparent foil. Samples are sectioned to a suitable size in a manner so as to avoid any Beilby damage (24). Afterwards, these samples are thinned to a thickness of about one mil prior to electropolishing. Here the common approach is to use chemical polishing provided, of course, that a proper solution is available for the sample under evaluation (11). When a sample has been properly prepared, large transparent areas are commonplace.

It is surprising to note that many metallographers are unaware of the traditional problem areas in the thinning of metals by electrolytical methods. These difficulties for the case of jet machining were studied in detail by DuBose and Stiegler (25). They report that if a sample is thinned with the proper current density, the cross-section of the disc sample should have almost parallel surfaces

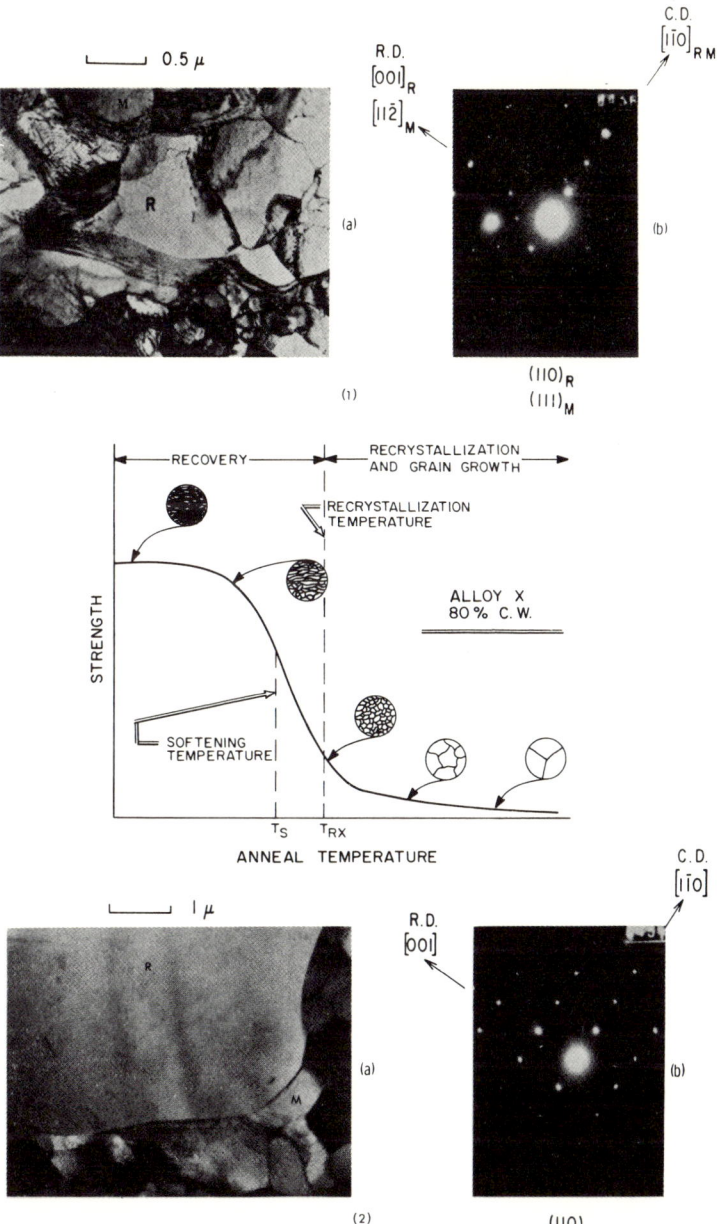

Figure 9. Evidence of subgrain coalescence to explain kinetics of softening (taken from Reference 17). In (1a), a recrystallized grain (R) is shown at an early stage of formation in the polygonized matrix (M). In (1b), the diffraction pattern is given for both (R) and (M). In (2a), a well-developed recrystallized grain exists in a polygonized matrix. In (2b), the diffraction pattern for the recrystallized grain is given.

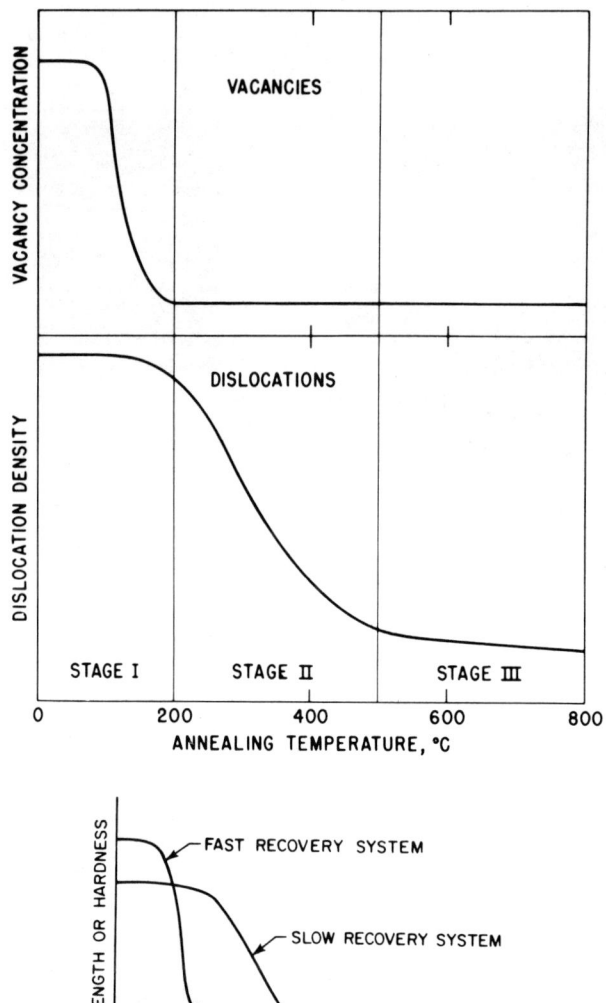

Figure 10. Softening studies in high purity iron (taken from Reference 18).

so that more than one hole will eventually occur. This condition is depicted in Figure 17; here the effects of current density are graphically given for the cases of low, nominal or high current. Also given therein is an indication that the normal low-resistance electrolyte current density-voltage curve does not hold for jet machining methods. Instead, thinning by jet machining offers a somewhat linear response for current density and voltage.

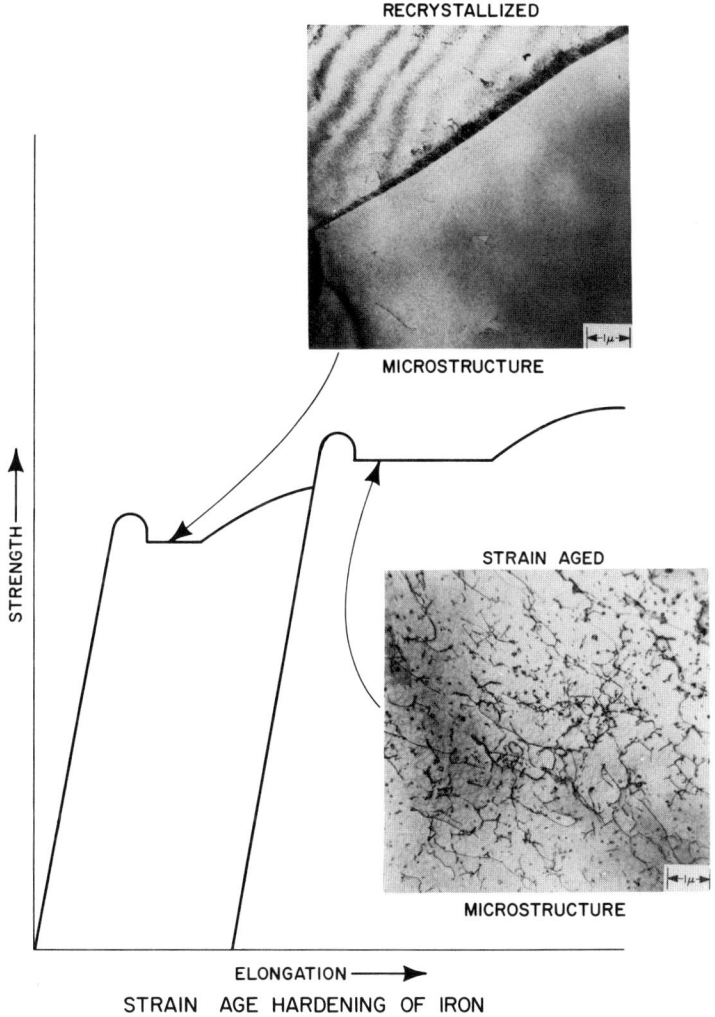

Figure 11. Role of precipitate in conventional strain aging processes (taken from Reference 18).

Unfortunately, the novice has come to view jet polishing as jet machining and vice versa. Because of this confusion, Figure 17 has become much misunderstood by uninformed metallographers who attempt to evaluate submerged-jet polishing appliances using pump cathodes. Figure 18, taken from Glenn (26), reveals that jet polishers do not normally exhibit a linear behavior for voltage versus current density plots; instead, the conventional low-resis-

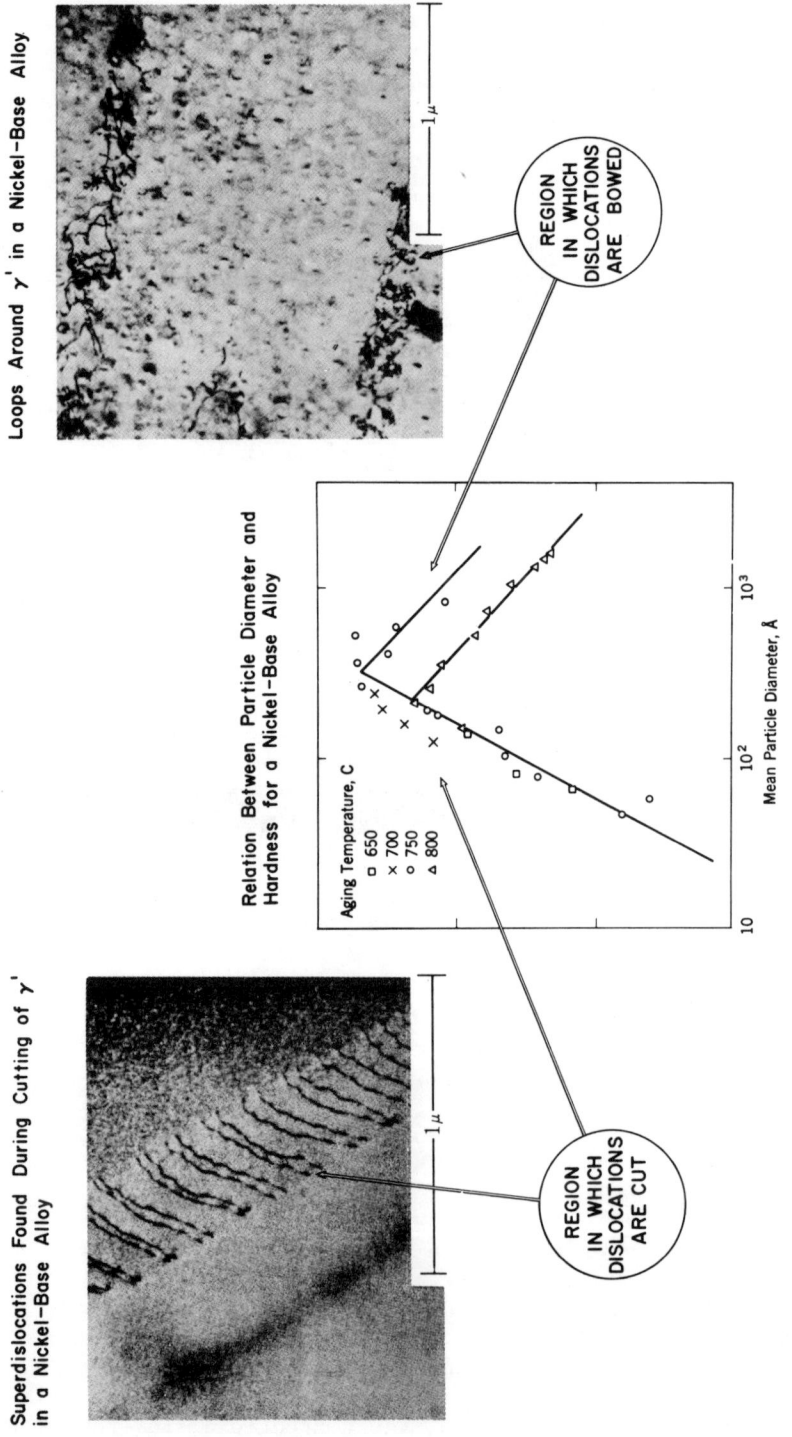

Figure 12. Foils can be used to evaluate aging treatments (taken from Reference 20).

ELECTROPOLISHING OF THIN METAL FOILS 321

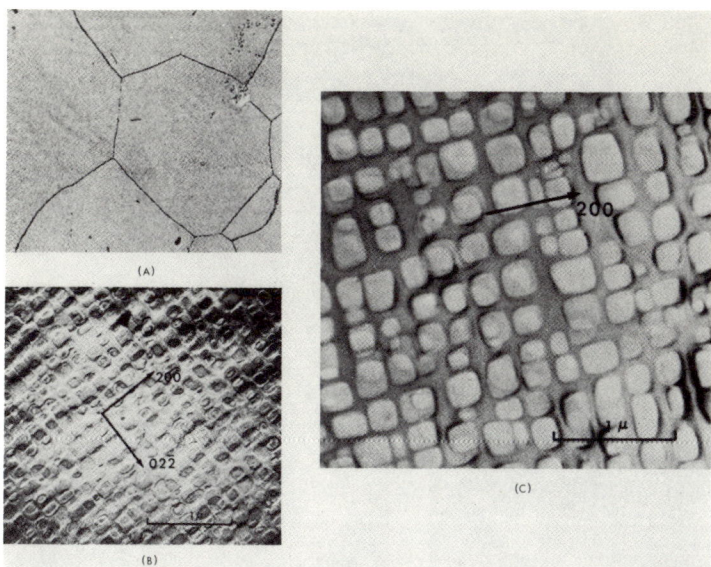

Figure 13. Effect of aging times for nickel-base superalloy; (a) optical micrograph at 500X, (b) electron micrograph of aged condition at 24000X and (c) electron micrograph of ripened γ' after an additional 16 hr. aging at 1800°F-24000X (taken from Reference 21).

Figure 14. Directional precipitation involving Ni$_3$Ti (taken from Reference 20).

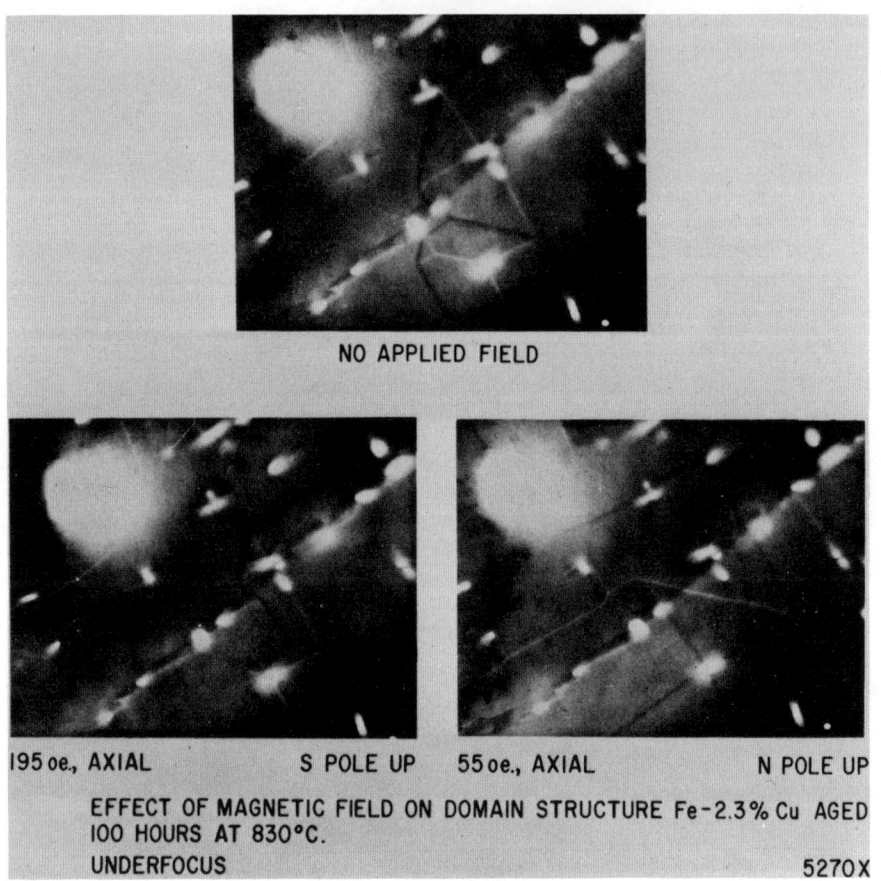

Figure 15. Use of foils to evaluate magnetic domain arrays (taken from Reference 21.

tance electrolyte curve is held to be valid. Also shown in Figure 18 are the traditional problem areas for this form of electropolishing method; viz., (1) the lowest voltage for polishing is optimum (not the highest voltage as some operators might conclude) and (2) to vary submerged jet velocities is to increase the current density at specific locations (this facilitates the control of hole formation). At this point it is appropriate to leave the subject of problems encountered in the making of electron transparent foils; let us, instead, direct our attention to process improvements which appear promising.

Table I offers a survey of research trends for the electropolishing appliances previously developed. Of special interest therein is the primary design motivation for each class of electro-

TABLE I. JUSTIFICATIONS FOR PAST RESEARCH TRENDS IN ELECTROPOLISHING APPLIANCES

Cathode Type or Shape	Design Motivation
Immobile	For fixed, predetermined and optimum anode-cathode distance
Pointed	To concentrate thinning in localized area
Vertical	To allow ample maneuvering space
Parallel	To achieve streamlined electrolyte flow
Sheet	For simplicity and large cathode/anode area ratio
Non-vertical	To facillitate *in situ* observations
Non-parallel	To vary anode-cathode spacing
Complex	To extract foils from specific areas or to avoid lacquering
Pump	To attain optimum anolyte layer action
Mobile	To achieve uniform field polishing

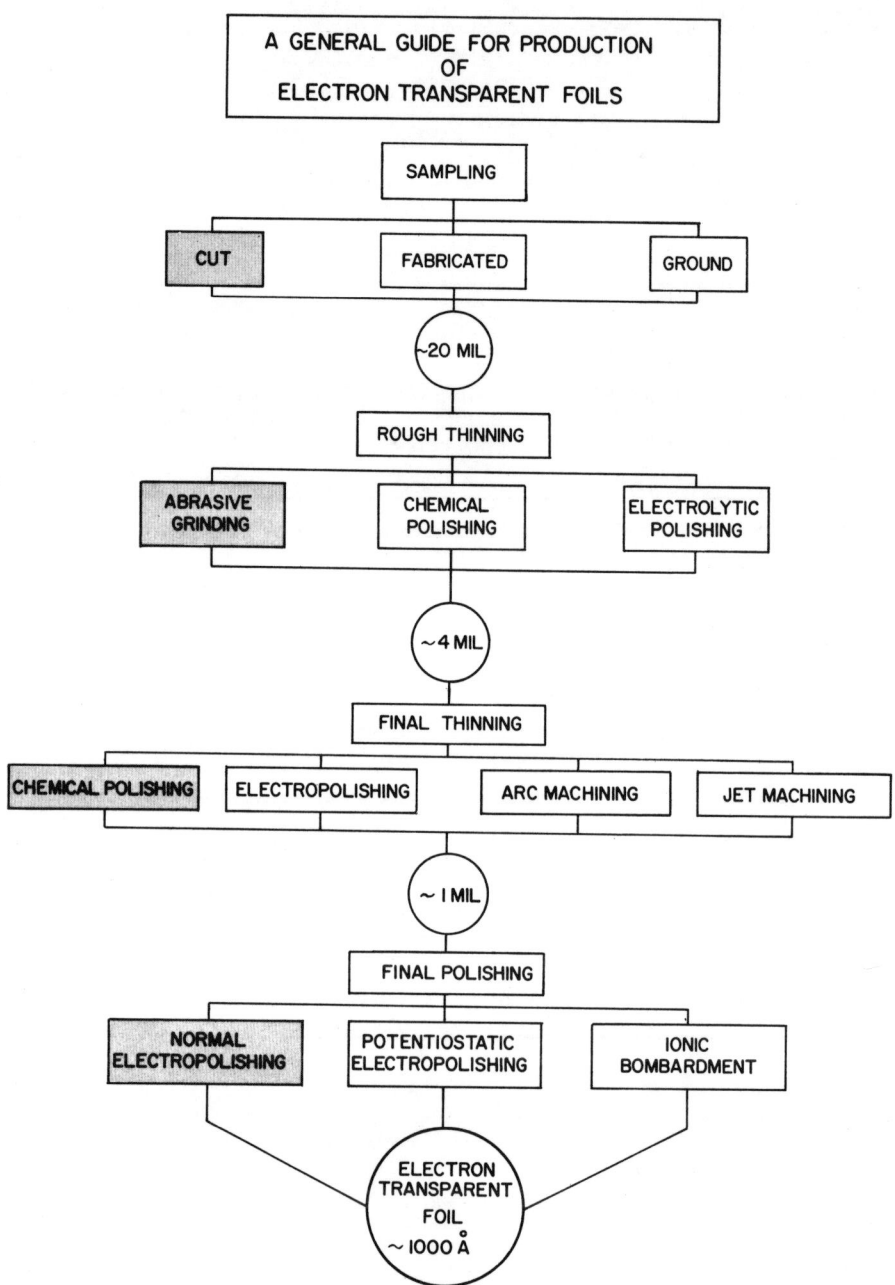

Figure 16. Schematic showing the general methods to produce an electron transparent foil (the shaded areas indicate the most popular method used).

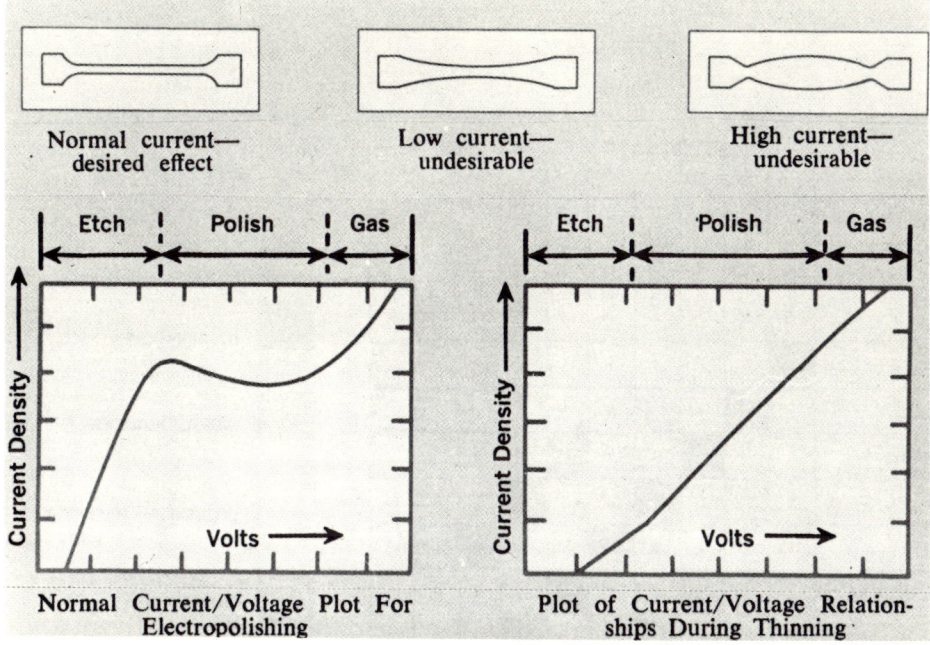

Figure 17. Traditional problem areas in the thinning of metal by jet machining (taken from Reference 25).

polishing appliance. It is quite clear that each class has a special advantage relative to the user. If a cross-plot is made for these different polishing appliances with respect to the prime investigators that are known, the results shown in Table II are thereby obtained. It is to be noted that with respect to just cathode type or shape, at least eight exciting variations are, as yet, presumed to be undeveloped devices. If variations other than just cathode type or shape are also considered, then many additional possibilities remain, as yet, undiscovered; note Table III. It is this challenge that I leave to you: "Much has been done yet more remains".

In the way of concluding remarks, look at Figure 19, a representation of an aerospace engine application which is so typical of the problems presented to a modern electron metallographer (32). The answer to this problem can only be obtained by a study of the substructure using electron transmission microscopy. I suggest that most of our material developments in the next decade will somehow involve substructure and, in this way, these improvements will each involve those who prepare specimens for transmission electron microscopy.

TABLE II. CROSS-PLOT OF DIFFERENT POLISHING APPLIANCES

CATHODE TYPE OR SHAPE	IMMOBILE	POINTED	VERTICAL	PARALLEL	SHEET	NON-VERTICAL	NON-PARALLEL	COMPLEX	PUMP	MOBILE
IMMOBILE	■	2	1	3	1	3	4	5	6	■
POINTED	2	■	7	▨	8	2	7	□	□	□
VERTICAL	1	7	■	9	1	■	4	5	6	□
PARALLEL	3	▨	9	■	3	3	■	10	6	14
SHEET	1	8	1	3	■	3	3	■	□	11
NON-VERTICAL	3	2	■	3	3	■	3	5	6	11
NON-PARALLEL	4	7	4	■	3	3	■	12	13	14
COMPLEX	5	□	5	10	■	5	12	■	□	□
PUMP	6	□	6	6	□	6	13	□	■	□
MOBILE	■	□	□	14	11	11	14	□	□	■

KEY
■ INVALID COMPARISON
▨ IMPRACTICAL DESIGN
(white filled) LITERATURE REFERENCE
□ UNDEVELOPED AREA

CROSS PLOT REFERENCE NO.	LITERATURE SOURCE	LITERATURE REFERENCE NO.
1	Heidenreich (1949)	1
2	Bollmann (1956)	3
3	Mirand and Saulnier (1958)	5
4	Nicholson, Thomas and Nutting (1958)	6
5	Yoshida (1960)	8
6	Glenn and Schoone (1964)	9
7	Strutt (1961)	27
8	Hays (1959)	28
9	Kelly and Nutting (1959)	29
10	Nagashima (1960)	30
11	Althof, Krings and Markworth (1965)	10
12	Seeger (1960)	7
13	Maurer and Schaffer (1967)	31
14	Markworth (1968)	2

TABLE III. AREAS OPEN FOR FUTURE RESEARCH TRENDS

● *Cathode Considerations*

Complex Pointed Cathodes
Pointed Pump Cathodes
Mobile Pointed Cathodes
Sheet Pump Cathodes
Complex Pump Cathodes
Mobile Pump Cathodes
Mobile Complex Cathodes

● *Anode Considerations*

Rotating Anode
Smaller Cathode/Anode Area Ratios
Variable Anode-Cathode Spacing
Better in situ observations
Improved Bubble Control

● *Electrolyte Considerations*

Instantaneous Current Density Monitoring
Increased Usage of Viscous Additives
Development of Improved Solutions
More Potentiostatic Applications

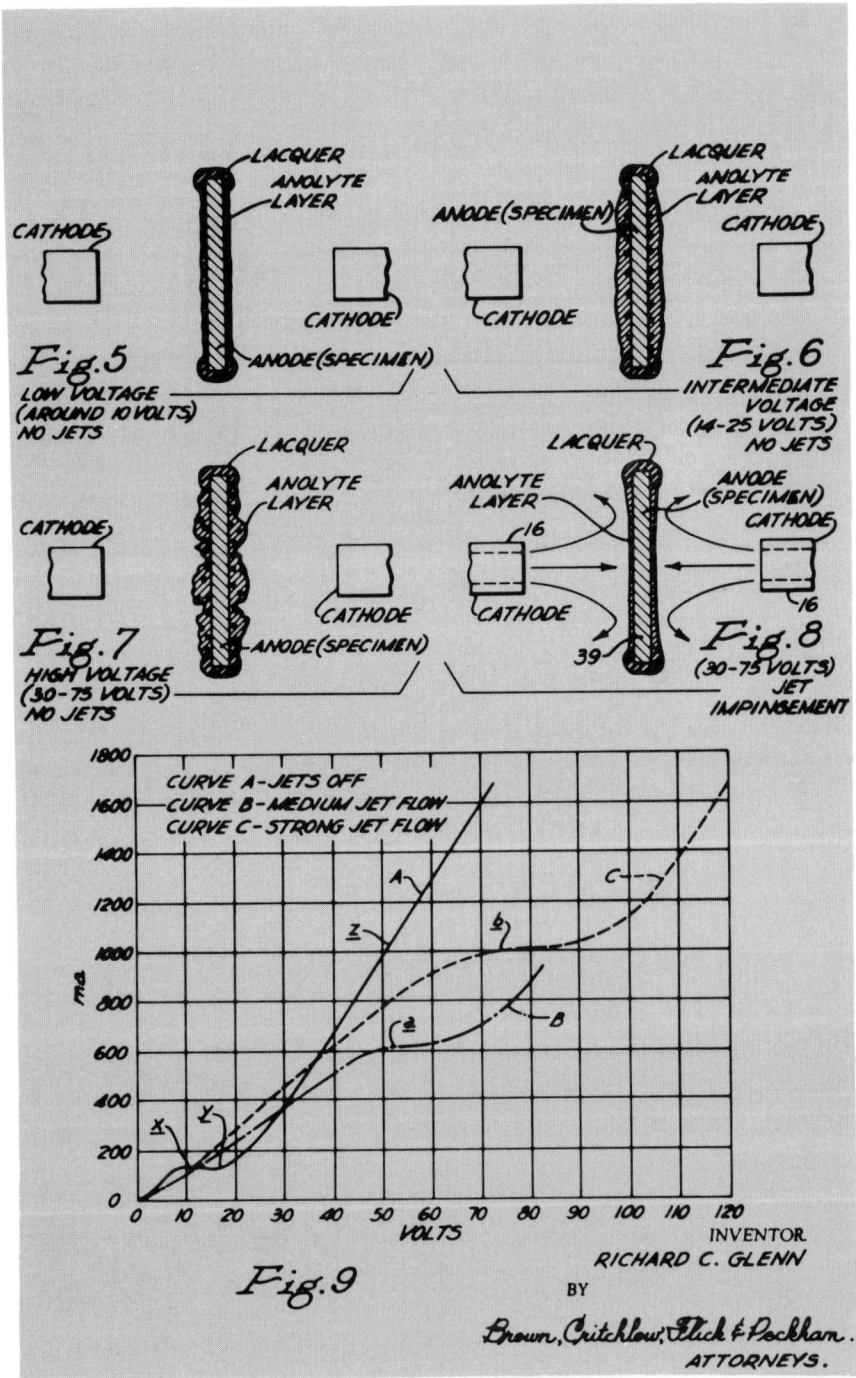

Figure 18. Traditional problem areas in the thinning of metal low voltage electropolishing (taken from Reference 26).

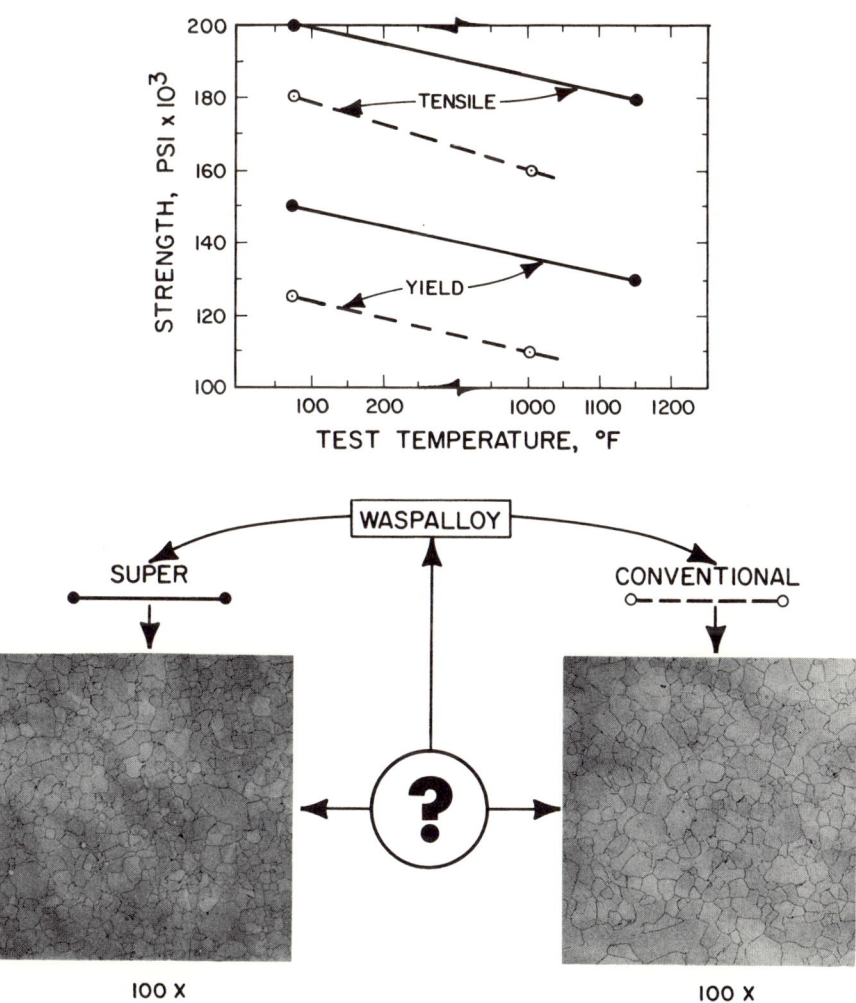

Figure 19. Examples of Waspalloy structure-property correlations which indicate that an improvement in strength and performance is caused by the substructure (taken from Reference 32).

REFERENCES

1. R. D. Heidenreich, J. Appl. Phys., $\underline{20}$, (1949) p. 993.
2. M. Markworth, Metallographie Praktische, $\underline{6}$,(1969), p. 323.
3. W. Bollmann, Phys. Rev., $\underline{103}$, (1956), p. 1588.
4. R. M. Fisher, as reported by P.M. Kelly and J. Nutting, J. Inst. Metals, $\underline{87}$, (1959) p. 385.
5. P. Mirand and A. Saulnier, 4th Intl. Conf. on Electron Microscopy-1958, Springer-Verlag, Berlin (1960) p. 390.
6. R. B. Nicholson, G. Thomas and J. Nutting, Brit. Jour. Appl. Phys., $\underline{9}$ (1958) p. 23.
7. A. Seeger, as reported in Techniques for Electron Microscopy, F. A. Davis Co., Philadelphia, 2nd Ed., (1965) p. 371.
8. S. Yoshida as reported by K. Shimizu and Z. Nishiyama, Jour. Elec. Microscopy, $\underline{12}$, (1963) p. 10.
9. R. C. Glenn and R. D. Schoone, Rev. Sci. Instr., $\underline{35}$ (1964) p. 1223.
10. F. C. Althof, H. Krings and M. Markworth, Metallographie Praktische, $\underline{2}$ (1965) p. 193.
11. W. J. McG. Tegart, The Electrolytic and Chemical Polishing of Metals in Research and Industry, Pergamon Press, London, 2nd Ed., (1959).
12. D. H. Kay, Techniques for Electron Microscopy, F. A. Davis Co., Philadelphia, 2nd Ed., (1956) p. 356.
13. R. C. Glenn, Private Communication, (1966).
14. C. Hays, R. C. Glenn, R. E. Swift and H. Conrad, Metallography, $\underline{4}$, (1), (1971) p. 51.
15. R. C. Glenn, G. Langford and A. S. Keh, United States Steel Report Memo 1114, (1968).
16. A. S. Keh and S. Weissmann, Electron Microscopy and Strength of Crystals, Interscience, New York, (1963) p. 231.
17. H. Hu, "Annealing of Silicon-Iron Single Crystals", in Recovery and Recrystallization of Metals, Interscience, New York, (1963) p. 311.
18. C. Hays, Unpublished Work, (1969).
19. W. C. Leslie, Acta Met, $\underline{9}$, (1961) p. 1004.
20. R. F. Decker, Steel Strengthening Mechanisms Symposium, Zurich, Switzerland, May 5, 1969.
21. J. M. Oblak and W. A. Owczarski, Trans. AIME, $\underline{242}$, (1968) p. 1563.
22. R. C. Glenn, ASTM Metallography Exhibit 19, as submitted to C. Hays, (1973).
23. R. C. Glenn and C. A. Johnson, United States Steel Research Report PR 227, (1964).
24. A. Szirmae and R. M. Fisher, ASTM Spec. Tech. Publi. 372, (1963) p. 3.
25. C. K. H. DuBose and J. O. Stiegler, Oak Ridge Nat'l Lab. Report, ORNL-4066, (1967).
26. R. C. Glenn, Private Communication, (1973).

27. P. R. Strutt, Rev. Sci. Instr., $\underline{32}$, (1961) p. 411.
28. C. Hays, Unpublished Work, (1959).
29. P. M. Kelly and J. Nutting, J. Inst. Metals, $\underline{87}$ (1959) p. 385.
30. S. Nagashima and Y. Matsuo, Kinzoku Butsuri, $\underline{6}$ (1960) p. 161.
31. K. L. Maurer and H. Schaffer, Metallographie Praktische, $\underline{4}$, (1967) p. 388.
32. Colin Stead, Private Communication, (1973).

PREPARATION OF SAMPLES FOR ELECTRON MICROSCOPY BY ION MICROMILLING

Ram Kossowsky and G. P. Sabol

Westinghouse Research Laboratories

Pittsburgh, Pennsylvania 15235

INTRODUCTION

Preparation of thin foils for transmission electron microscopy was initially limited to conductors that could be thinned by electropolishing. Chemical etching was used for non-conductors, but with only limited success. Uneven thinning, preferential etching of grain boundaries or second phases were a few of the problems encountered by chemical etching. This is not to say that complex, multiphase metallic alloys, could be easily thinned by electropolishing (1). These subjects are discussed in other chapters. A recently developed thinning technique applicable to both conductors and non-conductors is ion micromilling. This technique, which operates on the principle of ion bombardment, has added a wide variety of solids to the list of materials that can be studied by transmission electron microscopy.

Although the most glamorous use of the technique has been in the characterization of lunar rocks (2,3), use of the technique is now being extended to studies of systems of commercial interest. In this paper, we discuss briefly the ion milling instrument and the principle of its operation. We then proceed to discuss milling parameters and techniques for a few ceramic materials, i.e., hot pressed silicon carbide (SiC) and silicon nitride (Si_3N_4) and metal-oxide scales. Finally, a few micrographs illustrating the results of successfully prepared specimens are presented.

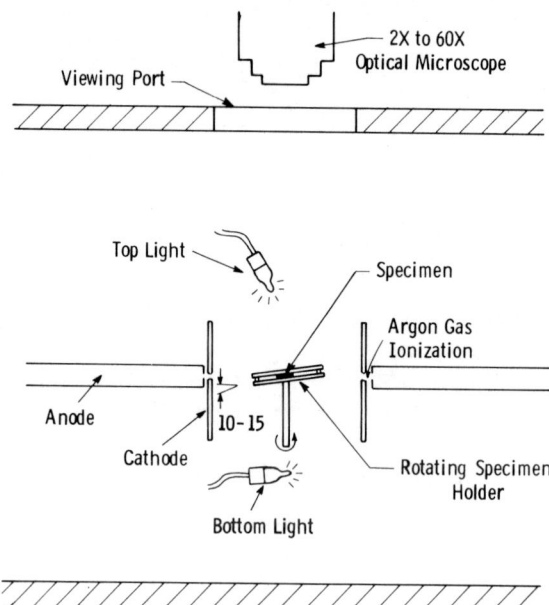

Figure 1. Schematic illustration of the milling chamber of the IMMI.

THE ION MICROMILLING INSTRUMENT (IMMI)

The ion milling instrument* consists of three main parts: (a) the work chamber with ion guns, specimen holder and a microscope, (b) power supply and control panel, and (c) vacuum system. A schematic drawing of the work chamber is shown in Figure 1. A 3 mm diameter, standard microscopy sample is mounted in a holder, Figure 2, which rotates, and can be tilted up to 30° from the horizontal position. Ion milling is accomplished by a sputtering effect. Dry argon gas is introduced into the previously evacuated chamber through the anode. The gas is ionized at the anode-cathode gap (about 2-3 mm) with a potential of 4-6KV. A jet of ionized gas, about 1 mm diameter, is impinged on the specimen with typical ion currents of 50-100 μA. To provide uniform milling, the sample is rotated at about 60 rpm. One can operate both guns, for a fast thinning toward the center of the specimen, or one gun only if one of the surfaces of the specimen should be preserved. Perforation is determined by viewing the specimen through the microscope with the bottom light turned on.

---

* Commonwealth Scientific Corporation, Alexandria, Virginia.

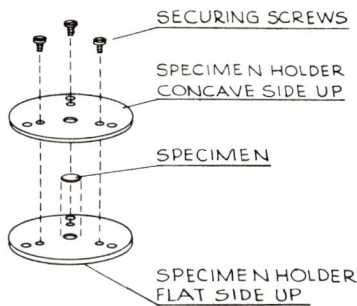

Figure 2. Diagram of the specimen holder of the IMMI.

SPECIMEN PREPARATION

Hot Pressed SiC and Si_3N_4

A piece of material about 1 mm thick is sliced off the bulk to be examined with a thin diamond cutting wheel. Figure 3 illustrates the common sources of specimens, i.e., as-received billets (usually 12 x 12 x 2.5 cm), tensile or creep specimens, usually with a gauge diameter of 6 mm, and flexural-bend specimens, 3 x 6 x 30 mm. In the case of the flexural specimens, the outer surfaces are of interest, since the maximum tensile or compressive stresses occur there. Therefore, the tested specimen was sliced in half on the thickness, and ground from the newly cut surfaces out. In all cases, the cut slices were glued with wax to a glass slide and ground carefully, using a precision wafering machine and a special 300-grit diamond wheel.* With careful grinding, one can achieve a slice 60-70 μm thick. After grinding, 3 mm diameter discs were cut from the wafers by use of ultrasonic cavitation tools, placed in the holder, Figure 2, and thinned to perforation.

The effects of ion current and angle of incidence on the rate of thinning and the shape of the cavity are shown in Figures 4 through 9. Cavity depths as a function of milling time, for three ion currents and two angles of incidence are plotted in Figures 4

---

* Supplied by the Norton Company.

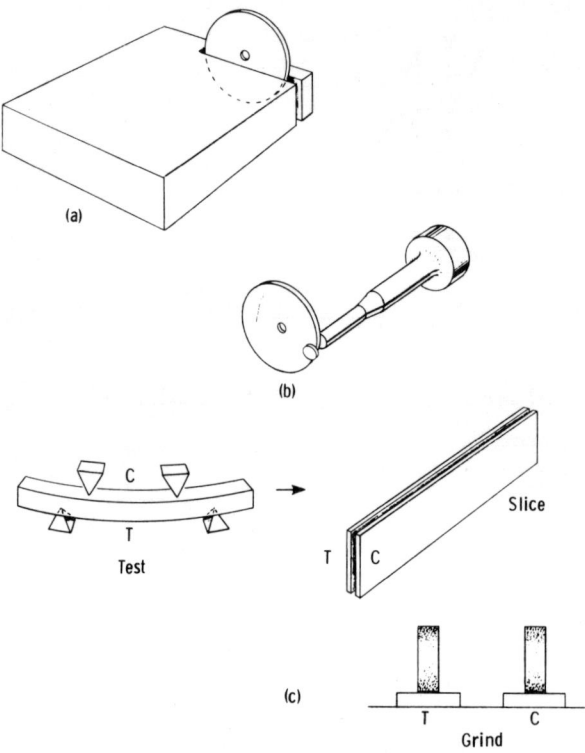

Figure 3. Common sources of specimens from hot pressed Si_3N_4 and SiC (a) as received billet (b) creep or tensile specimen (c) flexural-bend specimen.

and 5, for SiC and Si_3N_4, respectively. The data are summarized in Figure 6 from which it is derived that

$$R \simeq AI_i^3$$

where R is milling rate, I_i the ion current, and A is a constant depending upon the material and angle of incidence.

The effect of angle of incidence upon the shape of the cavity is demonstrated in Figures 7-9. As might be expected, a shallow angle of incidence produces a wider cavity with gradually rising walls and a narrow flat bottom. At a steeper angle of incidence, the cavity looks more like a dish, with steep walls and a wide flat bottom. The latter shape of a cavity provides a wider field of observation from the edge of a perforation, as compared to the cavity produced with a shallow angle. However, if milling is not stopped as soon as a perforation occurs, there is the risk of losing the

PREPARATION OF SAMPLES BY ION MICROMILLING

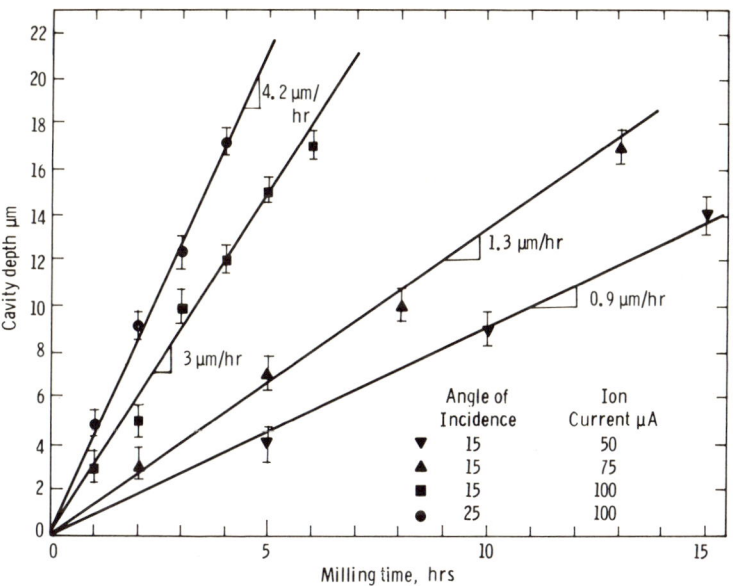

Figure 4. Cavity depth as a function of milling time, hot-pressed SiC.

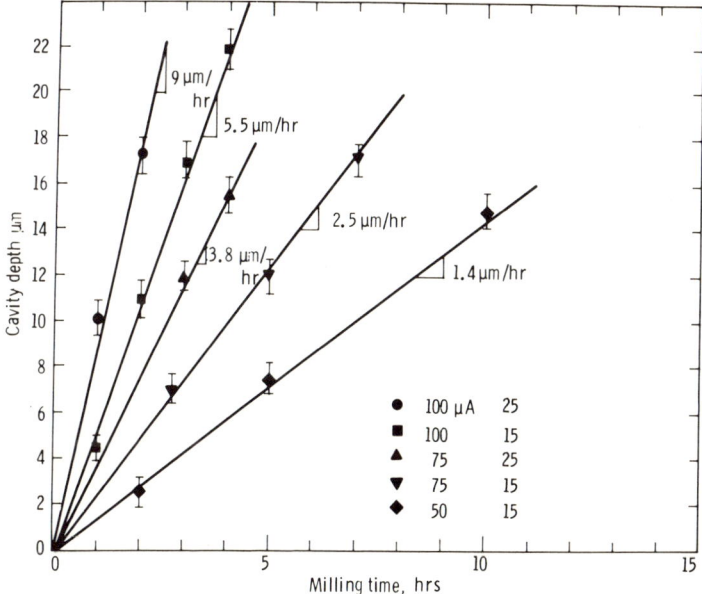

Figure 5. Cavity depth as a function of milling time, hot-pressed Si_3N_4.

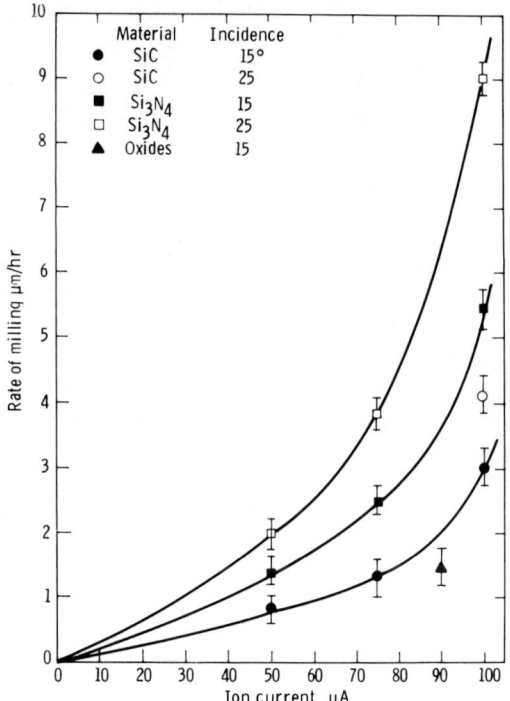

Figure 6. Rate of milling as a function of ion current.

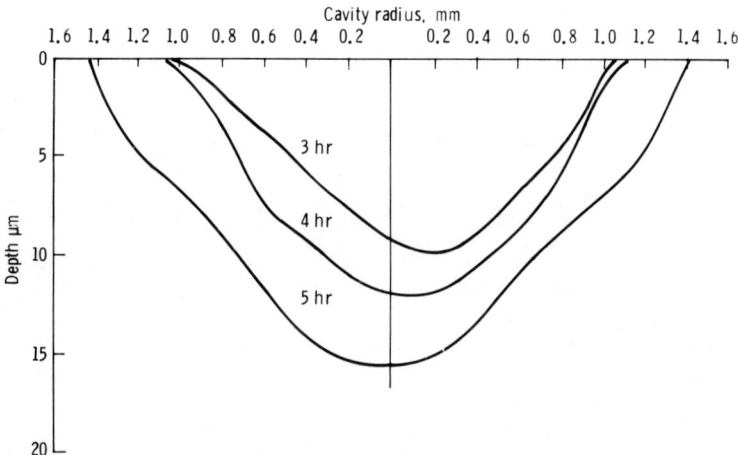

Figure 7. Cavity profile as a function of milling time. Hot-pressed SiC (NSC-472-7). 100 µA ion current, 15° tilt.

PREPARATION OF SAMPLES BY ION MICROMILLING

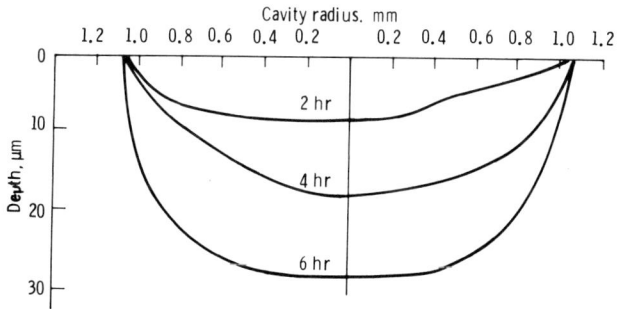

Figure 8. Cavity profile as a function of milling time. Hot-pressed SiC (NSC-472). 100 µA ion current 25° tilt.

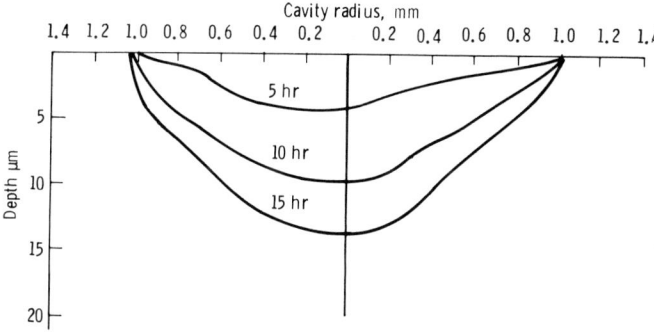

Figure 9. Cavity profile as a function of milling time. Hot-pressed SiC. 50 µA ion current, 15° tilt.

entire specimen, since the perforation will grow very fast to cover the entire area of the flat bottom. With a shallow angle of incidence, the field of view may be limited to an area a few microns away from the edge of the perforation, but there is less risk of losing the specimen if milling continues for some time after perforation has occurred.

Scanning electron micrographs of a cavity in a SiC specimen milled for 6 hours at 100 µA, and 15° incidence, are shown in Figure 10. The edge of the cavity shows concentric markings, while radial milling markings are seen along the walls of the cavity. The white particles are WC inclusions present in hot pressed SiC[4]. The flat bottom of the cavity is shown in Figure 10d. From the angle of tilt of the specimen, one can estimate the larger pertubation, P, to be about 6 µm diameter and about 1.5 µm deep.

Oxide Films

The techniques used for the preparation of TEM specimens of oxide films have been described previously, [5,6,7] but the procedures will be reiterated here for the sake of completeness.

Figure 10. Scanning electron micrographs of cavity hot-pressed SiC. 100 µA ion current, 15 deg. incidence. A: general view (foil tilted 25°), B,C: cavity rim, D: cavity bottom. White particles are WC inclusion, present in material.

It was found that the preferred method of isolating a metal-oxide film depends upon the film thickness. In general, for films of thickness less than 3-4 μm it was necessary to provide mechanical support to prevent damage. The sequence of steps leading to satisfactory separation of the oxide from the metal is shown schematically in Figure 11. A corroded specimen of any thickness is ground from the side opposite the desired oxide film until the thickness of the composite metal-oxide is approximately 0.13 mm. For corrosion specimens greater than about 3 mm thick, the coupons are surface ground to about 0.7 mm, then ground by hand on metallographic paper to 0.13 mm. Throughout this grinding, care must be taken to avoid oxide damage due to abrasion and also overheating of the specimen. Following grinding, discs of 3 mm diameter are punched from the metal-oxide composite; and, the entire disc, except for a central 1 mm diameter region of the metal side, is painted with an acid resistant stop-off lacquer. The exposed metal at the central portion of the disc is then removed either chemically or electrolytically, depending upon the alloy.

For zirconium and its alloys, the metal was easily removed by chemical attack in $HF-HNO_3$ solutions. For metal-oxide composites of titanium, 316 stainless steel and Incoloy 800, the metal was removed using standard electrolytic solutions and voltages normally used for electropolishing. Metal removal was stopped as soon as a light beam could be detected passing through the translucent oxide. This assured a relatively small area of exposed oxide with maximum support at the circumference. After the electropolish, the disc was washed in water, the stop-off lacquer dissolved in acetone, and the specimen finally rinsed in methanol.

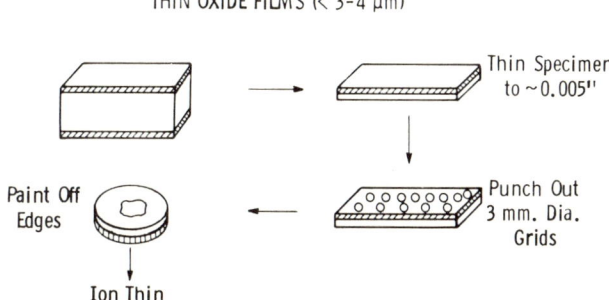

Figure 11. Schematic illustration of the technique used to prepare specimens of oxide of less than 3-4μm thickness for ion milling and subsequent examination by TEM.

For oxide films of thicknesses greater than about 3-4 μm a technique shown schematically in Figure 12 is used. A grid is scribed through the oxide film to the base metal, the grid spacing being approximately 3 mm. The oxide film is then detached from the metal by use of chemical dissolution of the metal for zirconium, or electrolytic dissolution for the other three alloys mentioned above. After the separation of the oxide from the metal, the segments of the oxide are washed several times in water and rinsed in methanol. Individual pieces of oxide are then sandwiched in folding copper grids of 3 mm diameter.

Specimens of both the disc-type and those sandwiched in the grids were thinned by ion milling at an ion current of 90-100 μA and 15° angle of incidence. At these conditions, the milling rate was approximately 1.5 μm/hour, independent of oxide composition or thickness. Similar to conventional techniques, milling was stopped shortly after noticeable perforation. The end point after milling of a sandwich-type specimen is shown in Figure 13. Perforation has occurred at several areas within the grid spacing.

Finally, it is significant to note that all non-conductive specimens, such as Si_3N_4 and the metal oxides, had to be coated with a thin layer of carbon prior to examination in the electron microscope. If an uncoated, non-conductive specimen is used, charge build-up is sufficient to divert the beam and interfere with normal microscopy work.

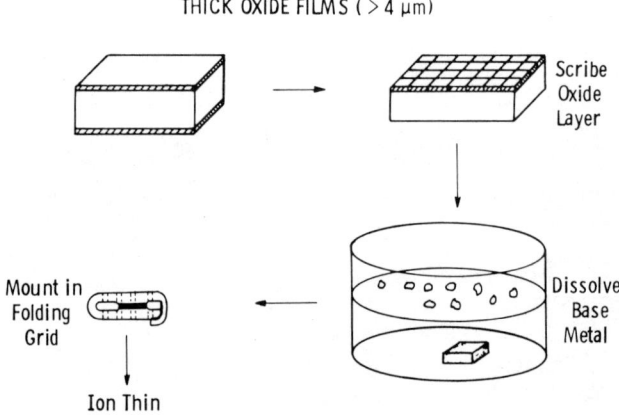

Figure 12. Schematic illustration of the method of preparation of samples having an oxide film thicker than 3-4μm.

Figure 13. A section of an oxide film sandwiched in a folding grid. Ion milling results in perforations within the grid spacings.

TRANSMISSION MICROGRAPHS

Typical micrographs of hot-pressed Si_3N_4 [8] and hot-pressed SiC (4) are shown, respectively, in Figures 14 and 15. The samples are suitable for standard transmission microscopy experiments. In most cases 100 KV was sufficient for good resolution. Figure 14a shows a large Si_3N_4 grain with dislocations and very fine inclusions. Figure 14b is a dark-field micrograph identifying second phase particles. SiC shows a more equiaxed structure, and larger average grain sizes, Figure 15. The typical defects in this material are second phase particles (not seen in the micrograph), dislocations, and a high density of stacking faults.

Electron micrographs of oxide films formed on Zircaloy-4 are shown in Figures 16 and 17. The flexibility of differential thinning is demonstrated in Figure 17; the initial oxide of a 3.5 μm thick film observed by milling from the opposite side with one gun

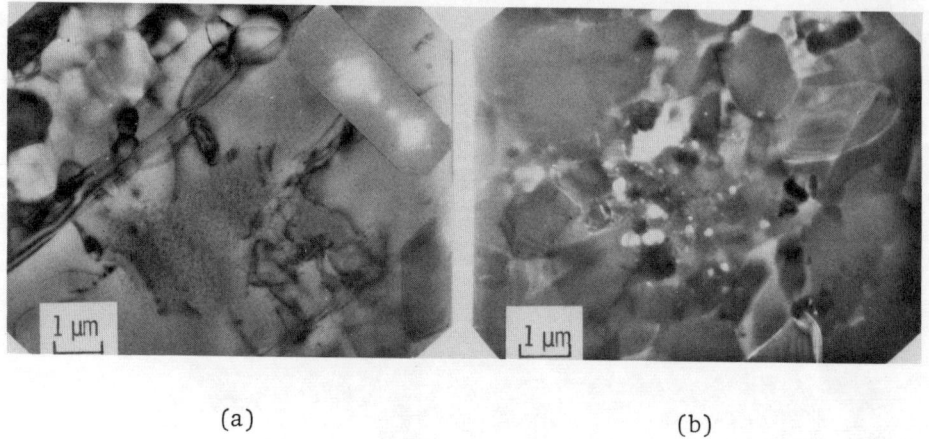

(a)   (b)

Figure 14. Typical transmission micrographs, as-received hot-pressed Si_3N_4. Milled at 100 µA, 15° incidence, carbon coated. 800 kV.
a. Typical microstructural features. b. Dark-field image showing inclusions.

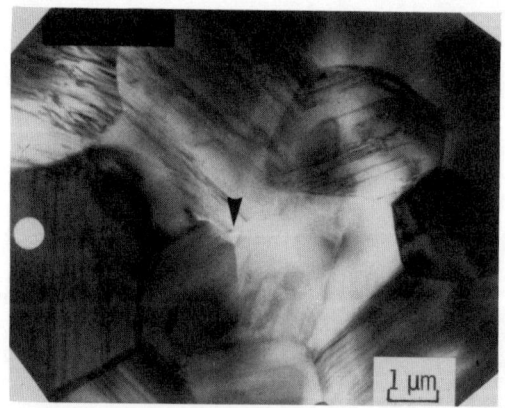

Figure 15. Transmission micrograph, hot-pressed SiC. Milled at 100 µA and 15° incidence. Equiaxed structure, showing dislocations and a high density of stacking faults. Arrow marks a void. 100 kV.

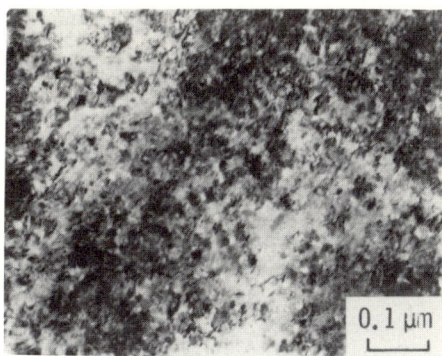

Figure 16. Electron micrograph of a 1.1 μm thick oxide film formed on Zircaloy-4.

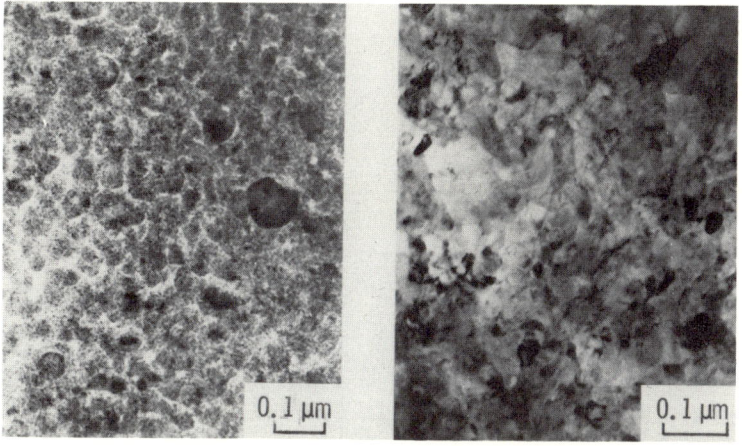

a) Initially formed oxide observed by milling with only one gun from the metal-oxide interface.

b) The mid-thickness location prepared by use of both guns.

Figure 17. Structure of 3.5 μm thick oxide formed on Zircaloy-4. Structure delineated at a) the oxide-environment surface, and (b) the mid-thickness of the oxide.

only consists of very fine grains of approximately 50 Å diameter, Figure 17a, whereas, at the mid-thickness (use of both guns) the grain diameter was substantially larger, approximately 300 Å, Figure 17b. An extremely fine-grained film is formed on titanium during oxidation in air at 700°C, Figure 18. On the other hand, the corrosion film formed on Incoloy 800 during autoclave exposure in 10% sodium hydroxide has a platelet morphology, Figure 19a, with many of the platelets possessing pores, Figure 19b. Other examples of the structures developed in oxide films have been reported elsewhere (5-7).

Extensive use of the ion-milling procedure has recently been applied in a study (8) of the mechanism of oxide film growth on zirconium-based alloys.

Some ambiguities in interpretation of electron micrographs may arise when voids are present. Some time, a hole created by extensive local thinning may be mistaken for a void. Generally, a thinning hole will exhibit a gradual change in contrast, Figure 20, small arrows, while a pre-existing hole or crack exhibits a sharp change from the white to dark contrast (large arrow, Figure 15). When natural cavities and cracks are sought it is advisable to examine the foil as far away from the edge of the foil as possible. In this case, high voltage microscopy presents a definite advantage.

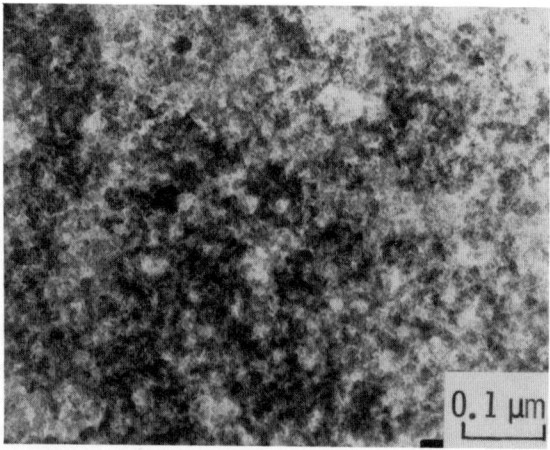

Figure 18. Structure of the oxide formed on titanium during oxidation in air for 15 min at 700°C.

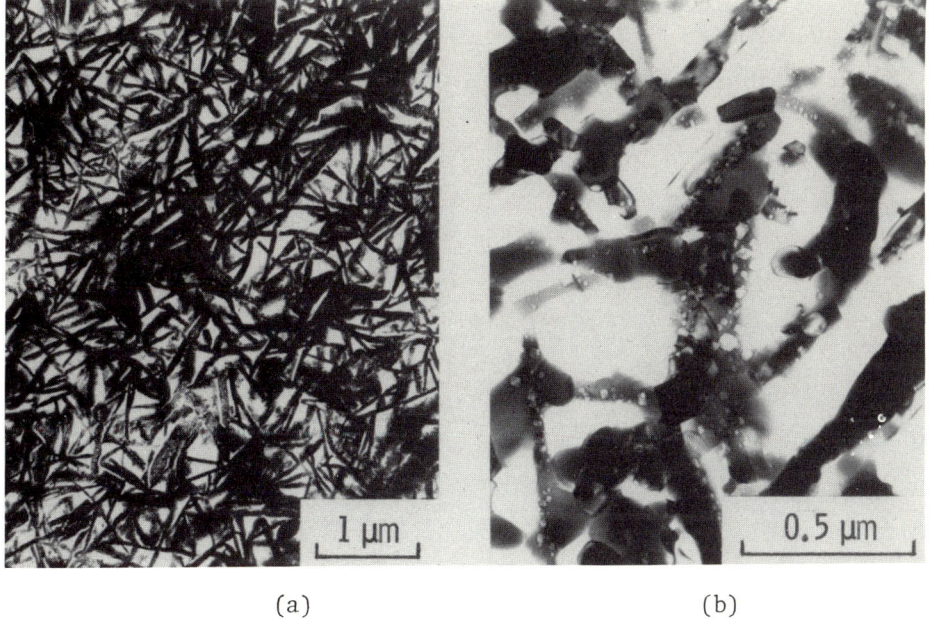

(a)   (b)

Figure 19. Micrographs of a 1.0 μm corrosion film formed on Incoloy 800 during autoclave exposure in 10% NaOH.

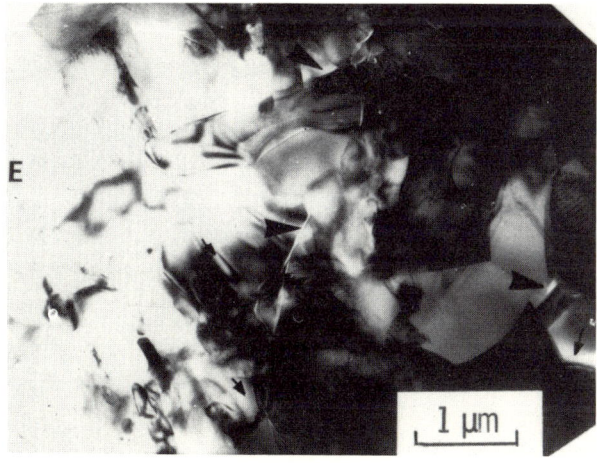

Figure 20. Transmission micrograph of hot-pressed Si_3N_4 creep specimen. Large arrows indicate real cracks and triple-point cavities. Small arrows indicate possible thinning effects. Edge of foil at (E). 100 kV.

REFERENCES

1. D. J. Barber, "Thin Foils of Non-Metals Made for Electron Microscopy by Sputter Etching", J. Materials Sci., $\underline{5}$, (1970) p. 1.
2. J. M. Cristie, D. T. Griggs, R. M. Fisher, J. S. Lally, A. H. Heuer, and S. V. Radcliffe, "Deformation of Lunar and Terrestrial Minerals", Electron Microscopy and Structure of Materials, G. Thomas, Ed., Univ. of Calif. Press, Berkeley, (1972) p. 1234.
3. P. E. Champness and G. W. Lorimer, "Electron Microscope Studies of Some Lunar and Terrestrial Pyroxene", Electron Microscopy and Structure of Materials, G. Thomas, Ed., Univ. of Calif. Press, Berkeley, (1972) p. 1245.
4. Ram Kossowsky, D. G. Miller and E. S. Diaz, "Microstructure and Strength of Hot Pressed SiC", paper presented at the Fall meeting of the Basic Science Div., Am. Ceram. Soc., Pittsburgh, September, 1973, to be published.
5. G. R. Airey and G. P. Sabol, "Transmission Electron Microscopy of Ion-Thinned Oxides Formed During Corrosion of Zircaloy-4", J. Nucl. Mater., $\underline{45}$, (1972/1973) p. 60.
6. C. W. Hughes, G. P. Sabol and G. P. Airey, "A Technique for the Examination of Thick Oxide Films by Use of Transmission Electron Microscopy", Microstructural Science, R. Gray and J. L. McCall, eds., Elsevier Publishing Co., New York, 1973.
7. G. P. Airey and G. P. Sabol, "A Technique for Thinning Corrosion Films Formed on Zirconium Base Alloys for Study by Transmission Electron Microscopy", Practical Metallography, $\underline{10}$, (1973) p. 282.
8. Ram Kossowsky, "Microstructure of Hot Pressed Si_3N_4", J. Mat. Sci., in press.
9. G. P. Sabol, S. G. McDonald, and G. P. Airey, "Microstructure of the Oxide Films Formed on Zr-Based Alloys", paper presented at the ASTM-AIME Symposium on Zirconium in Nuclear Applications, Portland, Oregon, August, 1973, to be published.

AUTHOR INDEX

A

Aeschliman, R., 143, 148, 152
Ahmad, I., 251, 273
Ahmed, W., 53, 55
Airey, G. R., 340, 346, 348
Albrecht, E. D., 95-107
Allen, M.D., 201, 206
Althof, F. C., 309, 330
Ambler, J.F.R., 207, 212, 231
Ang, C.Y., 147, 153
Armstrong, D., 207, 214, 215, 218, 231
Ashdown, F. A., 251, 252, 272
Avery, H.S., 156, 176

B

Bainbridge, J.E., 207, 224, 228, 231
Baker, C., 289, 295
Ball, R., 207, 220, 231
Bannister, G.H., 289, 295
Barber, D.J., 333, 348
Barranco, J.M., 251, 273
Barrett, J., 69-76
Bartell, H.F., 32, 36
Bates, L.G., 251, 252, 272
Beaman, D.R., 285, 295
Beardmore, P., 76
Beauchamp, R.H., 233-249
Belt, R.F., 143, 148, 152
Bender, J.H., 3, 34
Bertino, J.P., 15, 17, 34
Bertone, T.J., 62, 68, 251-273
Biechele, H.I., 41, 46, 54
Bierlien, T.K., 207, 228, 232
Bingman, F.O., 207, 212, 231
Bitter, F., 155, 176
Bollman, W., 309, 330
Bomar, E.S., 186, 205
Boumans, P.W.J.M., 207, 228, 232
Brammar, I.S., 76
Brandenburg, C.F., 3, 34
Braun, J.D., 201, 206
Bridges, W.H., 193, 206
Briers, G.W., 76
Broderick, S.J., 19, 35
Browning, C.E., 267, 273
Buchheit, R.D., 77-94
Bucklow, I.A., 76
Buoma, A.H., 123, 127
Butler, E.M., 286, 295

C

Cain, F.M., Jr., 207-232
Calabra, A.E., 1-41
Camahort, J.L., 251, 272
Campbell, J.J., 182, 205
Carter, G., 207, 232
Cathcart, J.V., 181, 182, 205
Cawthorne, C., 293, 296
Chalmers, B., 251, 272
Champness, P.E., 333, 348
Chang, J.V., 15, 18, 19, 35
Charvat, F.R., 95-107
Christie, J.M., 333, 348
Church, J.S., 3, 4, 34, 121, 127
Clausing, R.E., 186, 193, 205, 206
Clayton, C.V., 32, 36
Clemens, R.S., 45, 54
Cobine, J.D., 207, 232

Cochran, F.L., 15, 17, 18, 22, 35, 281, 282
Coffinberry, A.S., 275, 280, 282
Cole, M., 76
Colligan, J.S., 207, 232
Conn, G.K.T., 193, 206
Conrad, H., 311, 313, 330
Cook, E., 156, 176
Coons, W.C., 109-120
Costas, L.P., 22, 36
Cramer, E.M., 275-282
Crittendon, D.B., 66, 68
Crouse, R.S., 15, 16, 17, 18, 35, 179-206
Cummings, W.V., 285, 286, 288, 295

D

Dahlin, R., 4, 18, 22, 28, 34
Danksys, R.J., 267, 273
Dawe, C., 76
Decker, R.F., 313, 320, 321 330
Dewey, M.A.P., 76
Diaz, E.S., 340, 343, 348
Dobbins, A.G., 29, 31, 36
Donahue, D.A., 286, 295
Dragel, G.M., 291, 295
Drain, L.E., 76
DuBose, C.K.H., 186, 205, 316, 325, 330
Duffner, R.L., 19, 35
Duran, J.B., 207, 212, 223, 231

E

Ehman, M.F., 143-153
Ellis, O.W., 2, 19, 33
Elmore, W.C., 155, 156, 176
Engel, A. von, 207, 232
Everhart, J.L., 2, 3, 33, 34

F

Fallier, C.N. Jr., 56, 68
Faust, J.W. Jr., 148, 153
Feitknecht, W., 207, 232
Fisher, R.E., 207, 212, 224, 231
Fisher, R.M., 309, 316, 330, 333, 348
Fisher, T.F., 207, 229, 232
Fisinai, G.F., 156, 177
Fradetter, L.B., 3, 19, 22, 28, 33
Frazer, J.W., 207, 230, 232
French, P.M., 15, 18, 19, 35
Fulton, E.J., 293, 296
Furth, J.V., 15, 34

G

Gassmann, F., 143, 148, 152
Gehlbach, R.E., 201, 206
Gendron, N.J., 121-128
Gibbon, G.L., 15, 34
Glenn, R.C., 309, 311, 312, 313, 314, 319, 328, 330
Goddard, D.M., 267, 273
Godden, C.A., 22, 27, 35
Gray, R.J., 15, 17, 18, 34, 35, 146, 153, 155-177, 179-206 285, 295
Greco, J. P., 251, 273
Greeson, R., 281, 282
Gregory, T.G., 109, 120
Griggs, D.T., 333, 348
Grigson, C.W.B., 76
Gruber, W.J., 15, 18, 22, 34, 35, 202, 206
Gwathmey, A.T., 181, 205

H

Haddrell, V.J., 289, 295
Hamos, L. von, 155, 176
Hancock, J.R., 251, 272

AUTHOR INDEX

Hanke, J., 143, 152
Harvey, E.A.M., 156, 176
Haynes, W.D., 22, 35
Hays, C., 307-331
Hays, D.D., 280, 281, 282
Heidenreich, R.D., 307, 308, 309, 330
Heuer, A.H., 333, 348
Hewette, P.M. II, 15, 16, 35
Hill, J.H., 207, 230, 232
Hines, G.F., 289, 295
Hochschild, V.J., 19, 35
Hodgkin, N.M., 297-306
Hoffman, C.G., 29, 31, 36
Holland, I., 207, 228, 232
Holland, L., 207, 232
Holm, R.A., 289, 295
Homerberg, V.O., 5, 34, 156, 176
Hone, A., 192, 205
Horst, C.G. zur, 19, 35
Hoxie, E.C., 22, 36
Hu, H., 311, 317, 330
Hughes, C.W., 340, 346, 348
Hughes, J.P., 2, 15, 33
Hull, D., 76
Humphries, D.W., 2, 33
Hunt, M.D., 57, 68
Hurley, R.E., 207, 228, 232

I

Isasi, J.A., 285, 295
Iurosova, V.E., 207, 232

J

Jackson, R., 281, 282
Jacoby, W.R., 15, 18, 19, 35
Johns, W., 281, 282
Johnson, C.A., 314, 330
Johnson, G.W., 143-153, 180, 205
Johnson, K., 281, 282

K

Kaminsky, M., 207, 232
Kanaba, S.L., 77, 93
Katz, O.M., 291, 296
Kay, D.H., 309, 330
Kegley, T.M. Jr., 195, 205
Keh, A.S., 311, 314, 315, 316, 330
Kehl, G.L., 2,3,4,5,7,10, 13, 22, 33, 34, 121, 127, 180, 199, 204, 251, 272
Kelly, P.M., 309, 330, 331
Kirkpatrick, H.B., 22, 35
Klevtsov, P.V., 143, 152
Koppa, W., 32, 36
Koprowski, B.J., 286, 295
Kossowsky, R., 333-348
Krause, D.E., 3, 34
Krings, H., 309, 330
Krugler, O.L., 2, 4, 15, 33, 34
Kushner, J.B., 22, 25, 35

L

Ladroga, W.J. Jr., 7, 29, 31, 34
Lally, J.S., 333, 348
Langford, G., 311, 314, 330
Lannin, T.E., 291, 296
Laurenson, L., 207, 228, 232
Lauritzen, T.A., 285, 295
Leder, B., 77, 93
Leslie, B.C., 179-206
Leslie, W.C., 313, 330
Llewelyn-Jones, F., 207, 232
Lloyd, H.K., 76
Loeb, L.B., 207, 232
Long, E.L. Jr., 15, 18, 35, 146, 153, 193, 206, 285, 295
Lorimer, G.W., 333, 348
Luke, C., 76

M

Madsen, P.E., 207, 213, 215, 217, 231

Manley, A.J., 291, 296
Markworth, M., 308, 309, 330
Marlowe, M.O., 291, 293, 294, 296
Marshall, J.A. II, 267, 273
Martin, W.G., 207, 232
Mastel, B., 207, 228, 232
Matsuo, Y., 331
Maurer, K.L., 331
McCall, J.L., 77-94, 199, 200
McClung, R.W., 186, 205
McCutcheon, D.M., 207, 232
McDonald, S.G., 348
McKeehan, L.W., 155, 176
McLaughlin, H.B., 57-70, 255, 273
Meador, J.T., 146, 153
Medellin, D., 143-153
Metcalfe, A.J., 251, 272
Michels, L.C., 291, 295
Miley, D.V., 1-41
Miller, D.C., 143, 148, 152
Miller, D.G., 340, 343, 348
Miner, W.N., 280, 282
Mirand, P., 309, 330
Morris, C.J., 207, 230, 232
Mullaly, J.R., 207, 228, 232
Murr, L.E., 303, 305
Myers, J.R., 251, 252, 272

N

Nagashima, S., 331
Naish, J.N., 193, 206
Natesh, R., 286, 295
Nelson, J.A., 41-54
Nelson, R.C., 285, 286, 288, 295
Newkirk, J.B., 207, 232
Nicholson, R.B., 309, 330
Norris, E.S., 19, 35
Nutting, J., 309, 330, 331

O

Oblak, J.M., 314, 321, 322, 330

Oesterle, J.F., 3, 34
Oh, W., 289, 290, 295
Ondracek, G., 77, 93
Owczarski, W.A., 314, 321, 322, 330
Owens, C.M., 22, 36

P

Padden, T.R., 207, 216, 217, 218, 221, 222, 224, 227, 230, 231
Pahl, W., 207, 232
Palmberg, P.W., 186, 205
Pawel, R.E., 182, 205
Pearson, E.C., 192, 205
Pepper, R.T., 267, 273
Pepperhoff, W., 197, 206
Perryman, E.C.W., 192, 206
Peterson, G.F., 181, 205
Peterson, R.F., 285, 295
Picklesimer, M.L., 189, 205
Pierce, R.W., 303, 305
Pincus, I., 123, 127
Politis, C., 77, 93
Poulson, G.G., 303, 305
Pravdina, E.K., 207, 232
Prilizhaeva, I.N., 207, 232
Propst, F.M., 76
Puttick, K.E., 76

Q

Quarrell, A.G., 251, 272
Quon, H.H.D., 143, 148, 152

R

Radcliffe, S.V., 333, 348
Ragatz, R.A., 2, 3, 5, 22, 33, 34
Rahn, D.J., 22, 35
Ray, J.D., 267, 273
Reed, R.P., 251, 272
Reid, R.E., 77, 93

AUTHOR INDEX

Rexer, J., 207, 208, 224, 231
Reynolds, G., 289, 295
Richardson, J.H., 5, 7, 10, 13, 16, 17, 19, 22, 29, 31, 34, 267, 273
Richt, A.E., 15, 18, 35, 285, 295
Ritchie, E.E., 15, 18, 35
Robinson, R.K., 251, 273
Rosenbaum, H.S., 283-296
Roth, K.H., 129-142

S

Sabol, G.P., 333-348
Samuels, L.E., 4, 5, 7, 9, 10, 13, 15, 16, 17, 19, 22, 23, 29, 31, 34, 76, 131, 142
Saulnier, A., 309, 330
Schaeffler, A.L., 4, 22, 28, 34
Schaffer, H., 331
Schick, L.K., 143, 148, 152
Schleicher, H.M., 2, 3, 33, 34
Schmitz, F.J., 2, 33
Schneider, E.J., 79, 81, 94
Schonfeld, F. W., 276, 282
Schoone, R.D., 309, 330
Schuster, D.M., 251, 272
Schuyler, D.R., 109, 120
Seeger, A., 309, 330
Shalek, P.D., 15, 18, 19, 35
Shoemaker, H.E., 15, 17, 18, 35
Smith, C.S., 2, 33, 179, 204, 207, 232
Smith, F.M., 15, 18, 22, 35
Smith, G.V., 76
Smith, R.W., 57, 68
Smyth, H.D., 275, 282
Snide, J.A., 251, 252, 254, 272
Sognnaes, R.F., 304, 305
Sparks, C.J. Jr., 181, 205
Spiler, K., 77, 93
Spittle, J.A., 57, 68
Spivak, G.V., 207, 232
Sproule, G., 19, 35

Staub, R.E. Jr., 199, 206
Stead, C., 325, 329, 331
Stein, D.F., 186, 205
Stiegler, J.O., 197, 206, 316, 325, 330
Strong, J., 207, 212, 231
Strutt, P.R., 331
Swift, R.E., 311, 313, 330
Sykes, E.C., 207, 215, 216, 218, 231
Szaplonczay, A.M., 143, 148, 152
Szirmae, A., 316, 330

T

Taplin, D.M.R., 129-142
Tegart, W.J.M., 182, 205, 309, 330
Thiessen, P.A., 155, 176
Thomas, G., 309, 330
Thornton, H.R., 251, 273
Thurber, W.C., 186, 205
Tibbets, G., 76
Tighe, N.J., 22, 35
Titus, A.C., 123, 127
Tolansky, S., 81, 84, 96
Townes, C.H., 207, 232

U

Uzanas, R.A., 80, 96

V

Vogel, M., 207, 208, 224, 231
Vonnegat, B., 180, 204

W

Waldon, M.B., 275, 282
Wallace, W.P., 22, 35
Ward, J.W., 207, 224, 228, 231, 281, 282

Warren, P.C., 95-107
Warren, R.H., 76
Watts, E.C., 15, 18, 22, 35
Weber, C.E., 207, 230, 232
Weber, R.E., 186, 205
Wehner, G., 207, 232
Weinrich, P.F., 156, 176
Weissman, S., 311, 315, 316, 330
Weston, G.F., 207, 232
Westrich, R.M., 41-54
Wever, F., 2, 33
Williams, R.S., 5, 34
Williford, J.F., 233-249
Wolf, R.C., 291, 295
Wolff, U.E., 3, 19, 22, 28, 33, 289, 291, 292, 294, 295, 296
Woodbury, J.L., 51, 54
Woodman, T.P., 143, 152
Wyman, L.L., 1, 3, 4, 7, 10, 13, 33, 34

Y

Yakel, H.L., 166, 177
Yarnitsky, Y., 77, 94
Yoshida, S., 309, 330
Young, F.W., 181, 205

Z

Zamozhskii, V.D., 143, 152

SUBJECT INDEX

A

Abrasive cutting, 41-54
    conditions for, 43, 44
    consumable wheels, 41, 43-48
    cooling for, 45-46
    metal matrix diamond blades, 41-43, 48-49
    parameters for, 43, 53
    resin bonded diamond blades, 50-51
    rimlock wheels, 49-50
    wafering blades, 51-52
Abrasive particles, 43, 57-59
Abrasives
    final polishing, 109-120
Alumina abrasives, 95-107
    in metallographic polishing, 103-106
    preparation, 96-103
    quality control checks, 103
Anodizing, 192-198
Autoradiography, 201-202

B

Bakelite, 2,3,9
Boron filaments, 61-62
Boule powder, 101, 103
Brittle materials
    thinning of, 233-249, 333-347

C

Castable mounts, 15-19
Cathodic vacuum etching
    see vacuum etching
Composites, 251-270

D

Decorative etching, 179-204
Delta ferrite, 169, 175
Diamond blades
    metal matrix, 41-43, 48, 49,
    resin bonded, 50-51
    wafering, 51-52
Diamond compounds, 77-94
    characteristics, 89
    concentration, 80-81, 87, 91
    particle shape, 82
    particle size, 81-82, 83
    solubility in carriers, 80
Double mounts, 20-21

E

Edge retention, 3, 19, 22-29
Electric discharge machining, 56, 69-76
    zones, 72-73
Electrodeposited plating, 24-27
Electron microscopy of radioactive materials, 283-295

Electron probe microanalysis
    energy dispersive, 288-289
    of radioactive materials, 284-289
    specimens, 66-67, 286
    wave-length dispersive, 288
Electronic materials
    polishing, 143-152
Electropolishing
    thin metal foils, 307-347
Elemental printing, 199, 201
Epoxy resins, 121-123
Etching
    decorative, 179-204
    magnetic, 155-176

F

Ferrofluid, 155-176
Fiber reinforced composites, 251-270
    hard fiber-soft matrix, 252-266
    soft fiber-soft matrix, 267-270
Filler materials, 3-4, 27-29
Final polishing abrasives, 109-120
Fossils, 245, 246, 248

G

Gadolinium gallium garnet, 143-152

H

Honeycomb, 62-65

I

Ion micromilling, 333-347
    of oxide films, 340-343, 345, 346, 347
    of SiC and Si_3N_4, 335-340, 343-344, 347
Ion probe specimens, 66-67
Ion thinning and drilling, 228

L

Lapping, 123-127, 131
    gadolinium gallium garnet, 144-146
Low melting point alloys, 2, 37
Lunar basalts, 236-239, 247
Lunar breccias, 240, 241, 247

M

Magnetic domains, 156, 158, 160, 171, 174
Magnetic etching, 155-176, 202
Martensite, 161, 164, 165, 167, 171, 176
Mechanical mounts, 5-7
Meteorites, 242, 243, 247
Mounting methods
    see specimen mounts

O

Oxidation
    selective, 181-192
Oxide films, 340-343

P

Plastic overlay, 199, 200
Plated specimens, 24-27

SUBJECT INDEX

Plutonium, 275-281
Polishing, 124-127
    automated, 129-141
    for vacuum etching, 212
    gadolinium gallium garnet, 143-152
    irregular pieces, 140
    skid, 131
    vibratory, 143-152
Potted assemblies, 61
Pressure mounts, 7-9

R

Radioactive materials, 275-281, 283-295

S

Scanning electron microscopy of
    radioactive materials, 289-291
    specimen preparation, 297-305
Selective oxidation
    chemical, 186-192
    thermal, 181-185
SiC, 335-340, 343-344
Si_3N_4, 335-337, 343-344, 347
Sigma phase, 166, 168, 176
Spark machining
    see electric discharge machining
Spark planar, 74-75
Specimen mounting, 1-41, 121-127
    castable, 15-19
    double, 20-21
    filler materials, 3-4, 27-29
    for conductivity, 29-31
    for edge retention, 22-29
    for wire, tubing, sheet, 19-22
    for vacuum etching, 208-212
    history of, 1-4
    large sizes and quantities, 121-123
    marking, 32
    materials, 2-5, 6, 7-19
    mechanical, 5-7
    modern techniques, 4-32
    pressure, 7-9
    required properties, 5
    special techniques, 19-32
    standard sizes, 4
    thermoplastic resins, 10, 12-15
    thermosetting resins, 9-10, 11, 121-123
Staining, 186-192
Strain-induced martensite, 161, 164, 165, 167

T

Thermoplastic resins, 10, 12-15
Thermosetting resins, 9-10, 11, 121-123
Thin metal foils, 307-329
Thin sections, 62, 66, 233-249
Thin slabs, 62
Thin-walled tubing, 61
Transmission electron microscopy of radioactive materials, 291-294
Tungsten filaments, 61-62

U

Ultrathinning, 233-249

V

Vacuum deposition, 197, 199
Vacuum etching, 207-231
    alloys, 218-220
    ceramics, 224, 225
    complex materials, 218, 222
    deep, 224, 226, 227
    dissimilar metal couples, 218, 221

pure metals, 216-218
uranium and uranium alloys, 218, 223

W

Wire sawing, 55-68
    automated, 68
    parameters, 67
    wire/specimen force, 59
    wire tension, 59-60

X

X-ray diffraction specimens, 67